Babesiosis of Domestic Animals and Man

Editor

Miodrag Ristic
Professor
Department of Veterinary Pathobiology
University of Illinois
Urbana, Illinois

CRC Press
Taylor & Francis Group
Boca Raton London New York

CRC Press is an imprint of the
Taylor & Francis Group, an **informa** business

First published 1988 by CRC Press
Taylor & Francis Group
6000 Broken Sound Parkway NW, Suite 300
Boca Raton, FL 33487-2742

Reissued 2018 by CRC Press

Library of Congress Cataloging-in-Publication Data

Babesiosis of domestic animals and man.

Includes bibliographies and index.
1. Babesiosis. 2. Babesia. I. Ristic, Miodrag,
1918- . [DNLM: 1. Babesiosis. SF 780.6 B1135]
QR201 . B318B33 1988 636.089'6016 87-24256
ISBN 0-8493-4908-7

A Library of Congress record exists under LC control number: 87024256

Publisher's Note
The publisher has gone to great lengths to ensure the quality of this reprint but points out that some imperfections in the original copies may be apparent.

Disclaimer
The publisher has made every effort to trace copyright holders and welcomes correspondence from those they have been unable to contact.

ISBN 13: 978-1-315-89092-0 (hbk)
ISBN 13: 978-1-351-07002-7 (ebk)

Visit the Taylor & Francis Web site at http://www.taylorandfrancis.com and the
CRC Press Web site at http://www.crcpress.com

PREFACE

This book presents the state of the art information on basic and applied knowledge pertaining to various aspects of babesiosis, particularly bovine babesiosis. The emphasis on bovine babesiosis is inevitable because of its economic impact in tropical and semitropical regions where it remains one of the most important uncontrolled diseases of cattle. While most of the projected increases in livestock production are expected to take place in these regions of the world, the uncontrolled disease situation discourages investment in cattle industries by private and government sectors. Aside from this, substantial sections of the book are devoted to babesial infections in nonbovine hosts.

Although babesiosis as a disease entity has been known for nearly a century, it is only during the last 2 decades that a better understanding of the nature of the etiologic agents and the pathogenesis of the disease has been acquired. The major research accomplishment during this period has been the development of a method for continuous in vitro propagation of *Babesia bovis*, one of the most important agents of bovine babesiosis. This accomplishment provided, for the first time, an opportunity to study the agent away from its natural host and carefully examine its biological properties. Other important recent findings include the understanding of the cycle of development of bovine babesia species in their tick vector, the nature of protective immunity, and the development of more effective and safer chemotherapeutic and immunoprophylactic control measures. Among the direct applications of the in vitro propagation technique is the development of the first commercial protozoan vaccine against *B. canis* (Pirodog®) which serves as a model for other vaccination endeavors directed to intraerythrocytic parasites. In addition, an interesting aspect of bovine and other *Babesia* species, which has been recognized only during the last 2 decades, is their zoonotic potential, thus refuting the traditional concept of host specificity of these organisms.

In the taxonomic field it has been shown that *B. equi*, hitherto considered a member of the genus *Babesia*, should be reclassified in the genus *Theileria*, since it has been shown to have a lymphoid tissue stage in its cycle of development.

A comprehensive modern review of all these and other aspects of bovine, canine, equine, and human babesiosis by well-known and experienced scientists makes this book an exciting, highly useful and timely venture. The book should serve as a valuable source of information for research workers, graduate and undergraduate students of veterinary and agricultural sciences, field veterinarians, and allied professionals involved in animal production and disease control. In view of the zoonotic nature of babesias, the book also serves as an authoritative reference text for medical and public health workers.

THE EDITOR

Miodrag Ristic, D.V.M., Ph.D., is a Professor in the Department of Pathobiology and Associate Director, Center for Zoonosis Research, University of Illinois. He is Honorary Professor, College of Veterinary Medicine, Autonomous Metropolitan University, Mexico City, Mexico, has served as the President of the American Association of Tropical Veterinary Medicine, and is currently Secretary-Treasurer of that society. He is Honorary Doctor, Pan American Association of Veterinary Medicine and Zootechnics, is President (U.S. Office) of the Laveran International Foundation for research and training in malaria and babesiosis, is a charter member and diplomate of the American College of Veterinary Microbiologists, and a recipient of the Certificate of Merit from the U.S. Army Medical Research and Development Command.

Miodrag Ristic is author and co-author of more than 300 scientific papers, contributor to 40 scientific books, and editor of 8 books. He is a member of the team which made the first isolation of the causative agent of Potomac horse fever which was later named in his honor as *Ehrlichia risticii*.

CONTRIBUTORS

Masamichi Aikawa, M.D.
Professor of Pathology
Department of Pathology
School of Medicine
Case Western Reserve University
Cleveland, Ohio

Georg Fayet, D.V.M.
Department of Marketing
Rhone Merieux
Lyon, France

Karl T. Friedhoff, Dr.med.vet.
Professor
Department of Parasitology
School of Veterinary Medicine
Hannover, West Germany

B. V. Goodger, M.Sc., Ph.D.
Principal Research Scientist
Division of Tropical Animal Production
CSIRO
Brisbane, Australia

George R. Healy, M.S., Ph.D.
Research Parasitologist
Division of Parasitic Diseases
Centers for Disease Control
Atlanta, Georgia

Ikuo Igarashi, D.V.M., Ph.D.
Research Associate
Department of Pathology
School of Medicine
Case Western Reserve University
Cleveland, Ohio

Mark A. James, B.S., M.S., Ph.D.
Visiting Assistant Professor
Department of Veterinary Pathobiology
University of Illinois
Urbana, Illinois

Ibulaimu Kakoma, D.V.M., M.S., Ph.D., Dip.Bact.
Associate Professor
Department of Pathobiology
University of Illinois
Urbana, Illinois

Gottfried Kellermann, Ph.D.
President
ProtaTek International, Inc.
St. Paul, Minnesota

Julius P. Kreier, V.M.D., Ph.D.
Professor of Microbiology
Department of Microbiology
Ohio State University
Columbus, Ohio

Kenneth L. Kuttler, D.V.M., M.S., Ph.D.
Veterinary Medical Officer
Agricultural Research Officer
United States Department of Agriculture
Pullman, Washington

Serge Martinod, D.V.M.
Ciba-Geigy-SA
St. Aubin, Switzerland

Sonia Montenegro-James, D.V.M., M.S., Ph.D.
Visiting Assistant Professor
Department of Veterinary Pathobiology
University of Illinois
Urbana, Illinois

Yves Moreau, D.V.M.
Head of Development
Department of Parasitology & Bacteriology
Rhone Merieux
Lyon, France

Ingrid Reiter, D.V.M.
Institute of Comparative Tropical Medicine & Parasitology
University of Munich
Munich, West Germany

Miodrag Ristic, D.V.M., Ph.D.
Professor
Department of Veterinary Pathobiology
University of Illinois
Urbana, Illinois

Eberhard Schein, D.V.M.
Professor
Institute of Parasitology
Free University of Berlin
Berlin, West Germany

K. Roger Tsang, Ph.D.
Director of Laboratory
ProtaTek International, Inc.
St. Paul, Minnesota

Georg Weiland, D.V.M.
Institute of Comparative Tropical Medicine & Parasitology
University of Munich
Munich, West Germany

Ian Gordon Wright, Ph.D., D.V.Sc., B.A.
Senior Principal Research Scientist
Division of Tropical Animal Production
CSIRO
Brisbane, Australia

Alan S. Young, B.Sc., Ph.D., A.R.C.S.
Team Leader
Overseas Development Adminstration of U.K.
Kenya Agricultural Research Institute
Kikuyu, Kenya

TABLE OF CONTENTS

Chapter 1

WORLD-WIDE IMPACT OF BABESIOSIS

Kenneth L. Kuttler

TABLE OF CONTENTS

I. INTRODUCTION

Babesiosis, also known as piroplasmosis, tick fever, red water, Texas fever, splenic fever, tristeza, etc., is caused by various tick-borne, intraerythrocytic, protozoan parasites of the genus *Babesia*. These parasites are capable of inducing disease in a wide range of vertebrates. The disease is usually characterized by fever, hemolytic anemia, hemoglobinuria, and in severe cases, death. First recognized in cattle by Babes in 1888,[1] these blood parasites occur in many vertebrates including domestic and wild animals. A list of the more commonly encountered babesias, their distribution, usual vectors, and vertebrate host is presented in Table 1. All *Babesia* species are generally host specific, but exceptions are not uncommon. *Babesia bovis*, primarily a cattle parasite, has also been recovered from sheep and goats;[2] *B. microti*, a rodent parasite, is also pathogenic for man;[3,4] and *B. divergens*, also a cattle *Babesia*, produces a parasitemia in man, gerbils, and rats.[5-8]

In nature, the babesias are transmitted biologically by ixodid ticks, but other mechanical means of transmission, such as biting flies and fomites that transfer blood to a susceptible host, may theoretically induce infection. *Babesia* spp. are not only intraerythrocytic parasites of vertebrates, but also parasitize their tick vectors where they produce adverse or pathogenic effects in the course of multiplication.[9] The geographic distribution of different *Babesia* species often correlates with the tick vector in a given area.[10]

The different *Babesia* species often have characteristic morphology which is amply described in the literature, but the pleomorphic tendencies of the parasites make identification on the basis of morphology alone difficult. There are, however, usually two morphologically distinct forms of *Babesia* species. One is small, rounded or pear shaped, with paired forms occurring at an obtuse angle; the other is a large type, which also may be round, oval, or pear shaped, with the paired forms being at a more acute angle. Typically, there are at least one large and one small *Babesia* species found in each susceptible vertebrate species. In the case of cattle, several large and small forms have been identified (Table 1), and this may be the case in other animals. In recent years, there has been an increased awareness of babesiosis in companion animals as well as sheep and swine. The increasing number of babesiosis cases in man has stimulated public health awareness of these blood parasites.

II. BOVINE BABESIOSIS

Babesiosis in cattle has often been thought of as a disease of the tropics and subtropics, where it most often occurs, but in reality, different pathogenic babesias affecting cattle occur throughout the world, with widely different vectors.

A. Historical

In response to a persistent cattle disease, the Romanian government in 1887 established a commission headed by Dr. Victor Babes to investigate and determine the cause and solution to the problem. As a result of these studies, he concluded that the cause of the disease, termed enzootic hemoglobinuria, was a small intraerythrocytic organism which, in 1888, he called *Haematococcus bovis*. Starcovici, a member of Babes' research team, in 1893 renamed the etiologic agent *Babesia bovis*.[11]

Smith, in 1889, reported observing the etiologic agent of Texas fever, but did not name it at that time; rather he gave credit to Stiles for being the first to observe the organism 20 years earlier.[12] In 1893 Smith and Kilborne proposed the name *Pyrosoma bigeminum* (later *Babesia bigemina*) for the North American parasite, and more importantly, first established that a pathogenic protozoan could be transmitted by an intermediate arthropod host (*Boophilus annulatus*).[13] While Starcovici, Smith, and Kilborne agreed that they were dealing with two distinct but closely related species, others considered them synonymous.[14] Rees in 1934

Table 1
SOME OF THE MORE COMMON BABESIAS

Animal and babesia	Morphology (length/width)	Principal vectors	Distribution
Bovine			
B. bigemina	Large 4.5 × 2 μm	*Boophilus annulatus, B. microplus, B. decoloratus, B. geigyi*	Central and South America, Southern Europe, Africa, Asia, Australia
B. bovis	Small 2.0 × 1.5 μm	*B. annulatus, B. microplus, B. geigyi, Ixodes ricinus*	Same as *B. bigemina*
B. major	Large 2.6—3.7 × 1.5 μm	*Haemaphysalis punctata*	Europe and North Africa
B. divergens	Small 1.5—2.0 × 0.4 μm	*I. ricinus*	Northern Europe
B. jakimovi	Large 4.5 × 2 μm	*I. ricinus*	Siberia, northern Russia
B. ovata	Large 3.2 × 1.7 μm	*Haemaphysalis longicornis*	Japan
B. occultans	Large 2.9 × 1.2 μm	*Hyalomma m. rufipes*	South Africa
Equine			
B. caballi	Large 3.0 × 2.0 μm	*Dermacentor (Anocentor), nitens Hyalomma, Rhipicephalus sp.*	World-wide
B. equi	Small 2.0 × 1.0 μm	*Dermacentor (Anocentor), nitens Hyalomma, Rhipicephalus sp.*	World-wide
Canine			
B. canis	Large 5.0 × 2.5—3.0 μm	*Rhipicephalus, Dermacentor, Haemaphysalis* (?) and *Hyalomma* (?)	Southern Europe, North America, Africa, Asia
B. gibsoni	Small 1.9 × 1.2	*Rhipicephalus sanguineus, Haemaphysalis bispinosa*	Far East, Asia, but expanding its distribution to Europe and North America
Ovine/caprine			
B. motasi	Large 2.5—4.0 × 2.0 μm	*Haemaphysalis Rhipicephalus*	Europe, Asia, Africa
B. ovis	Small 1—2 × 0.5 μm (rounded)	*Rhipicephalus*	Southern Europe, Middle East, Africa
B. trautmanni	Large 2.5—4.0 × 1.5—2.0 μm	*Boophilus* (?), *Rhipicephalus* (?), *Dermacentor* (?), *Hyalomma* (?)	Southern Europe, U.S.S.R., Africa
B. perroncitoi	Small 1.2—2.6 × 0.7—1.9 μm	Unknown	Southern Europe, Africa
Feline			
B. herpailuri	Large 2.7 × 2.2 μm	Unknown	Africa
B. felis	Small 0.9 × 0.7 μm	Unknown	Africa and southern Asia
Man			
B. microti	Small (pleomorphic)	*I. dammini*	Northeastern states — U.S.
B. divergens	Small	*I. ricinus*	Northern Europe

determined that both *Babesia bigemina* and *B. bovis* were present in the U.S. and were involved with Texas fever.[15] He further suggested, after looking at Smith and Kilborne's drawings, that *B. bovis* had been present from the earliest days of *Babesia* recognition.

Lignieres, in studying the cattle disease tristeza in Argentina, recognized in 1903 that he was dealing with two strains of *B. bigemina*. One form was large, resembling the organism described by Smith and Kilborne, while the other was smaller and often difficult to find in blood smears, but more easily seen in capillaries of the kidneys and the meninges of the central nervous system.[16] In Lignieres' 1910 paper, he named the small organism *Piroplasma argentina,* later known as *Babesia argentina.*[17] The term *B. argentina* was used for many years in Australia and South and Central America. It was not fully agreed until the mid 1970s by a consensus of parasitologists that *B. argentina* and *B. bovis* were the same and that the designation *B. bovis* should take precedence.[18]

M'Fadyean and Stockman, in 1911, encountered a small intraerythrocytic organism in England, which they named *Piroplasma divergens* (later *Babesia divergens*).[19] The paired parasites had a wide angle of divergence and are often found on the periphery of the erythrocyte. European workers observed similar organisms which they assumed to be *B. bovis,* and for many years *B. divergens* was thought to be synonymous with *B. bovis.* The Yugoslavian workers Simic, Petrovic, and Rakovic in 1955 pointed out that the ticks which transmit the parasite are important elements in identifying specific *Babesia* species.[20] *B. bigemina* and *B. bovis* (*B. argentina* and *B. berbera*) are transmitted by *Boophilus ticks,* which require a relatively warm climate to complete their development; hence, these are not found in northern Europe. On the other hand, *Ixodes ricinus* and *Haemaphysalis punctata,* which generally inhabit the cooler northern and western areas of Europe, are adequate vectors for *B. divergens* and *B. major,* where these organisms occur.

During their work in 1911, M'Fadyean and Stockman made reference to a large *Babesia* which they assumed to be *B. bigemina,* but was probably *B. major.*[19] Sergent et al. in 1926 recovered a large *Babesia* which appeared different from *B. bigemina,* he named *B. major.*[21] The morphologic similarities between *B. major* and *B. bigemina* were such that it was not until 1968 that Zwart et al. confirmed its difference from *B. bigemina.*[22] *B. major* as now seen in the U.K. and northern Europe is only mildly pathogenic, and for this reason, has probably in the past been overlooked.[23]

Brocklesby suggested in 1976 that there were only four valid *Babesia* species in cattle, which are *B. bigemina, B. major, B. bovis,* and *B. divergens.*[18] This proposition has been validated by serological studies.[24,25] Since that time, three new bovine babesias, *B. ovata, B. jakimovi,* and *B. occultans,* have been described and determined by various serological techniques to be unique.[26-28]

It is thought that cases of babesiosis had been encountered in the U.S. as early as 1814. One of the early accounts (1868) of babesiosis in the U.S. involved an outbreak in Indiana and Illinois after the importations of apparently healthy cattle from Texas.[29] It is surmised that the replete *Boophilus* females that dropped off the Texas cattle, laid eggs, so that when the larval progeny fed on local cattle a disastrous outbreak of babesiosis occurred in which an estimated 15,000 head of cattle died with a mortality rate approaching 90%. Similar patterns of death losses were experienced in Australia, where it is estimated that up to a million cattle died from babesiosis before the disease became endemic and a level of enzootic stability acquired.[30] Many of the early epizootics of babesiosis, which occurred throughout the world, were associated with the rapid exposure of large numbers of mature, previously unexposed cattle to ticks carrying the *Babesia* pathogen. Early workers were well aware of the phenomenon referred to as enzootic stability and knew that calves being raised in tick and *Babesia* endemic areas were resistant to infection. This gave rise to early vaccination programs involving premunition of calves, a method which is still in use with various modifications.

B. *Babesia bigemina* and *Babesia bovis*

These two *Babesia* pathogens responsible for bovine babesiosis occur together in most areas of the world, with the greatest incidence between 32°N and 30°S latitude where their *Boophilus* tick vectors commonly occur. They are found in most of Africa, southern Europe, southern Asia, Australia, Central and South America, and many of the islands in the

Caribbean and South Pacific.[31-38] *B. bovis* rarely occurs in those areas of Africa where *Boophilus decoloratus* is the only *Boophilus* tick present; however, in Africa where *B. microplus* and/or *B. geigyi* occur, *Babesia bovis* has been reported.[39,40-42] *B. bigemina* and *B. bovis* are both transmitted transovarially. *B. bovis* is usually transmitted during the larval feeding period and *B. bigemina* during the nymphal and adult stages.[43-45] Ticks must be reinfected with *B. bovis* during the nymphal and adult feeding period since these stages are thought to no longer carry the infection obtained by the previous generation, whereas *B. bigemina* may remain in the tick for more than one generation without reinfection.[41] By utilizing this information, pure isolates of either *B. bovis* or *B. bigemina* can be obtained.[45] The need for reinfection of ticks by *B. bovis* explains the apparent ability of some drugs, such as imidocarb, to apparently sterilize ticks infected with *B. bovis,* but not ticks with *B. bigemina*.[46,47] The drugs are probably not influencing the tick *Babesia* stages, but prevent tick reinfection by eliminating *B. bovis* in the blood of the treated bovine during tick engorgement.

It is interesting to note that nearly all the so-called "developing countries" are areas of endemic babesiosis, anaplasmosis, and *Boophilus* tick infestation. Efforts by these countries to import cattle having greater production potential from western Europe and North America are nearly always complicated by the presence of endemic babesiosis which, when transmitted to these susceptible cattle, invariably creates animal health problems. Death losses in mature animals can be high if these infections are not controlled by chemotherapy, vaccination, or vector management. The prevalence of the cattle ticks, babesiosis, and anaplasmosis are among the most important factors hindering the introduction of more productive livestock and the upgrading of existing livestock in the developing countries.[48]

Probably the most successful single program for the control of babesiosis was conducted in the U.S. from 1907 to 1940 when the *Boophilus* ticks were essentially eradicated. With the exception of some areas in Florida, tick eradication was completed in 1940.[10] *Boophilis* ticks were eradicated from a 700,000 mi^2 area (1,813,000 km^2), extending from the Atlantic coast to Texas and north to Oklahoma, Arkansas, Tennessee, and Virginia, plus a large area in southern California. Eradication was accomplished by dipping all cattle and other tick-infested livestock in an arsenic solution every 2 weeks. Generally, the strictly enforced quarantine lines progressed from west to east and from north to south. It should be remembered that *Boophilus annulatus* was the only *Boophilus* tick present in the U.S., with the exception of a few counties in Florida and south Texas where *B. microplus* occurred. Eradication in these areas of Texas and Florida was the most difficult.[10] Another factor which contributed to the success of this program was that the large areas, which environmentally were only marginally suitable for these ticks, also had a substantial annual winter tick kill. In addition, there was a very low number of wild ruminants (deer) in these areas, which allowed eradication to occur by removing all cattle from the pasture for extended periods, during which time the ticks would die. Pasture rest may no longer be a viable method of tick eradication because in many areas, present deer populations are so high that ticks can maintain themselves indefinitely in the absence of cattle.[10,49] Another significant factor in the success of this tick eradication program was its popular support by the cattle producers. Efforts to upgrade existing cattle in the south by importing superior breeding stock from Europe or the northern states were often frustrated by the high death and production losses in these cattle when moved to a tick environment. It was not until ticks were eradicated that the cattle industry developed to its present level in the southern U.S.

In 1906 it was estimated that the annual losses associated with ticks and babesiosis was $130 million which in today's dollars would probably be in excess of $1 billion. The cost of tick eradication was very modest in comparison. From 1907 to 1938 it is estimated that federal appropriations for the tick program were approximately $46 million.[10,50]

This program, while desired by a majority of cattlemen, was not without opposition. Cattlemen knew that without ticks their calves would fail to acquire immunity, and as a result would be fully susceptible when mature. Also, the arsenic dips were on occasion toxic to cattle, and many thought that a modest number of ticks were not a problem. A number of unique enforcement procedures were evolved, some of which are presented in a review of this subject by Graham and Hourrigan.[10] The final result of tick eradication was the elimination of babesiosis in the U.S. Since tick eradication, a permanent quarantine line creating a buffer zone has been established along the Mexican border, south of which both ticks and babesiosis occur. For the most part, this has worked well; however, on occasion, ticks are introduced into and even beyond the buffer zone. Constant surveillance is required with a dipping program instituted when needed. There have been few, if any, reports of serious or significant outbreaks of babesiosis in the U.S. since tick eradication.

An ambitious program of tick eradication was begun in January 1956 in northern New South Wales, Australia, to eradicate *Boophilus microplus*. The area under control encompassed over 2000 mi^2 (5180 km^2), 1750 premises, and almost 200,000 head of cattle. All cattle were dipped in 0.5% DDT every 2 weeks for 15 months.[51] This was followed by 1 year of inspections and a 9-month surveillance period. By July 1959, it was obvious that the campaign had failed. Acaricide-resistant ticks were not the problem in this effort, but by 1970, tick resistance was encountered in Australia, and has remained as an ever-increasing problem. A similar tick eradication effort was made in 1963 in New Guinea, which also failed.[51] It may be significant that these eradication attempts were made against *B. microplus*, whereas the U.S. effort was primarily against *B. annulatus*. While *B. annulatus* is more tolerant of climate extremes, *B. microplus* appears more able to feed and adapt to the nonbovine hosts.

The failure of eradication efforts, plus the occurrence of acaricide-resistant ticks, have led the Australians to alternative methods of tick and babesiosis control. Following extensive research programs, attenuated vaccines for both *Babesia bigemina* and *B. bovis* have been developed.[52,53] Immunity is dependent on a replicating, nonclinical infection which results in extended periods of protection against both needle and natural tick challenge exposure to virulent organisms. Investigations in South Africa and Mozambique have shown the African babesias to be antigenically similar to the Australian strains.[54-56] The Australian *Babesia* spp. vaccines appear effective in Africa and have also been used in the Western Hemisphere. Of equal significance has been the discovery that *Bos indicus* cattle are more capable of developing a high level of tolerance to ticks, which, in turn, reduces tick numbers and subsequent *Babesia* exposure.[57] Extensive epizootiological investigations have established the level of tick numbers and adult infection levels required to maintain enzootic stability.[58,59] This situation is characterized by a high calf infection rate resulting in an immune population. Changes in tick numbers over a sufficient period of time can lead to enzootic instability with the danger of babesiosis occurring in nonexposed susceptible cattle.[60] Vaccination is indicated in this situation as well as in cattle being introduced into an area of endemic babesiosis. Vaccination of older cattle may require chemotherapy to moderate the vaccinal response. This approach is often referred to as chemoimmunization.[61-63]

Chemoprophylaxis has been shown to be effective under some conditions. Imidocarb given at 5 mg/kg prevents *B. bigemina* infections for approximately 90 days and *B. bovis* for nearly 60 days.[64] If exposure occurs during this time, or at a time when drug levels are low, then a non clinical infection may occur with subsequent protective immunity. This approach has been successfully used where tick challenge was continuous over a period of time.[65]

A number of nonviable *B. bovis* vaccines using a variety of purified antigens with adjuvants have proven effective in significantly reducing the severity of response to virulent homologous and heterologous challenge.[66-68] In no instance has such a vaccine prevented infection, but they have reduced severity of the subsequent reaction and prevented death losses for as long as 6 months after vaccination.[69] These and genetically engineered vaccines, while not presently available commercially, offer promise for future development.

The reliance on tick control for the prevention of babesiosis historically has proven highly effective, but complete reliance on one system may impose circumstances that could prove hazardous. Of interest in this respect has been the relatively recent reports from Zimbabwe of significant cattle losses occurring in association with the period of political unrest (1973 to 1979).[36] During this time the previously well-established cattle dipping program for the control of tick-borne diseases broke down with the subsequent rapid increase in *Boophilus, Rhipicephalus,* and *Amblyomma* ticks. This was rapidly followed by an increase in the major tick-borne diseases of babesiosis, theileriosis, anaplasmosis, and heart water. We may surmise that the cattle in Zimbabwe during the tick control period had lost much of their natural immunity to these diseases, and as a direct result of the tick build-up, it is estimated that, as of 1979, nearly a million head of cattle had died from tick-borne diseases. The economic and social impact of such staggering livestock losses are beyond the scope of this review to assess, but the potential of these tick-born diseases to disrupt cattle production when control measures are abandoned for whatever reason is clearly apparent.

The natural reservoirs, or sources of infection, are cattle and ticks. *Babesia bigemina* and *B. bovis* have also been reported in water buffalo (*Bubalus bubalis*), African buffalo (*Syncerus caffer*), and American bison (*Bison bison*).[70-72] It is doubtful whether these hosts significantly influence the course, prevalence, and maintenance of disease in cattle. Efforts to induce *Babesia* infection in intact and splenectomized white-tailed deer (*Odocoileus virginianus*) by tick feeding and blood inoculation were unsuccessful.[73]

C. *Babesia divergens* and *Babesia major*

Babesia divergens, for years linked to and thought to be synonymous with *B. bovis,* is now recognized and generally accepted as a distinct species which morphologically is slightly smaller than *B. bovis* and is located more frequently on the erythrocyte margin.[14,18,27] The pleomorphic tendencies of both *B. bovis* and *B. divergens* make it difficult to distinguish these two organisms solely on morphology. Unlike the widely distributed *B. bovis,* transmitted by the one-host *Boophilus* tick, *Babesia divergens* is apparently restricted in distribution to Europe in association with its three-host vector *Ixodes ricinus*. Its transmission, mainly in the spring, is transovarial, with all tick developmental stages being capable of transmission after having acquired the infection as adults of the previous generation that fed on either *B. divergens* carrier or acutely infected cattle.[27,74] In the absence of tick control, susceptible cattle can rapidly acquire the disease when brought in contact with infected ticks, resulting in explosive and damaging outbreaks of babesiosis.

Babesia divergens has been reported throughout northern Europe including Ireland, Scotland, and western England.[27] Recent estimates of annual costs associated with *B. divergens* infection in the Republic of Ireland approach $15 million (U.S.) when production and death losses are added to treatment costs. The incidence of clinical infection varies from 1.3 to 2.1% with about 117,870 cases occurring each year accompanied by a 10% mortality.[75] In northern Ireland, estimates have been made of animal losses approaching $320,000 (U.S.), with 400 cases a year and a 12% mortality.[76] Annual losses in England and Scotland are significant, but based on 1974 estimates, were only about $450,000 (U.S.).[77] There are numerous reports of *B. divergens* being responsible for disease and death losses in Norway, the Netherlands, Germany, France, Austria, etc., but a cost figure of losses sustained by these countries is hard to determine.[78-82]

Losses from *B. divergens* can be reduced by tick control with regular and frequent spraying or dipping with effective acaricides, by chemotherapy, or chemoprophylaxis.[74,83-84] Some endemic areas, in the absence of tick control, experience enzootic stability as the result of the early acquisition of infection by young animals with the development of resistance or immunity.[85] The same shortcomings for each system of control exist as described for babesiosis caused by *B. bovis* and *B. bigemina.* Young animals are generally quite resistant, whereas older cattle are very susceptible.[86] Tick eradication of a three-host tick is really feasible with existing methods, but tick control has been used successfully to reduce the incidence of babesiosis.[83] However, if and when nonexposed cattle are moved to an area or pasture where tick control has not been pursued, severe losses can and do occur. Recovery is the rule, if treated early with any one of several recognized babesiacides. Amicarbalide, quinuronium, diminazene, and imidocarb are all reported to be effective when given in the early stages of infection.[74] Even though recovery following treatment is expected, the disease aftermath is often associated with significant production losses reflected as poor weight gains and reduced milk production.[75] Imidocarb at 2 mg/kg has been observed to produce a period of extended protection. Following this treatment, susceptible cattle continuously exposed to ticks and infection will often acquire a subclinical infection and immunity.[84]

Carriers of infection can be accurately detected by several serologic procedures such as the complement-fixation test (CFT), indirect fluorescent antibody test (IFAT), indirect hemagglutination (IHA), and enzyme-linked immunosorbent assays (ELISA).[87-89] Carrier infection can be eliminated by chemotherapy, if that is the choice, but in so doing, the risk of increasing susceptibility over a period of time must be accepted. Experimental vaccines involving irradiated organisms, nonviable particulate, and soluble antigens with and without adjuvants have been tested and show promise in preventing clinical episodes, but appear ineffective in entirely preventing infection and the establishment of cattle reservoirs that have the potential of producing future disease outbreaks.[90-94] Efforts to develop an attenuated organism by rapid passage through splenectomized calves, as described for *B. bovis,* have been unsuccessful.[95]

It is not known to what extent *B. divergens* might reside in the nonbovine host or if such infection might prove a reservoir for cattle. There is evidence that *Ixodes ricinus* can maintain infection to the F2 generation without reexposure.[27] *B. divergens* can be grown and maintained in tissue culture and has been recovered from gerbils, man, chimpanzees, rats, splenectomized sheep, goats, and deer, where it is apparently able to maintain itself, suggesting that its growth requirements are less demanding than other babesias.[96] With the exception of man, *B. divergens* is not thought to pose serious economic or health hazards for vertebrates other than cattle.

It is interesting to note that the U.S. and Canada have not reported *B. divergens* even though they do have *Ixodes* ticks. There are numerous recognized species in the U.S. including *I. dammini, I. scapularis,* and *I. pacificus. Ixodes dammini* is known to transmit *B. microti* to man and rodents. There should be and is cause for some concern about the possible introduction of *B. divergens* into the U.S., where these potential vectors occur.

Babesia major occurs in essentially the same temperate zones where *B. divergens* is found, and in association with its three-host tick vector, *Haemaphysalis punctata.* In addition to Europe, it has been reported from Africa, Israel, South America, and the U.S.S.R.[27] Its close morphologic resemblance to *B. bigemina* has in the past led to some confusion relative to its identity. As early as 1902, *B. major* was probably observed, but at that time it was considered *B. bigemina.*[23] Sergent in 1926, working in Algeria, first identified *B. major* as a distinct species, but it was not until the work by Zwart in 1968 that the species designation was recognized to be distinctively different from *B. bigemina.*[21,22] It is probable that earlier reports of *B. bigemina* in northern Europe were in reality *B. major.* After the description of *B. divergens* by M'Fadyean and Stockman in 1911, no further reports of a large *Babesia* in the U.K. surfaced until 1969 when their original drawings were published showing the presence of both large and small babesias now thought to be *B. major* and *B. divergens.*[97]

The failure to recognize *B. major* and the ambiguity associated with it in the past is probably due in part to its low pathogenicity for cattle. It is pathogenic for splenectomized cattle, but frequently goes unnoticed in intact cattle, even though the potential exists for it to become pathogenic.[98] Both the larval and nymphal stages of *H. punctata* are capable of transovarial transmission, but some doubt exists about the ability of adult ticks to consistently transmit, although they acquire the infection at this stage by feeding on infected animals. No estimates of losses due to this pathogen have been made, even though it is known to have a fairly wide distribution. *B. major* is readily differentiated from *B. divergens* based on morphology and serology.[99]

Contrary to the established pattern of low pathogenicity for cattle, an outbreak of babesiosis caused by *B. major* in a group of American bison resulted in severe clinical disease.[100] Babesiosis occurred in the bison within 2 weeks of being moved to a pasture in southeastern England which was previously occupied by cattle or sheep and presumably infested with *H. punctata*. A total of three cows and a bull subsequently died. *B. major* was diagnosed and treatment with amicarbalide was successfully used for other sick animals.

D. Other Babesias of Cattle

There are at least three newly identified *Babesia* species recently described in cattle. They all appear to be large babesias that have vectors other than *Boophilus* ticks, and two of the three are only mildly pathogenic in intact cattle.

At one time, the babesias found in Japan were thought to be a different strain of *Babesia bigemina*, but as early as 1968, significant differences became apparent, indicating the possibility of a distinct species.[101] Minami and Ishihara proposed the name of *Babesia ovata* in 1980 for the piroplasm found in Japan, which appears different from *B. bigemina, B. major, B. bovis*, and *B. divergens*.[102] *B. ovata*, transmitted transovarially by *H. longicornis*, is a relatively large babesia most closely resembling *B. bigemina* and *B. major*. It is widespread in Japan, but is not highly pathogenic; however, in splenectomized cattle parasitemias do occur accompanied by anemia, hemoglobinuria, and in some instances, death. *B. ovata* appears to be serologically and immunologically distinct, but is reminiscent of *B. major* in other aspects such as the lack of pathogenicity, its vector, size, and epidemiology. *H. longicornis* readily transmits *B. ovata*, but does not transmit *B. bigemina* or *B. major*. *Boophilus microplus* does not transmit *B. ovata*.[103] This babesia, while widely distributed, does not appear to cause a serious health problem in Japanese cattle.

In 1981, a similar babesia was observed in South Africa and designated *Babesia occultans*.[28] Like *B. major* and *B. ovata*, this *Babesia* is not highly pathogenic in intact cattle, and does not appear to be transmitted by *Boophilus microplus*.[104] *Babesia occultans* is transmitted transovarially by the two-host tick, *Hyalomma marginatum rufipes*, by the nymphal and adult stages. Morphologically, the intraerythrocytic forms of *B. occultans* are slightly smaller than similar forms of *B. bigemina*, but the vermicules in tick hemolymph are significantly larger.[28] Serologically, *B. occultans* is distinguishable from *B. major, B. bigemina, B. divergens*, or *B. bovis*. Its wide distribution and high prevalence indicates a well-established host-parasite relationship, suggesting that this babesia may have originated from an African antelope and has been adapted in its present form to cattle.[28] This babesia does not appear to pose a health hazard to South African cattle.

Unlike *B. occultans* and *B. ovata, B. jakimovi* is pathogenic for cattle and is the etiologic agent for Siberian piroplasmosis which poses a threat as an emerging disease of cattle in eastern Europe. Purnell quotes Nikolskii's 1977 observations that this organism, a large babesia, closely resembles *B. bigemina*, but is transmitted by *I. ricinus*.[105] On the basis of cross-serology using CFT and cross-protection, *B. jakimovi* was determined to be distinct from *B. bigemina*. In addition to cattle, *B. jakimovi* infects Tartarian roe deer (*Capreolus capreolus*), elk (*Alces alces*), and reindeer. There is a suggestion that roe deer may be its natural host and as such is responsible for sporadic outbreaks in cattle where they come in contact. *B. jakimovi* shares a common vector *I. ricinus* with *B. divergens* and *B. capreoli* (in roe deer), but morphologically is quite different from these two small babesias. While *I. ricinus* is the most likely vector, mechanical transmission by gadflies is reported.[105] An

I. ricinus-engorging female will transovarially pass the infection to its progeny, and the adult ticks of the F1 generation will transmit the parasite to a susceptible host. Clinical signs are characteristic of severe babesiosis with the salient features being fever, anemia, and hemoglobinuria. Nervous involvement is not uncommon, and a significant mortality will occur in the absence of treatment with trypan blue or trypaflavin. In all probability, *B. jakimovi* would respond to most of the currently used babesiacides.

III. EQUINE BABESIOSIS (EQUINE PIROPLASMOSIS) (*BABESIA CABALLI* AND *BABESIA EQUI*)

Equine babesiosis is a febrile, tick-borne disease of horses and donkeys (and their hybrids) caused by *Babesia equi* and/or *B. caballi*. Intraerythrocytic hemoparasites resembling *B. bigemina* were first observed in a horse by Guglielmi in 1899.[106] Theiler in 1901 and 1902, working with *B. equi*, first described the disease entity we now know as equine babesiosis (or piroplasmosis), with Laveran in 1901 conferring the name "equi" on this small babesia.[107-109] *B. equi* was first observed in zebras by Theiler in 1905[110,111] and has been transmitted from zebras to horses by blood inoculations.[112] More recently, *B. equi* from zebras was found to infect *Rhipicephalus evertsi*, thus potentially creating a natural means of transmission to horses.[113]

Equine babesiosis is generally characterized by fever, anemia icterus, hemoglobinuria, and in some cases, death.[114] The severity of clinical response is variable, and in many cases, spontaneous recovery may occur without obvious signs of infection. In such cases, a careful monitoring of packed cell volume will usually show a transient, mild anemia. Infections, however, with either *B. equi* or *B. caballi* may produce clinical disease and death. Observations in South Africa suggest that a mortality of up to 10% may occur in fully susceptible horses, and in some instances approach 50%.[106] While reliable figures of numbers and economic losses are not readily available, reports of disease and death losses are not uncommon when *Babesia*-free adult horses are introduced into an endemic area. This was observed among European horses introduced into South Africa during the Boer War, in Australian horses taken to India, and in northern European horses moved to Italy during World War I.[114]

B. caballi more often is associated with sludging or a packing of parasites in capillaries and small vessels of different organs, resulting in dysfunction of these organs with a low parasitemia in the peripheral blood circulation. *B. equi* infections are more frequently characterized by higher parasitemias, red cell lysis, and death due to anemia.[115] *B. caballi* parasitemias rarely exceed 10%, whereas a 90% *B. equi* parasitemia is not uncommon.

B. caballi, a large babesia, morphologically similar to *B. bigemina*, is transmitted by ticks species of the genera *Dermacentor* (*Anocentor*), *Hyalomma*, and *Rhipicephalus*.[116,117] In the case of *D.* (*Anocentor*) *nitens*, the female becomes infected by feeding on the parasitized horse, and the F1 larvae, nymphs, and adults are all capable of transmitting the parasite.[118,119] A similar pattern of transmission probably occurs with other tick vectors. Intrauterine infections have been described with *B. caballi*, but these are thought to be the exception rather than the rule.[116]

B. equi, a small babesia, is also biologically transmitted by ticks species of the same genera as *B. caballi*.[116] *B. equi*, however, is not thought to be transmitted transovarially, but rather transstadially, and by interhost transfer of adult males. Tick transmission most often occurs after the nymphs feed on a carrier or acutely infected horse, with transmission by adults; however, larvae can acquire the infections with transmission occurring at the nymphal stage. Tick infection does not persist after transmission in subsequent stages; thus, ticks are not potential reservoirs of the parasite.[116] Intrauterine transmission appears to be more common than with *B. caballi*, and can occur in the fetus of either acutely infected or carrier mares devoid of clinical sign. The typical intraerythrocytic, babesia-like bodies found resemble the small babesias or *Theileria* sp. seen in other vertebrates, and characteristically are seen as "maltese cross" formation, which helps to distinguish it from *B. caballi*. In addition to these blood forms, lymphocytic macro- and microschizonts have been described.[120] From the microschizonts, the released merozoites invade the erythrocytes where further division produces the typical cross forms. The existence of schizonts, the lymphocytic phase, the absence of transovarial transmission (typical of most babesias), and the relative resistance of *B. equi* to most babesiacidal compounds point to the theilerial-like nature of *B. equi*. Efforts to infect laboratory animals, sheep, goats, pigs, dogs, and cats with *B. equi* have been unsuccessful.[121,122]

B. caballi and *B. equi* generally have the same distribution patterns. Both organisms are widespread, occurring in most tropical and subtropical areas of the world and, in some instances, extending well into temperate climatic zones.[123]

In Europe, equine babesiosis is endemic in France, the Iberian Peninsula, Italy, and the Balkans, and large areas of the U.S.S.R. It is not known how far north these infections are autochthonous, but they are not thought to occur in the British Isles or northern Europe, except in imported horses. A serological survey in the U.K. confirms that no horses born or continuously resident in the British Isles showed positive serological titers.[124] The only positive horses were those imported from endemic areas, with the exception of four foals born in England to seropositive mares from Portugal, which showed positive serological responses due to transient passive maternal antibody transfer.[125] In Russia, equine babesiosis occurs throughout the southern areas, and *B. caballi* extends north for the range of *Dermacentor reticulatus* as far as 58°N latitude.[116]

Africa, Madagascar, and western Asia are endemic areas of equine babesiosis. The status in the Far East is uncertain; however, Japan is generally considered free of the disease. In South Africa a recent survey indicated that equine babesiosis was a serious infectious disease of horses, with a yearly incidence of 1.88% among approximately 10,000 horses.[126] A higher incidence of disease was reported for the summer months, during the rainy season, which also corresponds with accelerated tick activity; however, cases were reported throughout the year.

Australia was considered free of equine babesiosis until 1976 when *B. equi* was diagnosed in a group of horses imported from Spain.[127] The disease produced experimentally by blood inoculation into intact horses was unexpectedly mild in contrast to reports of significant virulence elsewhere.[127] A similar finding has been reported from the U.S. where blood induced infections of a European *B. equi* in intact horses produced a mild, almost nonapparent clinical response.[128] An indirect fluorescent antibody test (IFAT) conducted on sera from 700 horses in Australia some years after the first recognition of *B. equi* revealed 30 reactors at levels indicative of present or past infections.[129] These reactions occurred among 33 Andalusian horses born in Spain and imported into Australia. The three horses that failed to react were 15 years old and had been in Australia 4 to 6 years. It is not known whether or not *B. equi* is transmitted by natural means to native horses in Australia, but these results indicate that during this limited time no evidence of natural transmission was apparent.[129]

With the exception of southern Argentina and Chile, equine babesiosis, caused by both *B. equi* and *B. caballi*, appears endemic in South and Central America as well as the Caribbean Islands. The U.S. was considered free of equine babesiosis until 1961 when *B. caballi* was identified in southern Florida.[130] It is presumed that the first introduction to Florida probably occurred in 1958 when 30 Cuban walking horses were imported.[115] At that time, both *B. equi* and *B. caballi* existed in Cuba. Within a year of its recognition, *B. caballi* was found in most of southern Florida, and the tropical horse tick *Dermacentor* (*Anocentor*) *nitens* was shown to be a vector.[131] Control measures, including tick control, serological testing, treatment, and quarantine, have contributed to its confinement to Florida. The limited distribution of the tropical horse tick probably also contributed to the failure of *B. caballi* to spread beyond Florida.[115,131,132] Recent studies have shown that *D. albipictus* (the winter tick) and *D. variabilis* (the American dog tick) are competent experimental vectors for *B. caballi.*[133,134] The widespread distribution of these ticks in the U.S. raises the question of why *B. caballi* is not more prevalent.

The first field case of *B. equi* infection was diagnosed in a Florida race horse in 1965.[135] Presently, little is known about the incidence and prevalence of these two babesias in Florida, but it is probable that both *B. equi* and *B. caballi* occur there. The natural vector of *B. equi* in the U.S. is yet to be determined. The introduction and recognition of these exotic equine babesias into Florida led to regulations requiring a negative *B. equi* and *B. caballi* complement fixation test (CFT) on all imported horses, which has probably contributed to the continued babesia-free status of horses in other areas of the U.S.

The CFT developed following the recognition of these agents in the U.S. accurately identifies horses with both acute and carrier *Babesia* infections.[136-138] There is little or no cross-CFT reactions between *B. caballi* and *B. equi.* Other tests, such as the IFAT, passive hemagglutination, bentonite agglutination, ELISA, and capillary-tube agglutination CA have also been developed and used with varying degrees of success.[139-142] Existing evidence suggests that the IFAT is more sensitive than the CFT, and in serums that are anticomplementary, the IFAT would be indicated. Comparisons of CFT and IFAT suggest that the CFT, while rarely giving a false positive, may on occasion react negatively with carrier horse serum, whereas the IFAT is more sensitive.[143,144] Following successful treatment to eliminate the carrier infection, the CFT will revert to a negative status in a variable period of time, whereas the IFAT may persist longer. Although treatment to eliminate carrier infections of *B. caballi* is usually successful, no consistently successful procedures are available to eliminate *B. equi.* Imidocarb presently appears to be the drug of choice.[116]

For the most part immunity is acquired in endemic areas by the young animals that undergo a nonapparent infection that renders them resistant to future clinical illness. There is some evidence that horses subjected to heavy training, poor nutrition, or other stress factors may develop relapsing clinical episodes, even after having previously acquired a nonclinical infection.[106] There is at least one report of a successful killed vaccine, utilizing a blood-based antigen, which reduces the severity of subsequent challenge infections.[145] Control programs based on serological diagnosis, treatment, and quarantine can be successful, but eradication from areas where the infection is established has not been achieved.

IV. CANINE BABESIOSIS (*BABESIA CANIS* AND *BABESIA GIBSONI*)

Canine babesiosis is a febrile, tick-borne disease of both domestic and wild Canidae caused by *B. canis, B. gibsoni,* and *B. vogeli.* The validity of *B. vogeli* is yet to be confirmed, and some suggest it may be a variant or strain of *B. canis.*[146] These canine babesias were at one time thought to cause disease problems primarily in the tropics;[27] however, as early as 1934, canine piroplasmosis was reported in Florida.[147] With the greater movement of dogs and the presence of competent tick vectors in the temperate climatic zones, more reports are surfacing to indicate that the disease is becoming increasingly widespread throughout the world.[148-151]

B. canis, a large *Babesia,* is generally recognized as being more widespread than *B. gibsoni* and is transmitted primarily by *Rhipicephalus sanguineus.* In addition, *Dermacentor marginatus, D. reticulatus,* and even species of *Haemaphysalis* and *Hyalomma* have also been suggested as possible vectors.[27,146,149] *B. canis* has been isolated from a variety of wild Canidae including fox, coyote, coy dog, wolf, jackal, etc., but the epidemiological significance of these potential reservoirs is not fully known.[152, 153] Reports from the U.S. and Europe suggest that *B. canis* is becoming an increasing problem.[149,154] Utilizing an IFAT on 181 serum samples from greyhounds located in Mississippi, West Virginia, Oklahoma, Texas, and Florida, approximately 41% were positive.[150]

The clinical disease usually reflects signs and pathology associated with a rapidly developing anemia and the accompanying hypoxia.[155] Contrary to the pattern observed in cattle, where the young are relatively resistant, young dogs and pups appear highly susceptible, with death often occurring in the absence of treatment.[150] Several babesiacides, including diminazene and imidocarb, are highly effective in the treatment of acute infections, but treatment does not always eliminate infections, and a carrier state usually follows.[156,157] Some young pups will show a parasitemia as early as 12 days after birth, strongly suggesting transplacental transmission.[146,150]

Immunity is usually associated with premunition. Studies utilizing amicarbalide, diminazene, and isometamidium to treat acutely infected dogs have resulted in a disappearance of the parasitemia, but then at varying intervals many of these dogs show a secondary relapsing parasitemia.[156] Those dogs showing this secondary response were generally immune to later virulent challenge. Those dogs not showing this relapsing infection were susceptible to challenge. A sterile immunity has been demonstrated, and efforts to produce a nonpremunizing vaccine have been promising.[158,159] Both soluble and particulate antigens combined with saponin adjuvants have induced a significant level of protection against a virulent challenge.[158]

B. gibsoni, a small babesia, was at one time thought to occur only in Africa, India, Sri Lanka, and Malaysia.[160] More recently, its distribution was expanded to include Europe, the U.S., and the Far East (Korea and Okinawa).[146,148,161] The fact that both *B. canis* and *B. gibsoni* may occur together contributes to the ambiguity regarding the prevalence of *B. gibsoni* which, being smaller, can easily be overlooked. In the U.S. and Germany *B. gibsoni* is usually seen only in dogs brought in from endemic areas. Transmission, however, occurs by the same ticks that transmit *B. canis*, principally *R. sanguineus* and *H. bispinosa*, so that once established, it is likely that native-born dogs will be increasingly at risk. The same treatments are effective against *B. gibsoni*; however, it is considered somewhat more tolerant to the usual babesiacides than *B. canis*.[146] Parasitemias are relatively low, and it is thought that *B. gibsoni* may not be as virulent as *B. canis*.[162] Older dogs are more often seen with clinical disease, and transplacental transmission is less likely to occur.[146] There is no cross immunity between *B. canis* and *B. gibsoni*, and protection is usually related to premunition.[146]

V. BABESIOSIS OF OTHER DOMESTIC ANIMALS

Numerous *Babesia* parasites have been observed in various vertebrates, some of which are of economic importance, but many are relatively nonpathogenic and only rarely represent a threat to animal health. Strain differences, as well as individual animal susceptibility, may alter virulence. Splenectomy almost always renders the individual more susceptible to infection.

A. *Babesia motasi* and *Babesia ovis*

Babesiosis in sheep and goats, caused by *B. motasi* (large) and *B. ovis* (small), while usually not considered highly pathogenic, is widespread throughout the world.[27] There are other *Babesia* species reported in sheep (*B. foliata, B. taylori,* and *B. crassa*) as well as the suggestion of strain differences which may influence pathogenicity.[27,163,164] While not recognized as a problem in the Western Hemisphere, sheep babesiosis is observed elsewhere in both temperate and tropical environments. Depending on the geographic location, these babesias may be transmitted by tick species of the genera *Haemaphysalis, Rhipicephalus, Ixodes,* and *Dermacentor*.[27,165,166]

On the island of Gotland, Sweden, *B. motasi* has been identified as the possible cause of anemia and reduced growth rates in spring lambs.[165] A serological survey, utilizing the

IFAT, revealed a 58 and 88% prevalence rate in lambs and adult sheep, respectively, suggesting the existence of conditions conducive to rapid and efficient transmission. *B. motasi* is more often the principal cause of sheep babesiosis in northern Europe.[165,167] It may occur as mixed infections with *B. ovis* in southern Europe and elsewhere.

B. motasi and *B. ovis* generally respond favorably to therapy with the usual babesiacidal compounds. Premunition is also an accepted method of vaccination and prevention.

B. *Babesia trautmanni* and *Babesia perroncitoi*

The babesias associated with swint babesiosis, *B. trautmanni* (large) and *B. perroncitoi* (small) may be pathogenic, but occur at lower prevalence rates than the babesias of grazing animals. *B. trautmanni* has been recorded from the U.S.S.R., southern Europe, and Africa.[168] *B. perroncitoi* was once thought to be limited to North Africa and Sardinia, but a recent report from Senegal describes an outbreak involving 800 cases.[168,169] A survey of slaughter pigs in Nigeria revealed both *B. perroncitoi* and *B. trautmanni*, with the latter showing an infection rate of 17.8% in 4- to 6-month-old pigs, and 2.2% in pigs over 9 months of age.[170,171] This work, based on blood examination, probably does not identify nonparasitemic carrier pigs. While the vectors of swine babesiosis are not proven, species of the genera *Boophilus*, *Rhipicephalus*, *Dermacentor*, and *Hyalomma* are suspected of being involved.[27,168] One report has suggested the possibility of mechanical transmission by hematophagus flies.[172] Treatment with diminazene, phemanidine, and quinuronium is generally successful if given early in the course of infection.

C. *Babesia herpailuri* and *Babesia felis*

Numerous feline babesias in both domestic and feral cats have been identified; however, the economic significance of these organisms is not well documented. *B. felis* was first identified in a Sudanese wildcat and was found to be pathogenic for domestic cats.[173] Most of the feline babesias have been isolated first from feral Felidae, some of which do not infect domestic cats. In addition to *B. felis* (small), *B. cati* (small), *B. herpailuri* (large), and *B. pantherae* (large) have been described.[174-178] *B. pantherae* and *B. herpailuri* differ in several respects. *B. herpailuri* is somewhat larger and readily infects both splenectomized and nonsplenectomized dogs, whereas *B. pantherae* is noninfective for dogs. No cross immunity exists with *B. canis* in dogs.[176]

Infections with *B. felis* in domestic cats are frequently severe, characterized by anemia and death.[179] *B. felis* is unresponsive to most of the commonly used babesiacidal drugs. Primaquine phosphate *per os* or as an intramuscular injection was found to be highly effective and the drug of choice when compared with other compounds.[180] *B. herpailuri* is successfully treated with diminazene and presumably could be treated with other babesiacides.[175]

VI. BABESIOSIS OF WILD ANIMALS

Babesias are seen in various species of wild animals. A common procedure to assist in the observation of these parasites is to splenectomize captured animals and then periodically examine their blood for the occurrence of hemoparasites.[181] The pathogenicity of these *Babesia* parasites in intact animals, under free-ranging natural conditions, has not been determined, but is probably insignificant. Available evidence suggests that these parasites, with possibly one or two exceptions, do not pose a threat for domestic livestock. One report suggested the possible transmission to sheep of *B. capreoli* from Scottish red deer in the U.K. by *I. ricinus*. The ensuing infection was mild and of low pathogenicity.[182]

Babesia odocoileus (*B. cervi*, Emerson and Wright, 1968) of white-tailed deer, morphologically similar to *B. divergens*, has been identified and described in Texas and Virginia.[183,184] While not highly pathogenic in nonsplenectomized deer, it does induce a high parasitemia and a life-threatening infection in splenectomized deer. Efforts to transmit this infection to cattle by blood inoculation was unsuccessful. Although the vector for this parasite is not known, *Amblyomma* and *Ixodes* ticks are suspected because of their presence on infected deer.[184]

Numerous other babesias from various species of wildlife have been described. Some of these are from wild rodents (*B. microti* and *B. rodhaini*), wild Felidae (*B. pantherae, B. herpailuri, B. felis, B. cati*), raccoons (*Procyon lotor*) (*B. lotori* and *B. procyoni*), skunks (*Mephitis mephitis*) (*B. mephitis*), the wild dog (*Lycaon pictus*) (*B. canis*), jackal (*Canis mesomelas*) (*B. canis*), roe deer (*Capreolus capreolus*) (*B. capreoli* and *B. jakimovi*), reindeer (*Rangifer*) (*B. jakimovi*), etc.[185-188]

VII. HUMAN BABESIOSIS

Human babesiosis in the past has been ascribed to several animal babesias, but current evidence suggests that most often it is probably caused by either *B. microti* in North America or *B. divergens* in Europe.[5] Both of these babesias occur in other vetebrate hosts which act as reservoirs of infection. In North America, *B. microti* resides in wild rodents and is transmitted by *I. dammini.*[189] Presently, human babesiosis is thought to be limited to those areas of the northeast U.S. where this tick occurs. *B. divergens* is primarily a pathogen of cattle, but may be transmitted by the tick *I. ricinus* to man where it is capable of inducing serious illness.[5] *B. divergens* can persist in *I. ricinus* and is also infective for some rodents, suggesting potential reservoirs of infections other than cattle.[27,96]

At one time, human babesiosis was thought to occur only in individuals who had experienced a splenectomy, or in those with compromised immune systems. In recent years, it has become apparent, especially in the northeast U.S., that it is no longer a medical rarity, and infection among the general population is significant. Both acute clinical disease as well as chronic nonapparent infections occur. A recent serological survey in New York indicated a 5.9% incidence in the resident population during a single tick season.[190] In these areas, it is now possible that both human and rodent reservoirs of infection occur, from which transmission is possible by the tick vector. The clinical course of babesiosis in splenectomized and immunosuppressed individuals is often severe with possible fatal consequences,[191] but in most nonsplenectomized patients, the infection can be controlled with pentamidine, clindamycin, or clindamycin plus quinine.[192,193] Most of the antimalarial drugs are ineffective, but severe side effects have been noted following the use of some animal babesiacides.[194]

It is apparent that *Babesia* spp. are numerous and that most of our domestic animals and many wild animals can acquire one or more of these hemoparasites. Many of these infections probably go unnoticed, but the potential for production and death losses are a constant concern. Techniques have been developed for the control of babesiosis, such as chemotherapy, vaccination, and vector control, but babesias still persist as significant health hazards and economic barriers to efficient livestock production.[195] The discovery and incorporation of new technology promises future control measures that will be both unique and more effective in further reducing or eliminating these problems. Comprehensive reviews on human babesiosis are available elsewhere[196] and in this book.

REFERENCES

1. **Babes, V.**, Sur l'hemoglobinurie bacterienne du boeuf, *C. R. Acad. Sci.*, 107, 692, 1888.
2. **Callow, L. L.**, *Babesia bigemina* in ticks grown on non-bovine hosts and its transmission to these hosts, *Parasitology*, 55, 375, 1965.

3. **Benson, G. D., Galdi, V. A., Altman, R., and Fiurmara, N. J.,** Morbidity and Mortality Weekly Rep. No. 18, Center for Disease Control, Public Health Service, U.S. Department of Health, Education, and Welfare, Atlanta, 1969, 277.
4. **Western, K. A., Benson, G. D., Gleason, N. N., Healy, G. R., and Schultz, M. G.,** Babesiosis in a Massachusetts resident, *N. Engl. J. Med.*, 283, 854, 1970.
5. **Brocklesby, D. W.,** Human babesiosis, *J. S. Afr. Vet. Assoc.*, 50, 302, 1979.
6. **Lewis, D., Young, E. R., Baggott, D. G., and Osborn, G. D.,** *Babesia divergens* infection of the mongolian gerbil. Titration of infective dose and preliminary observations on the disease produced, *J. Comp. Pathol.*, 91, 565, 1981.
7. **Lewis, D., Purnell, R. E., Shaw, S. R., and Revington, J. P.,** The isolation and characterization of human and bovine strains of *Babesia divergens* from Drumnadrachit, Scotland, *Parasitology*, 81, 145, 1980.
8. **Phillips, R. S.,** *Babesia divergens* in splenectomized rats, *Res. Vet. Sci.*, 36, 251, 1984.
9. **Gray, J. S.,** The effects of the piroplasm *Babesia bigemina* on the survival and reproduction of the blue tick, *Boophilus decoloratus*, *J. Inverteb. Pathol.*, 39, 413, 1982.
10. **Graham, O. H., and Hourrigan, J. L.,** Eradication programs for the arthropod parasites of livestock, *J. Med. Entomol.*, 13, 629, 1977.
11. **Starcovici, C.,** Bemerkungen uber den durch Babes entdeckten Blutparasiten und die durch denselben Hervorgerbrachten Krankheiten, die Seuchenhafte Hamoglobinurie des Rinder (Babes), das Texasfieber (Th. Smith) under der Carceag der Schafe (Babes), *Zentralbl. Bakt. Parasitkd.*, 14, 1, 1893.
12. **Smith, T.,** Preliminary observations on the microorganism of Texas fever, *Med. News Philadelphia*, 55, 689, 1889.
13. **Smith, T. and Kilborne, F. L.,** Investigations into the nature, causation and prevention of Texas or southern cattle fever, *Bull. Bur. Anim. Ind. U.S. Dep. Agric.*, 1, 1, 1893.
14. **Hoyte, H. M. D.,** The tick fever parasites of cattle, *Proc. R. Soc. Queensl.*, 87, 5, 1976.
15. **Rees, C. W.,** Characteristics of the piroplasms *Babesia argentina* and *Babesia bigemina* in the United States, *J. Agric. Res.*, 48, 427, 1934.
16. **Lignieres, J.,** La piroplasmose bovine. Nouvelles recherches et observations sur la multiplicite de parasites, leur evolution, la transmission naturelle de la maladie et vaccination, *Arch. Parasitol.*, 7, 398, 1903.
17. **Lignieres, J.,** La prophylaxie et la pathologie des maladies protozoaires (piroplasmoses, trypanosomoses, etc.) avec demonstration des parasites specifiques et des animaux transmetteurs (tigues, moustiques, etc,), *Trans. 9th Int. Vet. Congr.*, (The Hague, 1901) 1, (S.G. 7.3), 1910, 1.
18. **Brocklesby, D. W.,** Recent observations on tick-borne protozoa, in *Tick-Borne Diseases and Their Vectors*, Wilde, J. K. H., Ed., Centre for Tropical Veterinary Medicine, University of Edinburgh, 1976, 263.
19. **M'Fadyean, J. and Stockman, S.,** A new species of piroplasm found in the blood of British cattle, *J. Comp. Pathol.*, 24, 340, 1911.
20. **Simic, C. P., Petrovic, S., and Rakovec, R.,** Les especes de Babesiella du boeuf d'Europe, *Arch. Inst. Pasteur Alger.*, 33, 310, 1955.
21. **Sergent, E., Donatien, A., Parrot, L., Lestoquard, F., and Plantureux, E.,** Les piroplasmoses bovines dues aux *Babesiella* etude d'ensemble avec description d'une espece nouvelle, *B. major*, originaire de France, *Arch. Inst. Pasteur Alger.*, 4, 318, 1926.
22. **Zwart, D., Van Den Ende, M. C., Kouwenhoven, B., and Buys, J.,** The difference between *B. bigemina* and a Dutch strain of *B. major*, *Tijdschr. Diergeneeskd.* 93, 126, 1968.
23. **Brocklesby, D. W. and Barnett, S. F.,** Large *Babesia* species transmitted to splenectomised calves by field collections of British ticks (*Haemaphysalis punctata*), *Nature (London)*, 228, 1215, 1970.
24. **Leeflang, P. and Perie, N. M.,** Comparative immunofluorescent studies on 4 *Babesia* species of cattle, *Res. Vet. Sci.*, 13, 342, 1972.
25. **Goldman, M. and Rosenberg, A. S.,** Immunofluorescence studies of the small *Babesia* species of cattle from different geographical regions, *Res. Vet. Sci.*, 16, 341, 1974.
26. **Minami, T., Yamabe, K., Hayashi, S., and Ishihara, T.,** Serological relationship of a Japanese *Babesia* species and *Babesia bigemina* by the complement fixation and capillary-tube agglutination tests, *Vet. Parasitol.*, 5, 29, 1979.
27. **Purnell, R. E.,** Babesiosis in various hosts, in *Babesiosis*, Ristic, M. and Kreier, J. P., Eds., Academic Press, New York, 1981, 25.
28. **Gray, J. S. and De Vos, A. J.,** Studies on a bovine *Babesia* transmitted by *Hyalomma marginatum rufipes*, Koch, 1844, *Onderstepoort J. Vet. Res.*, 48, 215, 1981.
29. **Dolman, C. E.,** Texas cattle fever, a commemorative tribute to Theobald Smith, *Clio Med.*, 4, 1, 1969.
30. **Seddon, H. R.,** Diseases of Domestic Animals in Australia. IV. Protozoan and Viral Diseases, Service Publ. No. 4, Commonwealth Department of Health, Canberra, Australia, 1966; as cited by **McCosker, P. J.,** *Babesiosis*, Ristic, M. and Kreier, J. P., Eds., Academic Press, New York, 1981, 1.
31. **Payne, R. C. and Scott, J. M.,** Anaplasmosis and babesiosis in El Salvador, *Trop. Anim. Health Proc.*, 14, 75, 1982.

32. **Thompson, G. D., Medellin, J. A., Trevino, G. S., and Wagner, G. G.,** Bovine babesiosis in northern Mexico, *Trop. Anim. Health Prod.*, 12, 132, 1980.
33. **Applewhaite, L. M., Craig, T. M., and Wagner, G. G.,** Serological prevalence of bovine babesiosis in Guyana, *Trop. Anim. Health Prod.*, 13, 13, 1981.
34. **Banerjee, D. P., Prasad, K. D., and Samad, M. A.,** Seroprevalence of *Babesia bigemina* infection in cattle of India and Bangladesh, *Indian J. Anim. Sci.*, 53, 431, 1983.
35. **Knowles, R. T., Montrose, M., Craig, T. M., Wagner, G. G., and Long, R. F.,** Clinical and serological evidence of bovine babesiosis and anaplasmosis in St. Lucia, *Vet. Parasitol.*, 10, 307, 1982.
36. **Norval, R. A. I.,** Tick infestations and tick-borne diseases in Zimbabwe, Rhodesia, *J. S. Afr. Vet. Med. Assoc.*, 50, 289, 1979.
37. **Miller, D. K., Diall, O., Craig, T. M., and Wagner, C. G.,** Serological prevalence of bovine babesiosis in Mali, *Trop. Anim. Health Prod.*, 16, 71, 1984.
38. **Xu, P. R.,** An investigation on an outbreak of babesiosis in dairy cattle at an animal farm in the suburb of Guangzhou, *Chin. J. Vet. Med.*, 7, 6, 1981.
39. **Norval, R. A. I., Fivax, B. H., Lawrence, J. A., and Daillecourt, T.,** Epidemiology of tick-borne diseases of cattle in Zimbabwe. I. Babesiosis, *Trop. Anim. Health. Prod.*, 15, 87, 1983.
40. **De Vos, A. J.,** Epidemiology and control of bovine babesiosis in South Africa, *J. S., Afr. Vet. Assoc.*, 50, 357, 1979.
41. **Potgieter, F. T.,** The Life Cycle of *Babesia bovis* and *Babesia bigemina* in Ticks and in Cattle in South Africa, Ph.D. thesis, Rand Afrikaans University, Johannesburg, South Africa, 1977.
42. **Akinboade, O. A. and Dipeolu, O. O.,** Bovine babesiosis in Nigeria, the vectorial capability of *Boophilus decoloratus* and *Boophilus geigyi* for *Babesia bigemina* and *Babesia bovis, Acarologia*, 26, 235, 1985.
43. **Mahoney, D. F. and Mirre, G. B.,** A note on the transmission of *Babesia bovis* (syn. *B. argentina*) by the one-host tick *Boophilus microplus, Res. Vet. Sci.*, 26, 253, 1979.
44. **Callow, L. L.,** Some aspects of the epidemiology and control of bovine babesiosis in Australia, *J. S. Afr. Vet. Assoc.*, 50, 353, 1979.
45. **Dalgliesh, R. J. and Stewart, N. P.,** The use of tick transmission by *Boophilus microplus* to isolate pure strains of *Babesia bovis, Babesia bigemina* and *Anaplasma marginale* from cattle with mixed infections, *Vet. Parasitol.*, 13, 317, 1983.
46. **Kuttler, K. L., Graham, O. H., and Trevino, J. L.,** The effect of imidocarb treatment on *Babesia* in the bovine and the tick (*Boophilus microplus*), *Res. Vet. Sci.*, 18, 198, 1975.
47. **Gray, J. S. and Potgieter, F. T.,** The retention of *Babesia bigemina* infection by *Boophilus decoloratus* exposed to imidocarb dipropionate during engorgement, *Onderstepoort J. Vet. Res.*, 48, 225, 1981.
48. **McCosker, P. J.,** The global importance of babesiosis, in *Babesiosis*, Ristic, M. and Kreier, J. P., Eds., Academic Press, New York, 1981, 1.
49. **Park, R. L., Skov, O., Seaman, G. A., and Bond, R. M.,** Deer and cattle fever ticks, *J. Wildl. Manage.*, 30, 202, 1966.
50. **Shillinger, J. E.,** Deer in relation to fever tick eradication in Florida, Texas, *Trans. North Am. Wildl. Conf.*, 3, 882, 1938.
51. **Kuttler, K. L. and McWhorter, G. M.,** Review of Cattle Fever Ticks and Diseases They Transmit in New South Wales and Queensland, Australia, TAES, Depart. Info. Rep. No. 77-1, ITVM, 1977.
52. **Dalgliesh, R. J., Callow, L. L., Mellors, L. T., and McGregor, W.,** Development of a highly infective *Babesia bigemina* vaccine of reduced virulence, *Aust. Vet. J.*, 57, 8, 1981.
53. **Callow, L. L., Mellors, L. T., and McGregor, W.,** Reduction in virulence of *Babesia bovis* due to rapid passage in splenectomized cattle, *Int. J. Parasitol.*, 9, 333, 1979.
54. **De Vos, A. J., Bessenger, R., and Fourie, C. G.,** Virulence and heterologous strain immunity of South African and Australian *Babesia bovis* strains with reduced pathogenicity, *Onderstepoort J. Vet. Res.*, 49, 133, 1982.
55. **De Vos, A. J., Combrink, M. P., and Bessenger, R.,** *Babesia bigemina* vaccine: comparison of the efficacy and safety of Australian and South African strains under experimental conditions in South Africa, *Onderstepoort J. Vet. Res.*, 49, 155, 1982.
56. **Callow, L. L., Kanhai, G. K., and Vandenberghe, A.,** Serological comparisons of strains of *Babesia bovis* occurring in Australia and Mozambique, *Trop. Anim. Health Prod.*, 13, 79, 1981.
57. **Johnston, L. A. Y., Haydock, K. P., and Leatch, G.,** The effect of two systems of cattle tick (*Boophilus microplus*) control on tick populations, transmission of *Babesia spp.* and *Anaplasma spp.* and production of Brahman cross-bred cattle in the dry tropics, *Aust. J. Exp. Agric. Anim. Husb.*, 21, 256, 1981.
58. **De Vos, A. J. and Potgieter, F. T.,** The effect of tick control on the epidemiology of bovine babesiosis, *Onderstepoort J. Vet. Res.*, 50, 3, 1983.
59. **Smith, R. D.,** *Babesia bovis*, computer simulation of the relationship between the tick vector, parasite, and bovine host, *Exp. Parasitol.*, 56, 27, 1983.
60. **Mahoney, D. F. and Ross, D. R.,** Epizootiological factors in the control of bovine babesiosis, *Aust. Vet. J.*, 48, 292, 1972.

61. **Callow, L. L. and McGregor, W.,** Vaccination against *Babesia argentina* infections in cattle during chemoprophylaxis with a quinuronium compound, *Aust. Vet. J.*, 45, 408, 1969.
62. **Kuttler, K. L. and Aliu, Y. O.,** Chemotherapy of babesiosis, in *Malaria and Babesiosis*, Ristic, M., Ambroise-Thomas, P., and Kreier, J. P., Eds., Martinus Nijhoff, Dordrecht, Netherlands 1984, 151.
63. **Aliu, Y. O.,** Chemoimmunization of ruminants against *Babesia, Anaplasma, Theileria* and *Cowdria* species, *Vet. Res. Commun.*, 4, 99, 1980.
64. **Kuttler, K. L. and Johnson, L. W.,** Chemoprophylactic activity of imidocarb diminazene and oxytetracycline against *Babesia bovis* and *B. bigemina*, *Vet. Parasitol.*, 21, 107, 1986.
65. **Day, W. C. and Kuttler, K. L.,** Animal health considerations involved in the movement of U.S. cattle to Haiti, *Southwest. Vet.*, 28, 229, 1975.
66. **Smith, R. D., Carpenter, J., Cabrera, A., Gravely, S. M., Erp, E. E., Osorno, M., and Ristic, M.,** Bovine babesiosis: vaccination against tick-borne challenge exposure with culture-derived *Babesia bovis* immunogen, *Am. J. Vet. Res.*, 40, 1678, 1979.
67. **Goodger, B. V., Commins, M. A., Wright, I. G., and Mirre, G. B.,** *Babesia bovis*: vaccinations of cattle against heterologous challenge with fractions of lysate from infected erythrocytes, *Z. Parasitenkd.*, 70, 321, 1984.
68. **Mahoney, D. F., Wright, I. G., and Goodger, B. V.,** Immunization against babesiosis: current studies and future outlook, *Prev. Vet. Med.*, 2, 401, 1984.
69. **Kuttler, K. L., Levy, M. G., and Ristic, M.,** Cell culture-derived *Babesia bovis* vaccine sequential challenge exposure of protective immunity during a 6 month post vaccination period, *Am. J. Vet. Res.*, 44, 1456, 1983.
70. **Muraleedharan, K., Ziauddin, K. S., Gopalaswamy, K., Muraleedhar, T., and Seshadri, S. J.,** Some observations on clinical cases of *Babesia bovis* (Babes 1888) Starcovici, 1893, in Buffaloes (*Bubalus bubalis*), *Indian Vet. J.*, 61, 76, 1984.
71. **Karbe, E., Gootenhuis, J. G., Kelley, S., and Karstad, L.,** Experiments on the *Babesia bigemina* carrier state in East African buffalo and Eland, *Tropenmed. Parasitol.*, 30, 313, 1979.
72. **Zaugg, J. L.,** *Babesia bigemina* Infections in American Bison, personal communication, 1986.
73. **Kuttler, K. L., Graham, O. H., Johnson, S. R., and Trevino, J. L.,** Unsuccessful attempts to establish cattle *Babesia* infections in white tailed deer, *J. Wildl. Dis.*, 8, 63, 1972.
74. **Gray, J. S., Fitzgerald, E., and Stickland, K. L.,** Prevalence of clinical babesiosis in an area in North Co., Meath, Ireland, *Vet. Rec.*, 113, 537, 1983.
75. **Gray, J. S. and Harte, L. N.,** An estimation of the prevalence and economic importance of clinical bovine babesiosis in the Republic of Ireland, *Ir. Vet. J.*, 39, 75, 1985.
76. **Taylor, S. M.,** Assessment of prevalence of clinical babesiosis in cattle in Northern Ireland, *Vet. Rec.*, 112, 247, 1983.
77. **Barnett, S. F.,** Economical aspects of tick-borne disease control in Britain, *Bull. Off. Int. Epizoot.*, 81, 167, 1974.
78. **Moen, J. H.,** A case of piroplasmosis in housed cattle, *Nor. Vet.*, 92, 736, 1980.
79. **Schreuder, B. E. C., and Van Wijk, C.,** An outbreak of babesiosis in Eastern Flevoland, *Tijdschr. Diergeneeskd.*, 109, 549, 1984.
80. **Ullmann, B., Centurier, C., Weiland, G., and Boch, J.,** A contribution to the study of bovine babesiosis in Western Allgau, *Berl. Muench. Tierarztl. Wochenschr.*, 97, 265, 1984.
81. **Morel, P. C.,** Reflections on the epizootiology of bovine babesiasis in Europe due to *Babesia divergens*, *Bull. Acad. Vet. Fr.*, 53, 451, 1980.
82. **Hinaidy, H. K.,** Babesiasis of cattle in Austria. I. Distribution and prevalence, *Wien. Tieraerztl. Monatsschr.*, 68, 52, 1981.
83. **Hope-Cawdery, M. J.,** Some field observations on the use of cypermethrin to control bovine babesiosis, *Ir. Vet. J.*, 39, 174, 1985.
84. **Haigh, A. J. B. and Hagan, D. H.,** Evaluation of Imidocarb dihydrochloride against redwater disease in cattle in Eire, *Vet. Rec.*, 94, 56, 1974.
85. **Taylor, S. M., Kenny, J., Purnell, R. E., and Lewis, D.,** Exposure of cattle immunised against redwater to tick induced challenge in the field: challenge by a heterologous strain of *Babesia Divergens*, *Vet. Rec.*, 106, 385, 1980.
86. **Brocklesby, D. W., Harness, E., and Sellwood, S. A.,** The effect of age on the natural immunity of cattle to *Babesia divergens*, *Res. Vet. Sci.*, 12, 15, 1971.
87. **Weiland, G., Reif, L., Schmidt, M., and Boch, J.,** Serologische untersuchungen zum Nachweis der *Babesia divergens* Infektion des Rindes, *Berl. Muench, Tierarztl. Wochenschr.*, 93, 261, 1980.
88. **Adam, K. M. G. and Blewett, D. A.,** A serological survey for *Babesia* in cattle in Scotland. II. The occurrence of antibody in the population and the distribution of infected herds, *Ann. Trop. Med. Parasitol.*, 72, 417, 1978.
89. **Purnell, R. E. and Hendry, D. J.,** Microplate enzyme linked immunosorbent assay for antibody to *Babesia divergens* in cattle, *Vet. Rec.*, 99, 102, 1976.

90. **Purnell, R. E., Lewis, D., Brabazon, A., Francis, L. M. A., Young, E. R., and Grist, C.,** Field use of an irradiated blood vaccine to protect cattle against redwater (*Babesia divergens* infection) on a farm in Dorset, *Vet. Rec.*, 108, 28, 1981.
91. **Taylor, S. M., Kenny, J., and Mallon, T.,** The effect of route of administration of a *Babesia divergens* inactivated vaccine on protection against homologous challenge, *J. Comp. Pathol.*, 93, 423, 1983.
92. **Purnell, R. E., Brocklesby, D. W., and Stark, A. J.,** Protection of cattle against *Babesia major* by the inoculation of irradiated piroplasma, *Res. Vet. Sci.*, 25, 388, 1978.
93. **Lewis, D., Purnell, R. E., and Brocklesby, D. W.,** *Babesia divergens:* the immunisation of splenectomised calves using irradiated piroplasms, *Res. Vet. Sci.*, 26, 220, 1979.
94. **Purnell, R. E., Lewis, D., Brocklesby, D. W., and Taylor, S. M.,** Bovine babesiosis steps toward an irradiated vaccine, *J. S. Afr. Vet. Assoc.*, 50, 339. 1979.
95. **Taylor, S. M., Kenny, J., and Mallons, T.,** The effect of multiple rapid passage on strains, of *Babesia divergens:* a comparison of the clinical effects on juvenile and adult cattle of passaged and irradiated parasites, *J. Comp. Pathol.*, 93, 391, 1983.
96. **Lewis, D. and Williams, H.,** Infection of the mongolian gerbil with the cattle piroplasm *Babesia divergens*, *Nature (London)*, 278, 170, 1979.
97. **Broaklesby, D. W., and Irvin, A. D.,** The possibility that a large *Babesia* species exists in British cattle, *Vet. Rec.*, 84, 430, 1969.
98. **Brocklesby, D. W. and Sellwood, S. A.,** *Babesia major* in Britain: tick transmitted infections in splenectomised calves, *Res. Vet. Sci.*, 14, 47, 1973.
99. **Joyner, L. P., Donnelly, J., and Payne, R.,** The indirect fluorescent antibody test for the differentiation of infections with *Babesia divergens* and *Babesia major*, *Res. Vet. Sci.*, 13, 515, 1972.
100. **Findlay, C. R. and Begg, T. B.,** Red water in American bison caused by *Babesia major*, *Vet. Rec.*, 100, 406, 1977.
101. **Ishihara, T.,** Bovine piroplasmosis in Japan, *JARQ*, 3, 23, 1968.
102. **Minami, T. and Ishihara, T.,** *Babesia ovata* sp. n. isolated from cattle in Japan, *Natl. Inst. Anim. Health Q.*, 20, 101, 1980.
103. **Ishihara, T. and Minami, T.,** Bovine theileriosis and babesiasis in Japan, *Protozoan Dis. (JGAPD)*, 1, 201, 1977.
104. **Thomas, S. E. and Mason, T. E.,** Isolation and transmission of an unidentified *Babesia* sp. infective for cattle, *Onderstepoort J. Vet. Res.*, 48, 155, 1981.
105. **Nikolskii, S. N., Nikiforenko, V. I., and Posov, S. A.,** Epidemiology of piroplasmosis in Siberia, *Veterinariya (Moscow)*, 4, 71, 1977; as quoted by **Purnell, R. E.,** in *Babesiosis*, Ristic, M. and Kreier, J. P., Eds., Academic Press, New York, 1981, 25.
106. **Henning, M. W.,** *Animal Diseases in South Africa*, 3rd ed., Central News Agency Ltd., South Africa, 1956, 643.
107. **Theiler, A.,** Die pferde malaria, *Schwei. Arch. Tierhkd.*, 43, 253, 1901; as quoted by **Henning, M. W.,** *Animal Diseases in South Africa*, Central News Agency Ltd., South Africa, 1956, 643.
108. **Theiler, A.,** Equine malaria, *J. Comp. Pathol.*, 15, 40, 1902; as quoted by **Henning, M. W.,** *Animal Diseases in South Africa*, Central News Agency Ltd., South Africa, 1956, 643.
109. **Laveran, A.,** Contribution a'l'etude du Piroplasma equi, *C. R. Soc. Biol.*, 53, 385, 1901.
110. **Theiler, A.,** Notes on piroplasmosis in the horse, mule and donkey, *Rep. Gov. Vet. Bacteriol. Transv.*, 1903—1904, 171, 1905.
111. **Theiler, A.,** Further notes on piroplasmosis of the horse, mule and donkey, *Rep. Gov. Vet. Bacteriol., Transv.*, 1904—1905, 94, 1906.
112. **Dennig, H. K.,** The isolation of *Babesia* species from wild animals, *Proc.* 1st *Int. Congr. Parasitol. Rome*, Corradetti, A., Ed., *Pergamon Press*, London, 1966, 262.
113. **Young, A. S. and Purnell, R. E.,** Observations on *Babesia equi* in the salivary glands of *Rhipicephalus evertsi*, *Bull. Epizoot. Dis. Afr.*, 21, 377, 1973.
114. **Hutyra, F., Marek, J., and Manninger, R.,** in *Special Pathology and Therapeutics of the Diseases of Domestic Animals*, 4th ed., Greg, J. R., Mohler, J. R., and Eichhorn, A., Eds., Alexander Eger, Chicago, 1938, 858.
115. **Holbrook, A. A.,** Biology of equine piroplasmosis, *JAVMA*, 155, 453, 1969.
116. **Friedhoff, K. T.,** The piroplasms of equidae, significance for International Commerce, *Berl. Muench Tierarztl. Wochenschr.* 95, 368, 1982.
117. **Thompson, P. H.,** Ticks as vectors of equine piroplasmosis, *JAVMA*, 155, 454, 1969.
118. **Roby, T. O., Anthony, D. W., Thornton, C. W., and Holbrook, A. A.,** The heredity transmission of *Babesia caballi* in the tropical horse tick, *Dermacentor nitens*, Neumann, *Am. J. Vet. Res.*, 25, 494, 1964.
119. **Stiller, D. and Frerichs, W. M.,** Experimental transmission of *Babesia caballi* to equids by different stages of the tropical horse tick, *Anocentor nitens*, *Rec. Adv. Acarol.*, 2, 263, 1979.
120. **Schein, E., Rehbein, G., Voight, W. P., and Zweygarth, E.,** *Babesia equi* (Laveran 1901). I. Development in horses and in lymphocyte culture, *Tropenmed. Parasitol.*, 32, 223, 1981.

121. **Frerichs, W. M., Johnson, A. J., and Holbrook, A. A.,** Equine piroplasmosis: attempts to infect laboratory animals with *Babesia equi*, *Am. J. Vet. Res.*, 30, 1333, 1969.
122. **Malhotra, D. V., Gautam, O. P., and Banerjee, D. P.,** Attempts to transmit *Babesia equi* in laboratory animals, *Trop. Anim. Health Prod.*, 11, 95, 1979.
123. **Holbrook, A. A. and Frerichs, W. M.,** Equine Piroplasmosis, *Proc. 72nd Annu. Meet. USAHA*, 1968, 220.
124. **Joyner, L. P., Donnelly, J., and Huck, R. A.,** Complement-fixation tests for equine piroplasmosis (*B. caballi and B. equi*) performed in the U.K. during 1976 to 1979. *Equine Vet. J.*, 13, 103, 1981.
125. **Donnelly, J., Phipps, L. P., and Watkins, K. L.,** Evidence of maternal antibodies to *Babesia equi* and *B. caballi* in foals of seropositive mares, *Equine Vet. J.*, 14, 126, 1982.
126. **Littlejohn, A. and Walker, E. M.,** Some aspects of the epidemiology of equine babesiosis, *J., S. Afr. Vet. Assoc.*, 50, 308, 1979.
127. **Mahoney, D. F., Wright, I. G., Frerichs, W. M., Groenendyk, S., O'Sullivan, B. M., Roberts, M. C., and Waddell, A. H.,** The identification of *Babesia equi* in Australia, *Aust. Vet. J.*, 53, 461, 1977.
128. **Kuttler, K. L., Gipson, C. A., Goff, W. L., and Johnson, L. W.,** Experimental *Babesia equi* infection in mature horses, *Am. J. Vet. Res.*, 47, 1668, 1986.
129. **Callow, L. L., McGregor, W., Rodwell, B. J., Rogers, R. J., Fraser, G. C., Mahoney, D. F., and Robertson, G. M.,** Evaluation of an indirect fluorescent antibody test to diagnose *Babesia equi* infection in horses, *Aust. Vet. J.*, 55, 555, 1979.
130. **Sippel, W. L., Cooperrider, D. E., Gainer, J. H., Allen, R. W., Mouw, J. E. B., and Teigland, M. B.,** Equine piroplasmosis in the United States, *JAVMA*, 141, 694, 1962.
131. **Roby, T. O. and Anthony, D. W.,** Transmission of equine piroplasmosis by *Dermacentor nitens* Neumann, *JAVMA*, 142, 768, 1963.
132. **Bryant, J. E., Anderson, J. B., and Willers, K. H.,** Control of equine piroplasmosis in Florida, *JAVMA*, 154, 1034, 1969.
133. **Stiller, D., Frerichs, W. M., Leatch, G., and Kuttler, K. L.,** Transmission of equine babesiosis and *Bovine anaplasmosis* by *Dermacentor albipictus*, (Packard) (Acari: Ixodidae), *J. N.Y. Entomol. Soc.*, 88, 75, 1980.
134. **Stiller, D., Coan, M. E., and Goff, W. L.,** Experimental transmission of equine piroplasmosis by the American dog tick, *Dermacentor variabilis* (SAY), *Abstr. 5th Int. Congr. Parasitology in Molecular Biochemistry, Parasitol.*, Suppl., Muller, M., Gutteridge, W., and Kohler, P., Eds., Elsevier, Amsterdam, 1982, 273.
135. **Knowles, R. C., Mathis, R. M., Bryant, J. E., and Willers, K. H.,** Equine piroplasmosis, *JAVMA*, 148, 407, 1966.
136. **Frerichs, W. M., Holbrook, A. A., and Johnson, A. J.,** Equine piroplasmosis: complement-fixation titers of horses infected with *Babesia caballi*, *Am. J. Vet. Res.*, 30, 697, 1969.
137. **Frerichs, W. M., Holbrook, A. A., and Johnson, A. J.,** Equine piroplasmosis: production of antigens for the complement-fixation test, *Am. J. Vet. Res.*, 30, 1337, 1969.
138. **Holbrook, A. A., Cooperrider, D. E., Roby, T. O., Kirkham, W. W., and Martin, W. H.,** Protocol for the complement-fixation test for equine piroplasmosis, *Proc. 75th Annu. Meet. USAHA*, 1971, 617.
139. **Todorovic, R. A.,** Serological Diagnosis of Babesiosis: a review, *Trop. Anim. Health Prod.*, 7, 1, 1975.
140. **Sibinovic, S., Sibinovic, K. H., and Ristic, M.,** Equine babesiosis: diagnosis by bentonite, agglutination and passive hemagglutination tests, *Am. J. Vet. Res.*, 30, 691, 1969.
141. **Singh, B., Banerjee, D. P., Gautam, O. P., and Malhotra, D. V.,** A capillary tube agglutination test for equine babesiosis, *Indian Vet. J.*, 55, 677, 1978.
142. **Weiland, V. G., Aicher, B. M., and Boch, J.,** Serodiagnosis and therapy evaluation of equine piroplasmosis with the complement-fixation test and the indirect fluorescent antibody test, *Berl. Muench. Tierarztl. Wochenschr.*, 96, 341, 1984.
143. **Donnelly, J., Joyner, L. P., and Frank, C.,** Quantitive epidemiological studies on the prevalence of babesiosis in horses in Kuwait, *Trop. Anim. Health Prod.*, 12, 253, 1980.
144. **Donnelly, J., Joyner, L. P., Graham-Jones, O., and Ellis, C. P.,** A comparison of the complement-fixation and immunofluorescent antibody tests in a survey of the prevalence of *Babesia equi* and *Babesia caballi* in horses in the Sultanate of Oman, *Trop. Anim. Health Prod.*, 12, 50, 1980.
145. **Singh, B., Gautam, O. P., and Banerjee, D. P.,** Immunization of donkeys against *Babesia equi* infection using killed vaccine, *Vet. Parasitol.*, 8, 133, 1981.
146. **Farwell, G. E., LeGrand, E. K., and Cobb, C. C.,** Clinical observations on *Babesia gibsoni* and *Babesia canis* infection in dogs, *JAVMA*, 180, 507, 1982.
147. **Eaton, P.,** *Piroplasma canis* in Florida, *J. Parasitol.*, 20, 312, 1934.
148. **Dennig, H. K., Centurier, C., Gobel, F., and Weiland, G.,** Ein Beitrag zur *Babesiose* des hundes und ihrer Bedeutung in der Bundesrepublik Deutschland und Berlin-West, *Berl. Muench. Tierarztl. Wochenschr.*, 93, 373, 1980.

149. **Liebisch, V. A. and Gillani, S.,** Experimentelle Ubertragung der hundebabesiose (*Babesia canis*) durch Einheimische deutsche zeckenarten. I. Die braune Hundezecke *(Rhipicephalus sanguineus), Dtsch. Tier-aerztl. Wochenschr.*, 86, 149, 1979.
150. **Breitschwerdt, E. B., Malone, J. B., MacWilliams, P., Levy, M. G., Qualls, W., and Prudich, M. J.,** Babesiosis in the greyhound, *JAVMA*, 182, 978, 1983.
151. **Uilenberg, G., Top, P. D. J., Arends, P. J., Kool, P. J., Van Dijk, J. E., van Schieveen, P. B. E., and Zwart, D.,** Indigenous canine babesiosis in the Netherlands, *Tijdschr. Diergeneeskd.*, 110, 93, 1985.
152. **Van Heerden, J.,** The transmission of *Babesia canis* to the wild dog *Lycaon pictus* (Temminck) and black-backed jackal *Canis mesomelas* (Schreber), *J. S. Afr. Vet. Assoc.*, 51, 119, 1980.
153. **Roher, D. P., Anderson, J. F., and Nielsen, S. W.,** Experimental babesiosis in coyotes and coydogs, *Am. J. Vet. Res.*, 46, 256, 1985.
154. **Grogan, J. W.,** Piroplasmosis in a dog, *JAVMA*, 123, 234, 1953.
155. **Bansal, S. R. B. and Gautam, O. P.,** Reports of clinical cases of *Babesia canis* disease among the exotic breeds of dogs, *Indian J. Vet. Med.*, 1, 87, 1981.
156. **Stewart, C. G.,** A comparison of the efficacy of isometamidium, amicarbalide, and diminazene against *Babesia canis* in dogs and the effect on subsequent immunity, *J. S. Afr. Vet. Assoc.*, 54, 47, 1983.
157. **Klinefelter, M. R.,** Cause, diagnosis and treatment of canine piroplasmosis, *Vet. Med. Small Anim. Clin.*, October, 1982.
158. **Molinar, E., James, M. A., Kakoma, I., Holland, C., and Ristic, M.,** Antigenic and immunogenic studies on cell culture-derived *Babesia canis, Vet. Parasitol.*, 10, 29, 1982.
159. **Schindler, R., Schroder, G., Stieger, R., Wirahadiredja, S., and Kessler, W.,** Weitere Untersuchungen uber immunitat und serologische Reaktion bei hunder mach einer, Infektion unit *Babesia canis, Z. Tropenmed. Parasitol.*, 21, 182, 1970.
160. **Groves, M. G. and Dennis, G. L.,** *Babesia gibsoni,* field and laboratory studies of canine infections, *Exp. Parasitol.*, 31, 153, 1972.
161. **Anderson, J. F. and Magnarelli, L. A.,** Canine *Babesia* new to North America, *Science*, 204, 1431, 1979.
162. **Atilola, M. A. O. and Dipeolu, O. O.,** Studies on the degree of parasitemia and the morphological forms of *Babesia* during crisis in splenectomized dogs, *Zentralbl. Vet. Med. B.*, 30, 322, 1983.
163. **Hashemi-Fesharki, R. and Uilenberg, G.,** *Babesia crasia* n. sp (Sporozoa, Babesidae) of domestic sheep in Iran, *Vet. Q.*, 3, 1, 1981.
164. **Uilenberg, G., Rombach, N. C., Perie, N. M., and Zwart, D.,** Blood parasites of sheep in the Netherlands. II. *Babesia motasi* (Sporozoa, Babesidae), *Vet. Q.*, 2, 3, 1980.
165. **Christensson, D. and Thunegard, E.,** *Babesia motasi* in sheep on the Island of Gotland in Sweden, Vet. Parasitol., 9, 99, 1981.
166. **Moltmann, U. G., Mehlhorn, H., and Friedhoff, K. T.,** Ultrastructural study on the development of *Babesia ovis* (*Piroplasmia*) in the ovary of the vector tick *Rhipicephalus bursa, J. Protozool.*, 29, 30, 1982.
167. **Lewis, D., Holman, M. R., Purnell, R. E., and Young, E. R.,** Investigations on *Babesia Motasi* from Wales, *Res. Vet. Sci.*, 31, 239, 1981.
168. **Riek, R. F.,** Babesiosis, in *Infectious Blood Diseases of Man and Animals,* Vol. 2, Weinman, D. and Ristic, M., Eds., Academic Press, New York, 1968, 219.
169. **Vercruysse, J. and Parent, R.,** Observations on an epizootic of porcine babesiasis due to *Babesia perroncitoi* in Senegal, *Ann. Soc. Belge. Med. Trop.*, 61, 125, 1981.
170. **Dipeolu, O. O., Najaro, O. M., Akinboade, O. A., and Nwufor, K. J.,** Studies on the blood parasites of pigs in Ibadan, Nigeria, *Vet. Parasitol.*, 10, 87, 1982.
171. **Okon, E. D.,** Blood parasites of local pigs in Ibaden, *Trop. Anim. Health Prod.*, 8, 96, 1976.
172. **Dipeolu, O. O., Otesile, E. B., Adetunji, A., and Fagbemi, B. O.,** Studies on blood parasites of pigs in Nigeria: pathogenicity of *Babesia trautmanni* in experimentally infected pigs, *Zentralbl. Vet. Med.*, 30, 97, 1983.
173. **Davis, L. J.,** On a piroplasm of the Sudanese wildcat *(Felis ocreata), Trans. R. Soc. Trop. Med. Hyg.*, 22, 523, 1929.
174. **Futter, G. J. and Belonje, P. C.,** Studies on feline babesiosis. I. Historical review, *J. S. Afr. Vet. Assoc.*, 50, 105, 1980.
175. **Stewart, C. G., Hackett, K. J. W., and Collett, M. G.,** An unidentified *Babesia* of the domestic cat *(Felis domesticus), J. S. Afr. Vet. Assoc.*, 51, 219, 1980.
176. **Dennig, H. K. and Brocklesby, D. W.,** *Babesia pantherae* sp. nov., a piroplasm of the leopard *(Panthera pardus), Parasitology,* 64, 525, 1972.
177. **Mudaliar, S. V., Achary, G. R., and Alwar, V. S.,** On a species of *Babesia* in an Indian wildcat *(Felis catus), Indian Vet. J.*, 26, 391, 1950.
178. **Dennig, H. K.,** Eine unbekannte Babesienart bein Jaguarundi *(Herpailurus yaguarundi), Die Kleintier prax.*, 12, 146, 1967.

179. **Futter, G. J., Belonje, P. C., and Van Den Berg, A.,** Studies of feline babesiosis. III. Hematological findings, *J. S. Afr. Vet. Assoc.*, 51, 271, 1980.
180. **Potgieter, F. T.,** Chemotherapy of *Babesia felis* infection: efficacy of certain drugs, *J. S. Afr. Vet. Assoc.*, 52, 289, 1981.
181. **Kuttler, K. L., Robinson, R. M., and Rogers, W. P.,** Exacerbation of latent erythrocytic infections in deer following spelectomy, *Can. J. Comp. Med. Vet. Sci.*, 31, 317, 1967.
182. **Purnell, R. E., Lewis, D., Holman, M. R., and Young, E. R.,** Investigations on a *Babesia* isolated from Scottish sheep, *Parasitology*, 83, 347, 1981.
183. **Emerson, H. R. and Wright, W. T.,** The isolation of a *Babesia* in white-tailed deer, *Bull. Wildl. Dis. Assoc.*, 4, 142, 1968.
184. **Perry, B. D., Nichols, D. K., and Cullom, E. S.,** *Babesia odocoilei* Emerson and Wright, 1970 in white-tailed deer, *Odocoileus virginianus* (Zimmermann), in Virginia, *J. Wildl. Dis.*, 21, 149, 1985.
185. **Frerichs, W. M. and Holbrook, A. A.,** *Babesia* spp. and *Haemobartonella* sp. in wild mammals trapped at the Agricultural Research Center, Beltsville, Maryland, *J. Parasitol.*, 56, 130, 1970.
186. **Mudaliar, S. V., Achary, G. R., and Alwar, V. S.,** On a species of *Babesia* in an Indian wildcat *(Felis catus)*, *Indian Vet. J.*, 26, 391, 1950.
187. **Holbrook. A. A. and Frerichs, W. M.,** *Babesia mephitis* sp. n. (Protozoa: Piroplasmida), a hematozoan parasite of the striped skunk, *(Mephitis mephitis)*, *J. Parasitol.*, 56, 930, 1970.
188. **Anderson, J. F., Magnarelli, L. A., and Sulzer, A. J.,** Raccoon babesiosis in Connecticut, USA, *Babesia lotori*, sp. n., *J. Parasitol.*, 67, 417, 1981.
189. **Spielman, A., Clifford, C. M., Piesman, J., and Corwin, M. D.,** Human babesiosis on Nantucket Island, USA: description of the vector, *Ixodes* (*Ixodes*) *dammini*, N. sp. (Acarina: Ixodidae), *J. Med. Entomol.*, 15, 218, 1979.
190. **Filstein, M. R., Benach, J. L., White, D. J., Brody, B. A., Goldman, W. D., Bakal, C. W., and Schwartz, R. S.,** Serosurvey for human babesiosis in New York, *J. Infect. Dis.*, 141, 518, 1980.
191. **Zaino, E. C. and Amelkin, S.,** Babesiosis, a potentially grave disease for the immunosuppressed, N.Y. State, *J. Med.*, 81, 384, 1981.
192. **Francioli, P. B., Keithly, J. S., Jones, T. C., Brandstetter, R. D., and Wolf, D. J.,** Response of babesiosis to pentamidine therapy, *Ann. Intern. Med.*, 94, 326, 1981.
193. **Rowin, K. S., Tanowitz, H. B., and Wittner, M.,** Therapy of experimental babesiosis, *Ann. Intern. Med.*, 97, 556, 1982.
194. **Miller, L. H., Neva, F. A., and Gill, F.,** Failure of chloroquine in human babesiosis *(Babesia microti)*, *Ann. Intern. Med.*, 88, 200, 1978.
195. **Todorovic, R. A.,** Advances in control of bovine babesiosis (Texas Fever), in *Beef Cattle Science Handbook*, Baker, F. H. and Miller, M. E., Eds., Westview Press, Boulder, Colo. 1984, 510.
196. **Ristic, M. and Healy, G. R.,** Babesiosis, in *Handbook Series in Zoonoses*, Steele, J. H., Ed., CRC Press, Boca Raton, Fla., 1982, 151.

Chapter 2

TRANSMISSION OF *BABESIA*

Karl T. Friedhoff

TABLE OF CONTENTS

I. INTRODUCTION

Babesia are transmitted only by ticks; other vectors have never been identified with certainty. Obviously, mechanical transmission does not occur, neither by ticks nor by other hematophagous arthropods. Nonvector-borne transmission, i.e., prenatal or iatrogenic infections, are epidemiologically insignificant. Ticks are true hosts for the parasites, since the major part of the babesial life cycle takes place in these arthropods which inoculate infective stages, sporozoites, with their saliva during infestation. The sporozoites are the final stages of a complicated life cycle in the tick host. Since development may take place in each stage of a tick generation, eggs, larvae, nymphs, and adults, transmission and development of Babesia differ considerably from transmission of other protozoa by insects or annelids (Figure 1).

The tick vectors have been identified for approximately 17 of 73 *Babesia* species, and only a few of these vector-parasite systems have been studied in detail. We now know that *Babesia* do not arbitrarily infect any tick stage, organ or tissue, but follow a highly specific pathway of infection in certain tick stages depending on the species or strain involved. However, our knowledge in this field is incomplete, although many publications deal with transmission of *Babesia*.[1] Most of the published results have to be reexamined for various reasons: experiments have been insufficiently planned or poorly conducted, intercurrent infestations or infections cannot be ruled out, the identity of the tick or *Babesia* species remains uncertain, or the tick stages that acquire (susceptible stages) or transmit (transmitting stages) *Babesia* have not been clearly identified. In addition, some of the older literature has not been critically or correctly reviewed.

It is almost impossible to maintain all stages of the *Babesia* life cycle under laboratory conditions, but sporozoite stabilates can be made available. Far too often, the erythrocytic stages are propagated in their mammalian hosts by serial inoculation of infected red cells. Such "needle-passaged" strains may change rapidly in various aspects, e.g., infectivity for ticks, virulence, immunogenicity, drug resistance, morphology, etc., and may no longer reflect the situation in the field. Some of these changes may also occur in tick-transmitted strains maintained for several years under laboratory conditions. Thus sporozoite stabilates from field isolates are important for maintaining *Babesia* strains.

Herein, the terminology of transmission, the basic features of the babesial life cycle including transmission within the tick population, and data on *Babesia* species and their vectors are reviewed. Since the review is confined to the genus *Babesia*, transmission of other piroplasms will not be discussed here. However, some *Babesia* species are included that will probably be assigned to the genus *Theileria* in the near future because of their life cycle. Reference is made to recent reviews[1-4] to limit repetition.

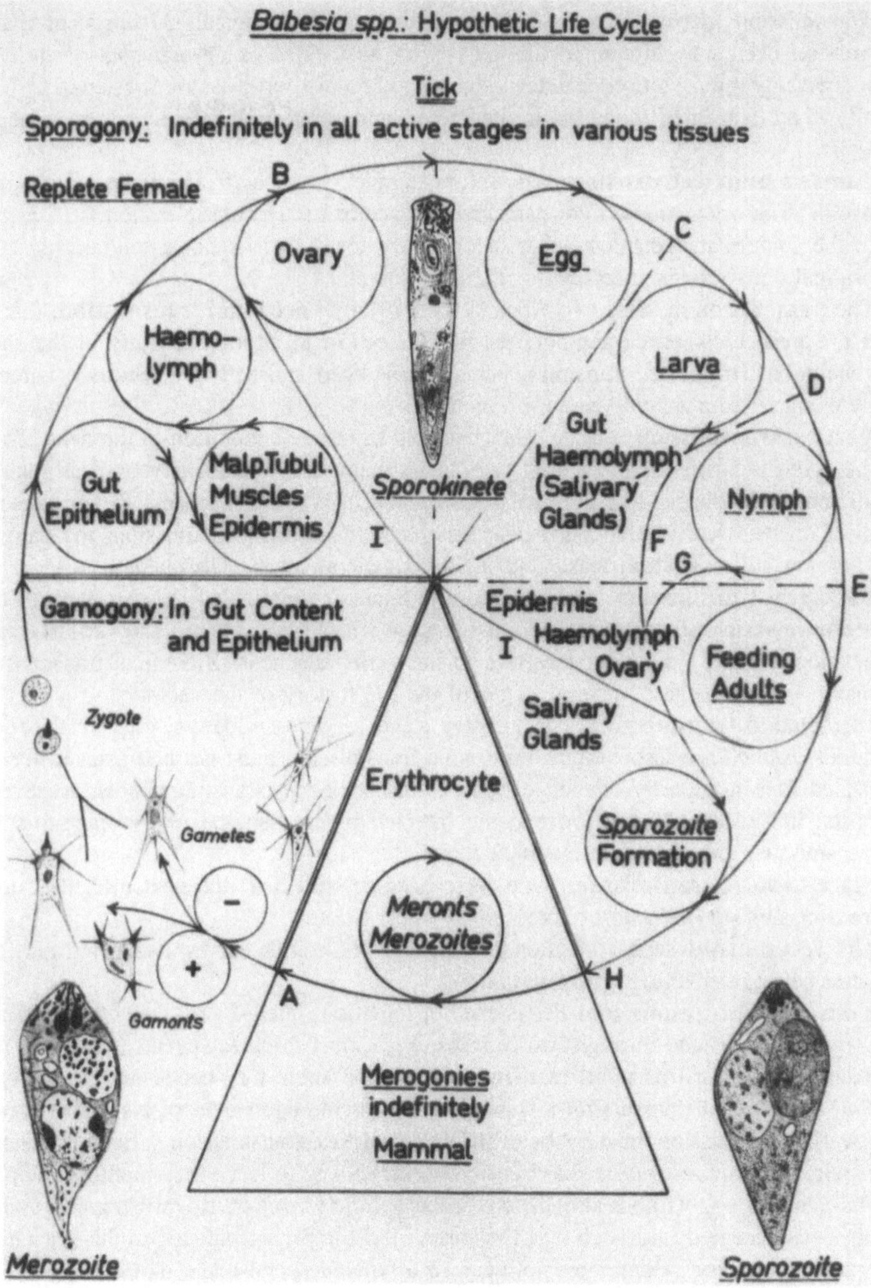

FIGURE 1. Life cycle of a *Babesia* species transmitted vertically in a three-host tick. (A) Alimentary infection. (B) Transovarial transmission. (C) Transstadial transmission from egg to larva, (D) from larva to nymph, (E) from nymph to adults. Transmission to mammalian host by larva (F), nymph (G), or adults (H). (I) Vertical transmission.

II. TERMINOLOGY OF TRANSMISSION

Unfortunately, different terms related to biological transmission of disease agents are used. Here, I apply essentially to terminology suggested by MacLeod:[5]

1. Transmission mechanisms are either mechanical or biological. **Mechanical transmission** occurs by direct inoculation (flying needle) as of *Trypanosoma evansi*, or indirect, by regurgitation or defecation of food as of sporocysts of *Sarcocystis*.[6] Usually, there is no multiplication in the vector, and duration of infectivity may be quite short.
2. **Simple multiplicative transmission**, for example, occurs by fleas infected with plague bacilli. The organisms do not parasitize the vector tissue: multiplication is incidental to the sojourn in the intestinal tract of the vector, and it is not essential, since the original units remain infective for the new host.
3. The transmission in these two types (1,2) is strictly **horizontal transmission**, that is, it is spread between contemporaries from one host to another, usually of the same generation. Horizontal transmission may be induced indirectly by vectors or directly by contact without involvement of vectors.
4. **Vertical transmission**, on the other hand, is from one generation to the next. Thus, the agent is carried over at least two host generations without any by-pass such as alimentary feedback. In horizontal transmission, it is not of immediate importance to the parasite whether or not the host be killed. In vertical transmission, the parasite must not kill its host, at least, until the next generation is established.
5. **Biological transmission** implies a more intimate association of the parasite with the vector: invasion of or at least close association with tissues occurs, and various stages develop that are no longer infective to the vertebrate host. Biological transmission may encompass one or several stages of the life history of the vector.
6. **Intrastadial transmission**. The infected stage becomes infective, e.g., in the mosquito-*Plasmodium* association. Intrastadial transmission must not be mistaken for repeated transmission by one stage, e.g., a male tick may transmit *Babesia* to several hosts; however, it may have become infected by transstadial or by stage-to-stage transmission already in the nymphal stage.
7. **Stage-to-stage transmission**. One stage becomes infected, the next infective, e.g., transmission of *Theileria* by three- or two-host ticks.
8. The vector may lose its infection or may become reinfected by its own strain, the latter being **reinfecting transmission**.
9. **Transstadial transmission**. Persistence of infection, latent or residual, throughout a complete stage, and through two ecdyses, e.g., most *Babesia* species (see below).
10. **Transovarial** (or transoval) **transmission**. This is merely a special instance of type 9 — transstadial transmission — since the egg is the equivalent of a stadium. Transovarial transmission must not be mistaken for vertical transmission in ticks. The latter requires continuous transstadial transmission from egg to larva to nymph(s) to female and again to egg without alimentary feedback. The term "transoval" regrettably has not been accepted, although it is the correct term implying passage to the egg rather than to the ovary, which does not necessarily include persistence into the egg.

III. DEVELOPMENT OF *BABESIA*

Development of *Babesia* implies transmission of these intracellular parasites from one host cell to the next, from one stage of the tick host to the next, from the tick to the mammalian host, and vice versa.

A. Erythrocytic Stages

Babesia do not parasitize any host cells other than the erythrocytes. The typical *Babesia* species multiply by the formation of only two daughter cells that completely incorporate the cytoplasm of their mother cell.[2] The piriform daughter cells are now called merozoites because of their morphogenesis and structure. Consequently, this cycle of asexual reproduction (Figure 1) should be termed merogony.[7]

Meront and merozoites of *Babesia* lie immediately within the red cell cytoplasm; there is no parasitophorous vacuole. This remarkable host-parasite relationship is unique among the intracellular parasites of the Apicomplexa in that all are located in a parasitophorous vacuole, at least during a certain phase of their intracellular development. Pigment, other iron-containing deposits, or residual bodies are not produced in red cells by *Babesia*. Gametocytes, as described for *Babesia microti*[8] have not been identified in the mammalian hosts of *B. bigemina*, *B. ovis*, and *B. canis*.[3,9-12]

B. Initial Stages in the Gut of Ticks

Babesia undergo morphologic, metabolic, and antigenic changes during their development from erythrocytic stages in the tick gut to the final stages in salivary glands of certain tick stages. The antigenicity and metabolism of *Babesia* stages occurring in tick tissues have not as yet been studied. There is, however, evidence that babesial antigens are stage-specific.[13] Also, the metabolism of *Babesia* developing in the hemolymph and ovary of the tick vector may be stage-specific, since these stages are not affected by babesicides.[14]

The intraerythrocytic stages in the gut of the engorged tick remain in host erythrocytes for some hours, during which time most of these stages deteriorate. Some of the spherical or ovoid stages without a pellicle or apical complex initiate further development. These ovoid stages are still morphologically identical to intraerythrocytic trophozoites in the mammalian host. Thus, gametocytes may not be identified in domestic animals. Within a few hours after detachment of the engorged tick, "strahlenkörper" or rayed bodies develop, some of which are obviously gametocytes.[3] These may give rise to two to four gametes which are probably isogametes.[3,4,15]

The final stage of the "strahlenkörper" or sexual forms is a spherical cell (= zygote) carrying a spike, a unique organelle produced by one of the two gametes before syngamy takes place. After reduction of the spike, the zygote infects selectively one special gut cell type, the basophilic epithelial cell, where multiple fission bodies develop. Typical, motile, club-shaped sporokinetes are then released (Figure 1). Thereafter, the host cells and their nuclei increase considerably in size.[16] The sporokinetes initiate cycles of multiple fission in various organs of the replete female tick. These cycles continue indefinitely until the tick dies. Oocytes also become infected, with transovarial transmission occurring. Similar cycles of multiple fission take place in the embryo, in various organs of the feeding, engorged, or moulting larvae, nymphs, and adult ticks.

A final cycle of multiple fission occurs in certain cells of the salivary glands of feeding ticks, as sporozoites are produced which may subsequently induce infection of the mammalian host (Figure 1).

IV. TRANSMISSION BY TICKS

A. General Aspects

The vectors of most *Babesia* species are still unknown for several reasons:

1. Most *Babesia* species are host specific or their host range is rather narrow. Thus, transmission trials have to be conducted with specific hosts, which is nearly impossible in many cases.
2. While many vectors have been incriminated, the results are biased by several factors. The identity of the *Babesia* or tick species is uncertain. Latent infections in the mammalian or tick host cannot be excluded, since experiments were conducted in endemic areas and, in many cases, the *Babesia* were not identified in the tick vector.

Table 1
KNOWN VECTORS OF *BABESIA* spp.

Babesia sp.	Tick species	Ref.
B. bigemina	*Boophilus microplus*	1, 29, 45
	Boophilus decoloratus	38, 43
	Boophilus annulatus (Syn: *B. calcaratus*)	17, 41, 194
	Rhipicephalus evertsi	20, 43
B. bovis	*Boophilus microplus*	48, 49
	Boophilus annulatus	41, 47, 52
B. divergens	*Ixodes ricinus*	54, 58, 63, 70, 195
	Ixodes persulcatus	51—53
B. major	*Haemaphysalis punctata*	66, 67, 73
B. occultans	*Hyalomma marginatum*	75, 76
B. ovata	*Haemaphysalis longicornis*	81—83
B. ovis	*Rhipicephalus bursa*	1, 91, 95, 96, 103
B. motasi	*Rhipicephalus bursa*	86, 92, 104
B. motasi (north Europe)	*Haemaphysalis punctata*	98, 100
B. caballi	*Dermacentor marginatus*	52, 94, 119, 122
	Dermacentor reticulatus (= *D. pictus*)	52, 108, 117, 120, 122
	Dermacentor silvarum	52, 114, 116, 122
	Dermacentor niveus (= *D. daghestanicus*)	122, 123, 196
	Dermacentor nuttalli	122, 123, 196
	Dermacentor (= *Anocentor*) *nitens*	110, 121
	Hyalomma marginatus (= *H. plumbeum*)	52, 107, 111, 112, 122
B. equi	*Dermacentor reticulatus* (= *D. pictus*)	52, 109, 122, 123, 131
	Dermacentor marginatus	52, 109, 130—132
	Hyalomma marginatum (= *H. plumbeum*)	52, 107, 122, 131
	Hyalomma anatolicum (= *H. a. anatolicum* = *H. a. excavatum*)	107, 128, 129
	H. detritum (= *H. scupense* = *H. uralense*[106])	94, 109, 122, 123, 131
	Rhipicephalus turanicus (= *R. sanguineus*?[107])	52, 85, 122, 128, 129
	Rhipicephalus bursa	52, 122, 197
	Rhipicephalus evertsi	124
B. canis	*Rhipicephalus sanguineus*	135—138
	Dermacentor reticulatus	137, 140—142
	Haemaphysalis leachi	138, 143
B. microti (Europe)	*Ixodes ricinus* (Germany)	145, 146
	Ixodes trianguliceps (Britain)	147, 148
B. microti (North America)	*Ixodes dammini* (= *I. scapularis*)	149, 151, 153
B. merionis	*Rhipicephalus sanguineus* (= *R. secundus*)	157—159
	Hyalomma anatolicum (= *H. a. excavatum*)	159—161
	Hyalomma dromedarii	159
B. meri	*Ornithodoros erraticus*	164, 165

3. There is some evidence for instability of *Babesia*-tick associations. Thus, certain strains of a tick vector may not be susceptible to infection by certain *Babesia* strains (see below).[17] In addition, the transmission mode may change, e.g., from vertical to non-vertical transmission (see below).

Taking into account these limiting factors, only a few vectors are known (Table 1). The vector role of many tick species listed in classical transmission charts[18] needs to be confirmed. Krylov[19] lists 90 *Babesia* and 51 "*Nuttallia*" species; Levine accepts only 71 *Babesia* species (this includes the obsolete genus *Nuttallia*). The newly described species *B. ovata, B. occultans,* and *B. crassa* need to be added to this list. Thus, the vectors of only 17 *Babesia* species are known, although 53 species have been incriminated.[19] Only those tick species that have been repeatedly confirmed as vectors by obviously unbiased laboratory and field studies are considered to be known vectors (Table 1). The bovine species have been studied most because of their economic importance.

B. *Babesia* Species Transmitted by Ticks

In general, two transmission patterns have been observed: (1) typical tick species acquire infection only in the female stage; transformation of *Babesia* erythrocytic stages into sexual stages and finally sporokinetes occurs in the gut of the replete female tick, always followed by transovarial transmission (Figure 1); (2) "Nuttallia"-like species, i.e., *B. microti, B. merionis,* and *B. equi,* are transmitted stage-to-stage only, thus closely resembling the *Theileria* transmission pattern.

1. Babesia bigemina

Babesia bigemina is transmitted by at least three *Boophilus* and one *Rhipicephalus* species. *Rhipicephalus bursa*[20] and *Rhipicephalus appendiculatus*[21] have not been confirmed as vectors. A reported vector, *Haemaphysalis punctata,* probably transmitted *B. major* rather than *B. bigemina,*[22] since the ticks were collected in an area (Norderney, Germany) endemic for *B. major.*[23] However, *R. bursa* should be reexamined, since convincing evidence has been shown by Sergent et al.[24] in Algeria that adults and probably nymphs of this two-host tick can indeed transmit *B. bigemina.* It is still uncertain whether *Boophilus kohlsi*[25] or *B. geigyi*[26] can transmit *Babesia bigemina* or *B. bovis* (see below). The epidemiological significance of *B. geigyi* and *R. evertsi* has not been examined. *Hyalomma* species are probably not vectors of *Babesia bigemina,* but may transmit *B. occultans* (see below).

Vertical transmission can be demonstrated (1) by identification of babesial stages (sporokinetes, etc.) in the ovary of the female tick before repletion, and also thereafter, before alimentary infection has proceeded to the ovary;[1] or (2) by prevention of alimentary infection. This can be achieved by adequate treatment of the mammalian host with highly effective babesicides during infestation.[1,20,27-29] The babesicides have no effect at all on stages developing in tick tissues, as has been shown by direct inoculation of drugs into engorged *Boophilus microplus* females infected with *Babesia bigemina.*[14] In this way, erythrocytic stages of *Babesia* can be eliminated without interfering with development in the vector tick. An additional indicator (3) for the absence of vertical infection is a delayed onset of transovarial transmission; eggs deposited during the first 12 hr[30] of oviposition are negative, since alimentary infection cannot proceed in time to the ovary so to infect the first oocytes developing after repletion.[1]

We could not demonstrate vertical transmission in a *Boophilus microplus* strain (Biarra) from Australia infected with a *Babesia bigemina* strain from Thailand by applying the three criteria mentioned above.[1,27,28,30] However, Büscher[31] could demonstrate vertical infection in the same ticks applying two of the criteria (1 and 3). This indicates a certain instability of the transmission modes in incompletely adapted vector-parasite systems. Vertical infection of *Boophilus microplus* (Australian strains) by *Babesia bigemina* was demonstrated by treatment of the tick-infested cattle with babesicides.[29] The progeny of ticks feeding on treated animals transmitted *B. bigemina.* By applying the same experimental procedure, Gray and Potgieter[32] demonstrated vertical infection by *B. bigemina* in *Boophilus decoloratus* (South African strain). Büscher,[20] using one stock of *Babesia bigemina* from Kenya and a *Boophilus decoloratus* strain from South Africa, could demonstrate vertical infection through two generations in unselected ticks by applying treatment of infested cattle and examination of ovaries and eggs. Thus, vertical transmission of *Babesia bigemina* undoubtedly exists in *Boophilus decoloratus.* Treatment of infested hosts with imidocarb dipropionate can safely prevent alimentary infection [20] with some *Babesia* species. However, sporozoites may turn into tick-infective stages before being eliminated by drug action. It has been shown that ticks may become infected alimentarily while feeding on refractory hosts.[31,33-35] Thus, raising ticks on such hosts is not a suitable method to demonstrate vertical infection. For these reasons, the ovaries of feeding and engorged female ticks should be examined before onset of alimentary infection.

Obviously, adaptation of *B. bigemina* to *Boophilus microplus* has occurred in endemic areas to the extent that the ovary and the salivary glands of the female tick become infected simultaneously, thus securing transovarial infection. There may have been insufficient adaptation of a *Babesia bigemina* strain from Thailand to infect an Australian *Boophilus* strain. This may explain the absence of vertical infection under these experimental conditions.[28] Strains of *Boophilus* spp. may be completely refractory to certain strains of *Babesia bigemina* originating from other endemic regions, e.g., *Boophilus annulatus* strains from Syria and Turkey did not transmit a *Babesia bigemina* strain from Tunesia, although it was transmitted readily by a Tunesian strain of the same tick species. Likewise, a *B. bigemina* strain from Venezuela was easily transmitted by *Boophilus microplus* from Venezuela.[17] However, selection for susceptibility has been successful. *Boophilus microplus* from Venezuela did not transmit a strain of *Babesia bigemina* from South Africa, but in about 1% of the female ticks, sporokinetes developed in the hemolymph. Obviously, the infection did not proceed in time to the oocytes. Consequently, transovarial infection did not occur, since the progeny of the infected female ticks did not transmit *B. bigemina*. When selection was continued through three generations, transmission occurred as up to 90% of the female ticks became infected alimentarily in the following generations. The same *B. bigemina* strain was still highly infective for *Boophilus decoloratus* from South Africa.[9] These results show that selection and adaptation probably account not only for the high susceptibility of vector ticks in endemic areas, but also for the mode of infection. It appears that vertical infection requires a higher degree of adaptation than transstadial infection. Whereas vertical infection is the rule in *B. decoloratus*, vertical and nonvertical transmission may occur in *B. microplus*.

There is evidence for vertical transmission of *Babesia bigemina* in *Boophilus annulatus*.[36] The absence of vertical transmission of *Babesia bigemina* in *R. evertsi* (both strains originate from different localities in Kenya) has been shown by Büscher.[20] However, whether this applies to all subspecies of *R. evertsi* and to well-adapted strains from other endemic areas remains uncertain.

R. evertsi transmits in the nymphal stage only. After transmission, the infection is eliminated from the nymph, and the following stages — males and females — are not infected. The engorging female tick becomes infected again only after uptake of infected erythrocytes (= alimentary infection; Figure 2).[20] Again, results obtained under laboratory conditions with nonadapted strains may not reflect the situation in the field. Thus, transmission of *B. bigemina* by *Boophilus annulatus* and *R. evertsi* should be studied using field isolates of infected ticks. Care should be taken to closely simulate local microclimatic conditions.

The route of infection (Figure 2) has been investigated in all known vectors of *Babesia bigemina*. Apparently, the engorged female tick is the only stage that can acquire the infection;[1] contradictory results have been encountered very rarely.[37] The *Babesia* are transmitted by nymphs, females, and males in all *Boophilus* species[20,38-43] except for *B. annulatus*, where transmission by female and male ticks has not been demonstrated. *R. evertsi* transmits in the nymphal stage only; thereafter, the infection is terminated until the replete female becomes reinfected alimentarily.

The various types of transmission of *Babesia bigemina* are summarized and schematically presented in Figure 2. Further details of transmission have been reviewed.[1] Alimentary infection can take place only if female ticks engorge on an infected host. Parasitemias as low as 10^{-6} (infected/uninfected red cells) may induce hemolymph infection rates of about 50% in *Boophilus microplus*.[1,44] However, only during the initial phase of parasitemia up to 4 to 7 weeks postinfection, the erythrocytic stages are capable of inducing infection in the tick.[1] There are no data on the persistence of vertical infection beyond two generations and on the effect of immunity on alimentary and vertical infection. The low infection rates in *Boophilus* larvae or nymphs in the field (see Chapter 5) as compared to *Babesia divergens* infections in *Ixodes ricinus* (see below) suggest that (1) alimentary infection is reduced in ticks feeding on immune cattle, and (2) vertical infection wanes within a few generations under field conditions.[1]

A certain additive effect of alimentary infection in vertically infected ticks resulting in higher infection rates of eggs has been shown for *B. ovis* in *R. bursa* (see below). There is little doubt that alimentary feedback is important for persistence of infection in a tick population. Vertical infection contributes only partially to this persistence[1] and is not essential. Vertical infection, however, is probably essential for euryxenous three-host ticks which feed on a variety of refractory hosts (see below).

2. *Babesia bovis*

Babesia bovis is transmitted by two *Boophilus* species, *B. microplus* and *B. annulatus* (Table 1), but not by *B. decoloratus*.[45] It appears that *B. geigyi* is also a vector,[26,46] but probably not *B. kohlsi*.[41] Besides the *Boophilus* species, only *R. bursa* has been incriminated.[47] The vector role of this tick species needs to be confirmed. Vertical infection does not occur in *B. microplus*[48] (Table 2). Only the larval stage can transmit *Babesia bovis* to the bovine host.[48,49] The larvae lose their infection after transmission has occurred.[48] Thus, infection of the tick population is based only on alimentary infection of the engorging female ticks followed by transovarial infection of a minor proportion of their eggs (Figure 2). This may account for the lower infection rates in tick populations in the field as compared to infection rates with *B. bigemina* in the same ticks.

Babesia bovis is also transmitted transovarially in *Boophilus annulatus*, and the larvae probably transmit.[41,47] Further studies are required to elucidate development and transmission of *Babesia bovis* in this important vector tick.

The different modes of transmission of *B. bovis, B. bigemina,* and *Anaplasma marginale* can be exploited to separate these disease agents and to eliminate *Theileria* and *Eperythrozoon* infections: (1) only *Babesia* spp. are transmitted transovarially by *Boophilus microplus*; (2) only larvae transmit *Babesia bovis*; (3) only nymphs and adults transmit (*B. bigemina*; (4) *A. marginale* is transmitted intrastadially by unengorged ticks of any stage.[39] Although lacking confirmation,[41] *Boophilus* ticks may transmit *Babesia bovis* and *B. bigemina* simultaneously, as was shown with field isolates of *Boophilus microplus* from Colombia;[50] however, the progeny of a single tick has not yet been tested.

3. *Babesia divergens*

Babesia divergens is almost exclusively transmitted by the three-host tick *Ixodes ricinus*. *Ixodes persulcatus* is also capable of transmitting *B. divergens*[51-53] (Table 1); however, not much is known of its vector role in the field.

Vertical infection has been demonstrated by feeding nymphs of *I. ricinus* on a rabbit and the ensuing adults on a sheep (Table 2).[54] The larval progeny then transmitted *B. divergens* to a calf. Although *B. divergens* is a rather euryxenous parasite, rabbits[55] and sheep[56] are apparently refractory, even though a splenectomized moufflon (*Ovis musimon*) was shown to be susceptible.[56] Thus, final proof for the existence of vertical infection is lacking, since infested animals were not treated with babesicides (see above).

The only stage that can acquire the infection is the engorging female tick,[54] whereas the larvae, nymphs, and females transmit[54,57] (Figure 2). A single nymph or female may transmit the infection,[54] however, it is not known whether males are capable of the same. *I. ricinus* males feed very little; therefore, it has to be shown whether this is sufficient for transmission to occur. However, males of *I. persulcatus* can transmit *B. divergens* (see below). *I. persulcatus* can transmit *B. divergens* in the larval, nymphal, and adult stages; one male or female tick can transmit.[51]

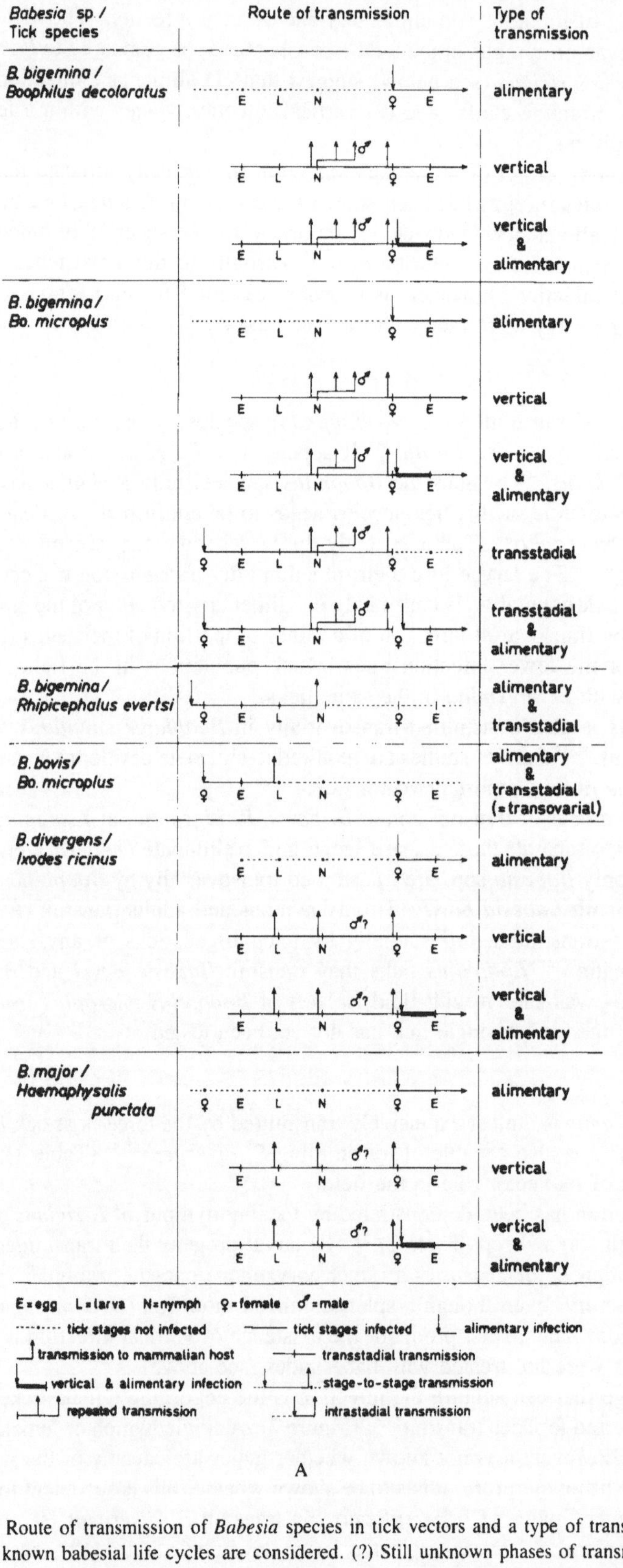

FIGURE 2. Route of transmission of *Babesia* species in tick vectors and a type of transmission are shown. Only known babesial life cycles are considered. (?) Still unknown phases of transmission.

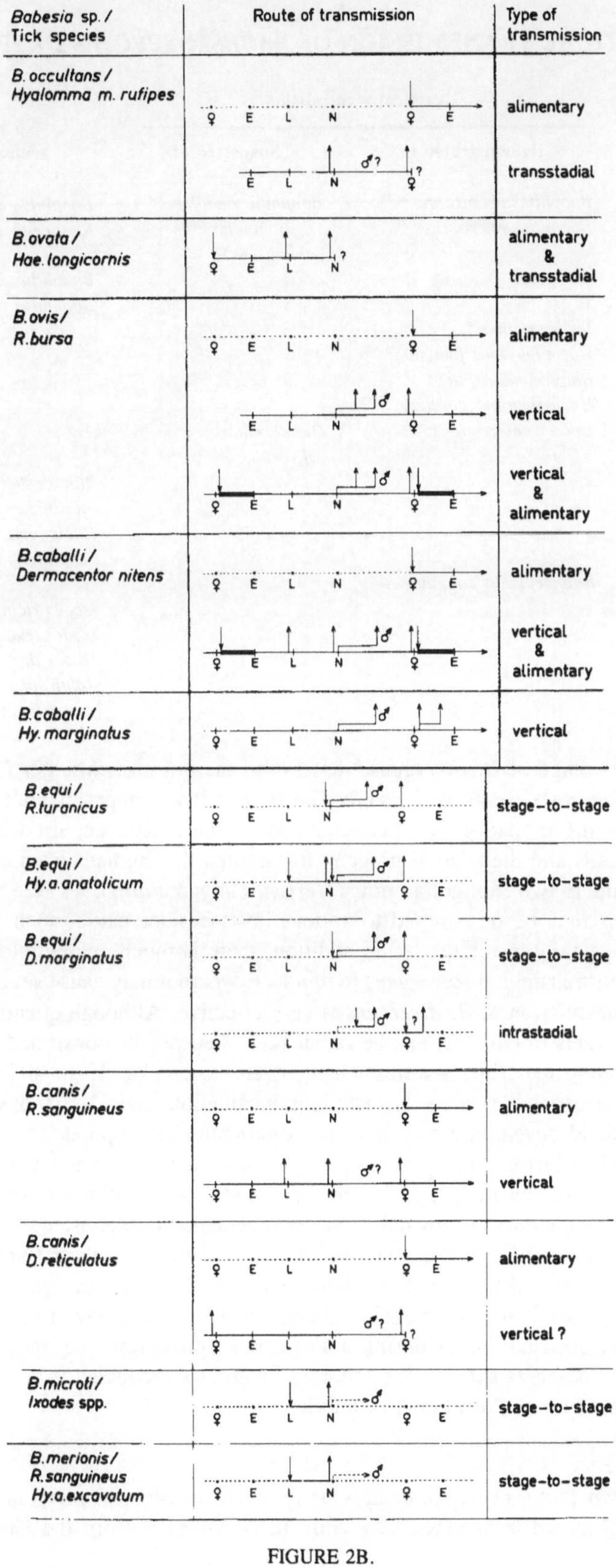
Babesia sp. / Tick species
Route of transmission
Type of transmission
B. occultans / Hyalomma m. rufipes
alimentary
transstadial
B. ovata / Hae. longicornis
alimentary & transstadial
B. ovis / R. bursa
alimentary
vertical
vertical & alimentary
B. caballi / Dermacentor nitens
alimentary
vertical & alimentary
B. caballi / Hy. marginatus
vertical
B. equi / R. turanicus
stage-to-stage
B. equi / Hy. a. anatolicum
stage-to-stage
B. equi / D. marginatus
stage-to-stage
intrastadial
B. canis / R. sanguineus
alimentary
vertical
B. canis / D. reticulatus
alimentary
vertical ?
B. microti / Ixodes spp.
stage-to-stage
B. merionis / R. sanguineus Hy. a. excavatum
stage-to-stage
♀ E L N ♀ E
♂?

FIGURE 2B.

Table 2
VERTICAL TRANSMISSION OF *BABESIA* SPECIES IN TICKS

Babesia sp.	Vertical transmission: Demonstrated in	Vertical transmission: Suspected in	No vertical transmission
B. bigemina	*Boophilus decoloratus*[20,43] *Boophilus microplus*[29,31]	*Boophilus annulatus*[36] (= *Boophilus calcaratus*)	*Boophilus microplus*[1,28] *Rhipicephalus evertsi*[20]
B. bovis			*Boophilus microplus*[48] *Boophilus annulatus*?
B. divergens	*Ixodes ricinus*[54]		
B. major	*Haemaphysalis punctata*[70]		
B. ovis	*Rhipicephalus bursa*[31,88]		
B. caballi	*Hyalomma marginatum*[111,112] *Dermacentor nitens*[110]	*Dermacentor silvarum*[113,114]	
B. equi			*Rhipicephalus turanicus*[128,129] *Hyalomma anatolicum*[128,129]
B. canis	*Rhipicephalus sanguineus*[135]		
B. microti			*Ixodes ricinus*[146] *Ixodes trianguliceps*[147] *Ixodes dammini*[153]
B. merionis			*Rhipicephalus sanguineus*[158,162]

The wide host range of *B. divergens*[56] has raised the question whether *I. ricinus* could transmit this *Babesia* species to various other hosts. A calf was experimentally infected after inoculation of 4 mℓ of parasitized human blood. Female ticks acquired infection while feeding on the calf, and their larval progeny transmitted this human strain of *B. divergens* to another calf and to two Mongolian jirds (*Meriones unguiculatus*).[58] These results indicate that *B. divergens* may be transmitted to various hosts besides cattle, such as red and roe deer,[56,59] and especially to reindeer.[60] In addition, there is much epidemiological evidence that *I. persulcatus* transmits *B. divergens* to reindeer, where it may cause severe disease.[61,62]

Obviously, transmission of *B. divergens* is very effective. Although quantitative data are scarce, infection rates in the field can be rather high. Petrov[63] demonstrated sporozoites in 25.6% of ticks collected from pastures where severe outbreaks (10 to 40% of the cattle) occurred. This corresponds to the extremely high inoculation rates (>1.0) observed in some herds.[50,64] The rapid development of infective sporozoites corresponds to the rather short infestation period of larvae and nymphs. Nymphs inoculate sporozoites within 24 hr after attachment.[65] This may also apply to larvae and females. Adaptation of *Ixodes,* especially the Taiga tick *I. persulcatus,* to low temperatures is remarkable. Persistence of *B. divergens* in these tick hosts under the climatic conditions that prevail on the northern Eurasian continent from Karelia to Kamchatka[52,61] is even more amazing. The babesia can proliferate only during the short parasitic and postparasitic phases of the tick, but have to persist for 2 to 6 years as obscure, dormant stages during a single tick generation (see Section III above). This phenomenon deserves special study that also could be conducted in *I. ricinus* or other tick vectors adapted to similar climatic conditions.

4. *Babesia major*

The only known vector of *B. major* appears to be *Haemaphysalis punctata* (Table 1). In 1914, Knuth[66] observed transmission by adult ticks collected from the German Friesian

Island of Norderney. In 1970, Brocklesby and Barnett[67] confirmed this observation. While *Boophilus annulatus* (= *B. calcaratus*) has been incriminated as a vector, however, the identity of the *Babesia* species in question remains uncertain.[52,53,68,69]

There is little doubt that Zeller and Helm[22] transmitted *B. major* rather than *B. bigemina* by feeding adults of *H. punctata* on a heifer. These ticks developed from adults collected on the Friesian Island of Norderney, where *B. major* has been demonstrated;[66] it is still endemic on the Dutch[70] and German Friesian Islands.[71]

In the Russian literature, *H. punctata* is considered to be a vector of *B. bigemina* in areas north of the *Boophilus annulatus* distribution.[53] This is reported to apply also to Czechoslovakia.[72] Again, there is evidence that in these areas *Babesia major*, rather than *B. bigemina*, is transmitted by *H. punctata*.

Vertical transmission of *B. major* probably occurs in *H. punctata* (Table 2). All instars of the progeny (F1) of infected adults (P) from the field were fed on rabbits.[70] Larvae of the next (F2) generation did not transmit, but the ensuing nymphs did.

Again, the rabbits were not treated with babesicides nor were the *Babesia* identified in ovaries and eggs, but alimentary infection is unlikely to occur via rabbits (see below). The only stage that acquires the infection is the engorging female tick; all three stages transmit *B. major* to calves (Figure 2). Infectivity also has been demonstrated in homogenated nymphs.[73,74]

5. Babesia occultans

This recently described species[75] is transmitted by *Hyalomma marginatum rufipes* in South Africa[75,76] (Table 1). Sporokinetes similar to those of *B. occultans* were found in engorged females of five *Hyalomma* species (*H. rufipes*, *H. truncatum*, *H. impressum*, *H. marginatum*, *H. impeltatum*) in West Africa (Nigeria).[77] Although there is some evidence for *B. occultans* infections in these *Hyalomma* species, the identity of the babesia remains uncertain.

Transmission of large babesia to cattle by *Hyalomma* has been observed elsewhere. *Hyalomma a. anatolicum* reportedly transmitted "*B. bigemina*" in the nymphal stage to 1 of 12 calves, but not in the larval (9 calves) and adult stages (5 calves).[78] Since transmission could not be achieved in the other calves, an intercurrent infection cannot be ruled out. Russian authors reported transmission of either *B. bigemina*[79] or the newly described species *Piroplasma* (= *Babesia*) *beliceri*, Abramov and D'Yakonov, 1974, by *H. anatolicum*.[80] The validity of *B. beliceri* should be ascertained; if the species is identical with *B. occultans*, then *B. beliceri* should have priority. It is unlikely that *Hyalomma* species are able to transmit *B. bigemina*. Büscher,[20] in our laboratory, was not able to infect *H. a. anatolicum* with a stock of *B. bigemina* that was infective for *Boophilus decoloratus*, *B. microplus*, *B. annulatus*, and *R. evertsi*. Vice versa, *Babesia occultans* was not transmitted by *Boophilus microplus*.[75]

Vertical transmission of *Babesia occultans* by *H. m. rufipes*, a two-host tick, has not been demonstrated. The female acquires infection and the nymph and the adults (males and females?) transmit, but not the larvae.[75] Since larvae/nymphs normally do not feed on cattle, females must acquire and adults must transmit the babesia in the field[75] (Figure 2). This interpretation is supported by the highly efficient alimentary infection observed: 59 to 90% of the engorged female ticks developed hemolymph infections after having engorged on blood with parasitemias as low as 0.01 to 0.003%.[75]

6. Babesia ovata

This newly described species is transmitted by *Haemaphysalis longicornis* in Japan[81,82] (Table 1). The same tick strain did not transmit *B. major* and *B. bigemina*.[81] Vertical infection was not demonstrated. The female ticks acquire the infection,[83] and the larvae[81,83] and probably the nymphs[82] transmit.[81-83] Apparently, transmission by adults has not been studied (Figure 2).

7. Babesia ovis

Rhipicephalus bursa is the only known vector of *B. ovis* (Table 1), although there is evidence that other ticks transmit this small, pathogenic sheep *Babesia*. Sporokinetes of *B. ovis* have been demonstrated in the hemolymph of engorged *R. turanicus* females.[84] We demonstrated *B. ovis* in hemolymph, ovaries, and eggs of experimentally infected females of *R. turanicus* and *Hyalomma a. excavatum*.[50] In addition, there is epidemiological evidence for the vector role of *R. turanicus*.[85]

Ixodes ricinus and *I. persulcatus* also have been incriminated as vectors.[52,86] However, the identity of these sheep *Babesia* which occur up to 60°N in the Leningrad area[86] are perhaps identical with the small sheep *Babesia* observed in Scotland,[87] but certainly not with *B. ovis*.

Vertical transmission of *B. ovis* in *R. bursa* has been demonstrated by applying all three criteria mentioned above[1,31,88] (Table 2). Both the tick and the *Babesia* strains used in our studies originated from Turkey (Ankara area). After vertical infection, approximately 95% of the female ticks produced infected eggs. However, less than 30% of the eggs were infected as compared to about 50% after alimentary infection, and about 67% after vertical and alimentary infection.[1,88] The additive effect of alimentary to vertical infection is only partial and sometimes almost impaired by resistance to alimentary infection induced by vertical infection. It appears that an existing vertical infection may protect the tick against heavy alimentary infection that is harmful to the tick.[13] These results correspond to field observations where infection rates in ticks feeding on cattle,[89,90] goats, or horses[90] were consistently lower than those in ticks feeding on sheep, e.g., 25.4% on sheep as compared to 10.5% on horses.[90] Markov and Abramov[91] followed vertical infection with *B. ovis* through 44 generations of ticks during 20 years; all instars of *R. bursa* were fed on rabbits. The ticks producing infected eggs were selected for maintenance of the infected tick colony. This selection explains the persistence of vertical infection through so many generations, although only a minor proportion of the tick progeny in each generation became infected (see above). Without selection, the infection rate in a tick population relying on vertical infection only will decrease rapidly. A gradual decline to an infection rate of 45.5% was observed in ticks fed for seven generations on cattle.[92] However, vertical transmission even of *Babesia* species highly adapted to their tick vectors is still insufficient to maintain infection in the field. Despite high transmission rates, the infection rate in the tick population would probably drop to 0% within a few generations, according to the fitness criterion of Fine.[93]

The female tick acquires the infection, both male and female transmit[1] (Figure 2). A single male or female tick is capable of transmission.[94] Males can infect secondary hosts.[50] Reportedly, a single female may induce a fatal infection.[95] Rarely, nymphs can acquire the infection, and rarely larvae and nymphs are capable of transmission.[95] In this case, transmission occurs only when thousands of ticks are used. Apparently, transmission by nymphs is uncommon, as is alimentary infection of nymphs. This view is supported by field observations in the U.S.S.R.[95]

8. Babesia motasi

Unfortunately, the taxonomy of the *Babesia* spp. of sheep and goats is not well-defined. It appears that the large *Babesia* sp., commonly referred to as *B. motasi*, is transmitted by *R. bursa* (Table 1) and does not occur north of the distribution area of this vector. *Babesia motasi* is said to be transmitted also by *Haemaphysalis otophila*[96] and *H. punctata*.[97] Schein[98] isolated *B. motasi* from a batch of mixed *H. punctata* and *H. otophila*.

In northern Europe (Germany,[99] the Netherlands,[98] Britain,[100] Sweden[101]), a large *Babesia* species has been identified as *B. motasi*. This species differs clearly from *B. ovis* and from the small *Babesia* spp. found in Scotland.[87] For practical reasons, I herein refer to it as *B. motasi* (north Europe). It is transmitted by *H. punctata*,[98,100] and perhaps by *I. ricinu*.[102]

Not much is known of the transmission of *B. motasi* and *B. motasi* (north Europe). Vertical infection has not been demonstrated, but transovarial transmission does occur.[96,98] *B. motasi* is probably transmitted by adults of *R. bursa*.[96,103,104] *B. motasi* (north Europe) is transmitted by adults of *H. punctata*.[98,100]

9. *Babesia crassa*

This recently described species[105] is distinct from *B. ovis* and *B. motasi* (isolates from Turkey and the Netherlands) by serological and morphological criteria; the vectors are unknown. A wide variety of ticks belonging to the genera *Haemaphysalis, Rhipicephalus,* and *Hyalomma* are found on sheep and goats in Iran besides a few *Ixodes* and *Dermacentor* species.[105]

10. *Other Babesia Species in Sheep and Goats*

The identity and the vectors of *B. foliata* and *B. taylori* remain uncertain.

11. *Babesia caballi*

Babesia caballi is transmitted in Eurasia by at least five *Dermacentor* species (*D. marginatus, D. silvarum, D. niveus, D. nuttalli, D. reticulatus*) and by *Hyalomma marginatum*. In the New World, the only known vector is *Dermacentor* (= *Anocentor*) *nitens* (Table 1), although *D. alblipictus* has been shown to be an experimental vector. Furthermore, transmission by *H. dromedarii, H. anatolicum* (= *H. a. anatolicum*), *H. detritum* (= *H. volgense*[106]), *R. sanguineus* (= *R. turaniccus?*) and *R. bursa*[107-109] also has been reported. Until these findings are confirmed, it remains unknown whether these tick species are important vectors in the field. The mode of transmission of *B. caballi* in the tropics and the Southern Hemisphere remains unclear.

Vertical infection (Table 2) has been demonstrated in *D. nitens,* where babesia were identified in the pregonadal tissues of nymphs and in the ovaries of unfed adults.[110] Transmission by the progeny of such ticks after inhibition of alimentary infection has not been studied. Vertical infection also has been observed in *H. marginatum*; the infected ticks retained infection through 11 generations, although all instars had fed exclusively on rabbits.[111,112] The infection was not monitored in the ovaries or eggs. However, it is unlikely that alimentary infection via rabbits can be as efficient as to secure transmission through 11 generations. *Dermacentor silvarum* retained an infection through three generations when fed on hosts other than horses,[113] or it was capable of maintaining infection only through one generation.[114] After having been maintained for two generations on rabbits, sheep, and cattle, the adults of the next generation did not transmit.[114] These authors[111,113,114] did not give information on selection of infected progeny, but persistence of vertical infection through 11 generations without selection is unlikely to occur, since infection rates are reduced during vertical infection (see above). In the field, infection rates of *Dermacentor* spp. are reduced rapidly on pastures where horses are substituted with cattle, sheep, or other animals.[115-117] When pastures are used only by cattle or sheep, *D. silvarum* ticks become negative within 3 years.[116]

Apparently, the only stage that acquires the infection is the female tick.[110,113] Infection of larvae or nymphs has been reported, but never has been demonstrated conclusively.[118] In the field, the adult ticks transmit.[119] Nymphs are also capable of transmission[119] as has been shown for *D. reticulatus*[120] and *D. silvarum*.[113] All feeding stages of the one-host tick *D. nitens* are capable of transmission, i.e., larvae, nymphs, females, and males. Males transmit even after delayed transfer, and after being withheld from the nymphal stage in vitro.[121] Single males or females of *D. silvarum,*[116] *D. marginatus,*[119] and *H. marginatum*[119] may transmit. Transmission of *B. caballi* by a single female of *D. marginatus* to two horses also has been demonstrated;[119] 2 to 4 days of attachment of a single adult tick may be sufficient for transmission to take place.[119] Infection rates of *D. marginatus* and *D. reticulatus* (= *D. pictus*) in endemic areas range from 3 to 35% (egg examination).[117] Infection rates are higher in ticks collected from horses than in ticks derived from cattle. Infection rates in ticks from horses were reportedly higher in endemic zones (8%) than in "latent" zones (3.4%), as compared to infection rates in ticks from cattle, which were higher in "latent" zones (2%) than in endemic zones (0.6%). [117]

12. Babesia equi

Babesia equi is transmitted by *Dermacentor* (*D. reticulatus, D. marginatus*), *Hyalomma* (*H. anatolicum, H. detritum*) and *Rhipicephalus* species (*R. evertsi, R. bursa, R. turanicus*) (Table 1). Russian authors[122,123] claim that "all"[122] *Dermacentor* species may transmit, including *D. nuttalli*.[52,123] *D. silvarum*,[52,123] and *D. niveus*.[123]

Little information is available on transmission by *R. bursa*[109] and *R. evertsi*;[124] however, epidemiological evidence supports experimental findings. *R. evertsi* appears to be the main vector in Ethiopia. Infection was observed in salivary glands of three *R. evertsi* males,[125] but the *Babesia* were not identified by transmission. Possibly *R. sanguineus*[107] as described in older papers may be identical with *R. turanicus*.[126]

Although *B. equi* is highly endemic in Latin America,[127] the vector of this important parasite is not known in the New World.

The absence of vertical infection has been demonstrated for *R. turanicus* and *H. anatolicum anatolicum*[118,128,129] (Table 2). *H. a. anatolicum* acquires the infection in the larval and nymphal stages, whereas *R. turanicus* is infected in the nymphal stage only. Infected nymphs of *H. a. anatolicum* that engorged on a rabbit were reported to become negative during development to adults. Both tick species did not transmit *B. equi* to their offspring[118,119,128,129] (Figure 2). Thus, stage-to-stage transmission was demonstrated, which is typical for *Theileria* spp. This, too, supports the view that *B. equi* is a *Theileria* rather than a *Babesia* species (see Chapter 12). These recent findings also support the view of Russian authors who doubt the postulated transovarial transmission of *B. equi*. Budnik[130] demonstrated transovarial transmission of *B. equi* in *D. marginatus* in only one of ten transmission trials. He concluded that transovarial transmission "is a very rare phenomenon" and *D. marginatus* is not "the main reservoir of *N. equi* in nature." Abramov et al., in a series of papers,[119,131,132] could not demonstrate transovarial, but only stage-to-stage transmission in *Dermacentor, Hyalomma*, and *Rhipicephalus* ticks.[119] *D. marginatus* acquires the infection in the nymphal stage and transmits it in the adult stage[131,132] (Table 2). The same is probably true for *D. reticulatus, H. detritum, H. marginatum*,[131] and *R. evertsi*.[124] *Hyalomma detritum* (= *H. scupense*) behaves as a single-host tick under the climatic conditions found in some areas of southern U.S.S.R., whereas in other areas (North Africa, Kazakstan, etc.) *H. detritum* is reported as a two-host tick.[106] Nevertheless, the one-host tick *H. detritum* (= *H. scupense*) transmits *B. equi* stage to stage. Even a single male may transmit.[131]

Budnik[130] and Abramov[131] demonstrated the only yet known intrastadial transmission of a piroplasm (Figure 2). Uninfected males of *D. marginatus* were placed on a parasitized horse for 10 days, then removed and incubated at 26 to 27°C for 17 days. The ticks then were transferred to an uninfected horse, where transmission occurred. This was shown in six trials with varying intervals for feeding and incubation.[130] This phenomenon was also observed by Abramov[131] who exposed males and females of *D. marginatus* for 2 to 4 days to an infected horse, incubated them for 28 days at 18 to 22°C, and then achieved transmission with 15 adults (6 females and 9 males).

Babesia equi is highly infective for nymphs of *H. a. anatolicum* and *R. turanicus*. Of *H. a. anatolicum* nymphs, 83% (48/58) became infected with parasitemias ranging between 0.005 and 0.1%, as well as 92% (34/37) of *R. turanicus* nymphs with parasitemias ranging from 0.05 to 0.1%. On the average, 17 acini of the salivary glands of an activated adult tick were infected. Infection rates and infection grades did not differ significantly between males and females.[128,129] Nymphs from both tick species acquired infection while feeding on a horse that was negative as assessed by microscopic examination and by CFT.[128,129]

13. Babesia canis

Babesia canis is transmitted by *R. sanguineus, D. reticulatus,* and *Haemaphysalis leachi* (Table 1). Transmission under experimental conditions has been demonstrated for *D. andersoni* (= *D. venustus*)[133] and *Hyalomma marginatum*;[134] however, these potential vectors are not of epidemiological significance. Vertical transmission has been shown for *R. sanguineus*. Ticks retained infections when fed on hedgehogs through five generations[135] (Table 2).

There is much evidence, that all known vectors acquire the infection only in the female stage,[136,137] although larvae and nymphs have also been incriminated.[134,138] Only in one case[136] was care taken to secure the negative status of the exposed tick stages. Thus, the other findings have to be confirmed. *R. sanquineus* is capable of transmission in the larval, [136-138] nymphal, [136-138] and adult stages[137,138] (Figure 2). However, infected larvae may not transmit,[136] and extremely large numbers are needed to achieve transmission.[136,137] This applies also to some extent to nymphs.[137] Again, the *instability of transmission patterns* becomes evident.

Adults appear to be the most important vectors in the field. It is not known whether both male and female stages transmit. Futhermore, no attempts have been made to study alimentary and vertical infection separately. *Babesia canis* isolates from France, where they are transmitted by *D. reticulatus,* are not infective for *R. sanguineus.*[137,139] This phenomenon of selective adaptation to a certain vector has also been observed with *B. bigemina* (see above). This should not, however, be used as a taxonomic criterion.

Dermacentor reticulatus transmits only in the adult stage, as larvae and nymphs normally do not feed on dogs.[137,140-142] Even when large numbers of infected larvae or nymphs are forced to attach to dogs, they do not develop, and transmission does not occur.[140] Thus, the immature stages are probably also incapable of acquiring infection. It is not known whether both adult stages transmit or whether vertical infection exists (Figure 2).

Haemaphysalis leachi was examined only recently by Shortt[138] who reported that larvae, nymphs, and adults transmit. Lounsbury[143] achieved transmission with adults only. In a large series of transmission experiments, he demonstrated that larvae and nymphs do not transmit and that both male and female adult stages are capable of transmission.[143]

14. Babesia gibsoni

Little is known of the transmission pattern of this canine parasite. Apparently, there have been no reports since the review of Neitz.[18] *Haemaphysalis bispinosa* and *R. sanguineus* are reported to be vectors.[18] *R. sanguineus* probably transmits *B. gibsoni* stage to stage, i.e., the larvae become infected and the ensuing nymphs transmit.[144]

15. Babesia Species of Felidae

The vectors of all feline *Babesia* species are unknown.

16. Babesia microti

Babesia microti is transmitted by *Ixodes* species. The vectors in Europe are *I. ricinus* in Germany[145,146] and *I. trianguliceps* in Britain[147,148] (Table 1). In Germany, *I. ricinus* transmits *B. microti* only to the field vole (*Microtus agrestis*), although tick transmission to the bank vole (*Clethrionomys glareolus*) is readily achieved in the laboratory.[145] The golden hamster (*Mesocricetus auratus*) is considered a suitable host for tick transmission trials.[146] In Britain, the only known vector is *I. trianguliceps,*[147,148] and besides the field vole, the bank vole is a common mammalian host.[148] *I. ricinus* and *I. hexagonus* are not susceptible to British isolates of *B. microti.*[147]

The only known vector of *B. microti* in North America is *I. dammini* (= *I. scapularis*).[149,150] Here, the mammalian hosts are deer mice (*Peromyscus leucopus*)[151,152] and the American field vole (*Microtus pennsylvanicus*),[152] whereas humans are accidental hosts (see Chapter 13). Transmission has been demonstrated also for *I. pacificus,*[153] but *D. variabilis*[151,153] and *D. andersoni*[153] are not susceptible for *B. microti*; an earlier report[154] could not be confirmed.[155] Hamster and gerbils are suitable hosts for tick transmission studies.[153]

There is no vertical, transovarial, or transstadial transmission, but only stage-to-stage transmission of *B. microti* in *Ixodes* spp. (Figure 2). The larvae become infected and the nymphs transmit.[146-148,151,153] It should be noted that this simple transmission pattern corresponds to transmission of *B. equi* (see above) and *Theileria,* with the exception that stage-to-stage transmission of *B. microti* obviously does not occur from the nymphal to the adult stage.

Apparently, adaptations to synecological situations are further developed in this rodent *Babesia* species than in *Babesia* species of domestic animals that fluctuate between countries and continents. This may explain the adaptations to various tick and mammalian hosts in isolated endemic areas (see *B. canis* and *B. bigemina* above). Whether the separation of reproduction has induced the development of valid species is an open question. *Babesia microti* has been detected recently in *Apodemus* species in East Asia,[156] but the vectors there are unknown.

17. Babesia Species of Rodents Other Than Babesia microti

Babesia merionis (= *Nuttallia danii, N. adleri, N. tadzhikistanica, Luhsia merionis*) is transmitted by *Rhipicephalus sanguineus* (= *R. secundus*),[106,157-159] *Hyalomma anatolicum* (= *H. excavatum*),[65,159-161] *H. dromedarii,* and *Hyalomma.* sp. (= "*H. aegyptium*")[159] to *Meriones tristrami* in Israel[159] and to *Meriones erythrourus* in the southern parts of the U.S.S.R.[160] *Haemaphysalis otophila,*[158,159] *H. cretica,* and *I. ricinus* are not susceptible to *B. merionis.*[159]

Vertical, transovarial, or transstadial transmission of *B. merionis* in its vector tick does not occur. This has been convincingly demonstrated (Table 2).[158,162] Always, the larvae acquire the infection and the nymphs transmit it.[65,158-161] Thus, the same stage-to-stage transmission occurs as was shown in ticks infected with *B. microti* (see above).

An unidentified *Babesia* species was experimentally transmitted to *M. oeconomus* by nymphs of *I. angustus* that were infected in the larval stage. Transovarial transmission did not occur; only the stage-to-stage type was observed.[163] The authors[163] also give epidemiologic evidence that *I. angustus* is a vector in Alaska. This tick could be a vector of rodent babesia in northern North America and northeastern Eurasia.[163]

B. meri is transmitted by *Ornithodoros erraticus* to the fat sand rat *Psammomys obesus.*[164,165] Other ticks (*O. tholozani, Hyalomma anatolicum excavatum, R. sanguineus*) were not susceptible when exposed on infected *Psammomys* or hamsters.[164] In total, three attempts to transmit *B. meri* with ticks collected in the field were successful. The ticks were placed on three sand rats 2, 39, or 81 days after collection. In addition, homogenates of ticks collected in the field 8 or 13 days before were infective.[164,165] Thus, *Ornithodoros meri* can be considered to be a vector, the only known argasid vector of a piroplasm. This phenomenon certainly deserves further research.

Vertical or transovarial transmission has not yet been demonstrated. There is some evidence for transstadial and repeated transmission. When negative "ticks" (nymphs or adults) were fed on infected sand rats or hamsters, sporozoites could be demonstrated at "different times after repletion" in the salivary glands.[166] Female ticks transmitted the infection 2 days after being collected in the field. Furthermore, a single adult tick (homogenate) was infective.[164] Unfortunately, further conclusions cannot be drawn, since nymphs and adults were used simultaneously in the transmission trials.[164,165] Alimentary infection as ascertained by examination of salivary glands occurred in 34% of the ticks fed on *Psammomys* and in 18% of the ticks fed on hamsters.[166]

The vectors of other *Babesia* species infective for rodents are unknown.[167] Research on these species is not only hampered by the problems in breeding potential vector ticks and some rodents, but also by the confused state of the taxonomy of *Babesia* species. It is unlikely that all these species belong to the genus *Babesia*. Essential criteria of the genus, i.e., transovarial transmission and absence of exoerythrocytic development in the rodent host have not been shown for a single species. On the contrary, transovarial transmission is absent in two species (*B. microti, B. merionis*) and has not been shown in another species (*B. meri*). The absence of exoerythrocytic stages has not been shown conclusively in any species. Many species of *Babesia* have been named from rodents, mostly because they were found in different hosts or because of structural differences in their erythrocytic stages. Cross immunity also has been used as a criterion. All three criteria, however, are not valid since many *Babesia* species are not host-specific and their morphology may be influenced by the host. Lack of cross immunity may occur between strains of the same species.

Levine[168] synonymized a number of piroplasms of rodents and suggests that the number of valid species may be as low as three species: (1) *B. muris*, a large, bigemina-type species, (2) *B. hylomysci*, a highly polymorphic and polyhostal species, and (3) *B. quadrigemina*, a small, *Nuttallia*-type species. *Babesia meri* has to be added as the only species transmitted by an argasid tick. *B. microti* and *B. rodhaini* are synonyms of *B. quadrigemina* as would most of the other rodent species be, with the exception of *B. muris, B. hylomysci*, and *B. meri*. However, Killick-Kendrick,[167] listing 54 *Babesia* spp. or isolates, maintains that a sweeping synonymy in the absence of knowledge of life cycles may lead to valid species being doubted, or even being later overlooked. Thus, identification of vectors is important, since they are essential for studies on the life cycle. A different tick host range would indicate definite isolation of reproduction, the often neglected, but still the most important, criterion for the validity of a species.

18. Babesia Species of Swine

The vectors of *B. trautmanni* and *B. perroncitoi* are not definitively known. Since the review of Neitz,[18] no further evidence has been given. The only experimental study appears to be that of Kurchatov and Markov[169] who fed larvae and nymphs of *Rhipicephalus turanicus* on rabbits; the ensuing adults transmitted *B. trautmanni* to swine. *Hyalomma marginatum, R. sanguineus, R. rossicus* and *Hyalomma* spp. (*H. aegyptium*) were not susceptible.[169] Markov[52] also listed *R. turanicus* as a vector of *B. trautmanni*. *B. decoloratus* in East Africa, *R. sanguineus, Hyalomma* spp. (*H. aegyptium*), and *D. reticulatus* in Europe also have been incriminated.[18] *Hyalomma aegyptium* is not a vector of *Babesia*, but only of *Haemogregarina* spp. of land tortoises. Thus *Hyalomma aegyptium* should not be incriminated as a vector of parasites of mammals or birds.[18,106]

The vectors of *B. perroncitoi* are unknown. *R. sanguineus, D. reticulatur*, and *Hyalomma* spp. (*H. aegyptium*) have been incriminated, however.[171] *Boophilus decoloratus, Amblyomma variegatum*, and *Rhipicephalus* spp. were found on sick pigs infected with *B. perroncitoi* during a recent outbreak at a large pig farm near Dakar, Senegal.[170]

19. Babesia Species of Cervidae

Several *Babesia* species have been described in Cervidae. *Babesia capreoli* (Enigk and Friedhoff, 1962) appears to be a valid species. It is reported to be transmitted by *I. ricinus*.[172] In this study, female ticks were infected with *Babesia* as shown by hemolymph examination.

The adult progeny of these females were fed on three roe deer (*Capreolus capreolus*). One animal was splenectomized 3 days after infestation and developed a parasitemia 2 days later. After another 2 days, parasitemia had reached 80% and the roe deer died. The second deer remained negative for 42 days, was then splenectomized, and became positive 1 day later. The third animal remained negative in spite of splenectomy 2 days after infestation.[172] Unfortunately, these results are not conclusive, since latent infections may relapse after splenectomy. The animals should have been splenectomized before infestation. There is, however, epidemiological evidence that *I. ricinus* is a vector of *B. capreoli* or other deer *Babesia*.[59]

In reindeer (*Rangifer tarandus*), probably one *Theileria* species and at least one *Babesia* species occur, although Levine[168] concluded that only *Theileria tarandi* (Syn.: *Piroplasma tarandi rangiferis, Franaiella tarandi rangiferis, Theileria rangiferis*) exists. However, a *Babesia* species (*F. tarandi rangiferis*) has been clearly described.[173] Also, concurrent *Theileria* and *Babesia* infections have been reported.[174] *Ixodes persulcatus* has been shown to transmit *B. divergens* (see above),[51] a *Babesia* pathogenic to reindeer inducing a disease associated with hemoglobinuria. It remains uncertain whether *B. tarandi rangiferis* is identical with *B. divergens*.[60] In either case, there is circumstantial evidence that *I. persulcatus* transmits a *Babesia* species to reindeer. There are several reports of *Babesia* infections in cervids in the New World;[175] however, the identity of the *Babesia* and the vectors remains uncertain. One species, *B. odocoilei,* has been described and accepted as a valid species.[168] *Amblyomma americanum* and *I. scapularis* could be potential vectors.[175]

20. Other Babesia Species

The vectors of *Babesia* species of mammals other than domestic animals and rodents are virtually unknown. *Dermacentor niveus* and *D. silvarum* are said to transmit *Babesia ninensis* (= *B. ninense*) in the larval and nymphal stages to hedgehogs (*Erinaceus rumanicus*). (For review, see Reference 123.) The larvae or nymphs of *D. silvarum* acquire the infection; in the next generations, the same stages transmit. (For review, see Reference 123.) The unusual transovarial transmission of a *Babesia* species from a small mammal indicates a relationship of this particular species to typical *Babesia* (see above). Thus, this parasite could perhaps serve as a model for research on *Babesia*.

Several tick species have been incriminated to transmit *Babesia* spp. to various mammals; however, no convincing evidence has been given. The piroplasms of birds — if they are piroplasms at all — cannot be considered as true *Babesia* species.

V. FACTORS INFLUENCING TRANSMISSION

Transmission can be strongly affected by various factors as has been reviewed before.[1] Here, only general and some additional information will be given.

A. Dynamics of Infection

Multiplication of *Babesia* in the tick host occurs only during the active phases of the tick that encompass only the minor part of the life span of one tick generation, especially in ticks adapted to temperate climatic zones. During the inactive phases of the tick host, *Babesia* remain in a dormant stage, so-called obscure stage, that has yet to be adequately characterized.[50] As soon as an infected tick is activated, the cells in various organs proliferate, and after 1 to 2 days, sporokinetes or sporozoites develop.[1] The main factor inducing activation of the tick and *Babesia* is a rise in ambient temperature (see below) affected by the body temperature of the mammalian host. Similar effects can be achieved by incubation of certain tick stages (see below).

Once development has been initiated, the *Babesia* follow a highly adapted pathway of infection by invading and transforming selected host cells in specialized organs throughout all (for typical *Babesia* spp.) or some tick stages resulting in exact timing of transmission of sporozoites and of alimentary infection by erythrocytic stages (see above, Section III).[1] Transovarial infection of eggs is affected by the mode of transmission. Alimentary infection proceeds to the ovary later than onset of egg formation. Thus, during the initial phase of oviposition, the eggs are not infected. This may involve as much as one half of the total egg mass. Vertical infection is established in the ovary before onset of egg formation. Thus, eggs become infected throughout the whole oviposition period. However, even in well-adapted systems, less than one third of the eggs become infected as compared to one half after alimentary infection. Alimentary infection is more efficient, but may be pathogenic to the tick host (see below). Combined vertical and alimentary infection has a certain additive effect on transovarial infection. However, the existing vertical infection may partly inhibit alimentary infection (see above). Thus, the degree of infectivity after combined infection can be erratic, but it has been shown that about 90% of tick eggs may become infected.[1]

B. Temperature

Unlike intrastadial transmission of most protozoa by insect vectors, stage-to-stage, transstadial, and especially vertical transmission of *Babesia* by ticks is not confined to climatic zones. *Babesia* transmission occurs as far north as Karelia or Alaska (see above). The dormant, obscure stages persist in inactive ticks during the long winters with temperatures below −30°C. However, during the active phase of the tick host, when the *Babesia* proliferate, higher ambient temperatures are necessary. Thus, alimentary and transovarial transmission of *B. bigemina* may be completely inhibited in *Boophilus microplus* at 20°C.[1] High temperatures have erratic effects. Persisting high temperatures, e.g., 35°C, inhibit or even eliminate infections.[161] Conversely, various stimulative effects of higher temperatures, especially of a sudden and pronounced nature, have been observed. *Babesia bovis* develops infective stages in *Boophilus* larvae within 2 to 3 days, and *Babesia bigemina* in nymphs within 8 to 10 days after infestation. Stimulation by feeding can be simulated by heat shock. Unfed infected larvae become infective after incubation at 37°C for 2 to 3 days; however, heat shock alone is not as effective as the complex stimulus provoked by feeding.[176] Development of infective stages can be enhanced by the prior storage of infected larvae. Storage of larvae for 3 to 5 weeks at 14°C consistently resulted in higher infection rates than storage for 1 to 5 days at 14°C, as was shown by microscopic examination of ticks 1 day *(B. bovis)* or 4 days *(B. bigemina)* after attachment. Storage of larvae infected with *B. bovis* at 27°C resulted in consistently lower infection rates than storage at 14°C. The effect was not as pronounced with *B. bigemina.*[176] From the results, it can be concluded that besides temperature, quiescence after hatching is important. The effect of low temperature after hatching is more critical for *B. bovis* than for *B. bigemina.* For the latter parasite, duration of quiescence is the most important factor.

The infective *Babesia* stages that develop after experimental heat shock have not been characterized. Such stages may develop even in tick instars where they do not normally occur (see above). Infective stages of *B. bovis,*[177,178] *B. bigemina,*[177] and *B. caballi*[121] may develop in tick eggs, or infective stages of *B. bigemina* in larvae.[177] In the case of *B. caballi,* this phenomenon occurred even without heat shock at 26°C. This again indicates a certain instability in the pattern of development and transmission. Obscure stages and infective stages induced by heat shock must be characterized for further elucidation of these important phenomena.

C. Relative Humidity and Light

These factors appear to be important only as far as the development of the tick vector is concerned.[1]

D. Age of Ticks

As shown above, a certain phase of quiescence after hatching is important; however, infections in ticks decrease proportionately with storage time. *Babesia bovis* remains viable in *Boophilus microplus* larvae stored at 14°C in 95% rH for 65 days, but not for 76 days. Adults of *R. bursa* infected with *B. ovis* become noninfective after 7.5 months.[1]

E. Elimination of the Sexual Cycle

Typical *Babesia* species can be propagated in cycles of either indefinite merogony or sporogony without gametogony being involved (Figure 1). Vertical transmission through many tick generations without alimentary feedback implying continuous cycles of sporogony (see above) apparently does not greatly affect the characteristics of a strain.[1] Propagation of merogony stages by subinoculation of blood into susceptible mammals ("needle-" or "syringe-passaged" strains), however, may change the characteristics of a strain.[1] Since it is a convenient method of maintaining strains, subinoculation of blood without the arthropod vector is a usual practice in many laboratories working with vector-borne protozoa. These strains, however, may not reflect the true nature of the parasite in question. These strains may change or lose the infectivity for ticks,[179-181] change their virulence for ticks,[180] alter the parasitic protein and antigen profiles,[182] and vary in the response to babesicides.[50] Whereas large *Babesia* species may lose their infectivity after a few blood passages, e.g., *B. major* after two passages, small *Babesia* species may retain their infectivity for longer periods, e.g., *B. divergens* for 11 years of direct serial passage.[1] This also applies to *B. equi*[118] and *B. merionis*.[161] The various methods of maintenance have been compared recently by studying the development and morphology of *B. bigemina* in the gut of *Boophilus microplus* females. Tick transmission to intact calves had no apparent effect, but tick transmission to spenectomized calves or syringe passage in splenectomized calves at intervals of 5 days or less had obvious deleterious effects on the gamogony.[183] Needle passage at 6-week intervals in intact calves produced similar results. Infectivity of *Babesia* for ticks appears to be a sensitive characteristic that can be irreversibly impaired by irradiation.[184]

F. Pathogenicity of *Babesia* for the Tick Vector

The pathogenic effects of *Babesia* on the tick vector are confined to alimentary infection. The gut, hemolymph, and perhaps the ovary are mainly affected. The severity of these effects is directly related to the degree of parasitemia and to the susceptibility of a tick strain for a certain *Babesia* strain.[1,75,185,186] The effects range from slight depression of egg production to death of the engorged female ticks before onset of oviposition.[1] Virulence of *B. bovis* for *Boophilus microplus* may be reduced after rapid blood passage in splenectomized calves.[180]

G. Interspecific Interaction Between Parasites of Ticks

Rickettsia-like, rod-shaped, or pleomorphic organisms measuring about 0.25 × 0.5 to 0.8 μm have been observed in epithelial cells of the *R. bursa* tick gut, in gut contents, and in the hemolymph of female ticks infected with *Babesia ovis,* 2 to 20 days after repletion.[187] The ticks showed typical signs of disease and died within a few hours. The hemolymph of some ticks was infected by apparently pathogenic bacterial organisms. Infection of ticks with both microorganisms could be prevented almost completely by antibiotic treatment of the tick host during the engorgement period.[187]

These pathogenic organisms may interfere with concurrent development of *Babesia*. Moreover, pathogenic effects may wrongly be attributed to *Babesia* infections. The interaction of *Babesia* with various symbionts in ticks has not been studied.

VI. OTHER MEANS OF TRANSMISSION

There is general agreement that *Babesia* are transmitted only by ticks. However, there are a few other possibilities to be considered.

A. Hematophagous Insects

Stomoxys and Tabanidae have been incriminated as mechanical vectors.[188] In one instance, 136 females of *Tabanus nipponensis* were fed daily for 12 days for a 1-min period/day on a calf infected with *B. ovata*, then transferred to another calf that was subsequently splenectomized. *Babesia ovata* was demonstrated in this calf 50 days after infection. However, unlike in *Anaplasma* infections, there is no convincing evidence for *Babesia* transmission by tabanids or by other insects.

B. Intrauterine Infections

Intrauterine infections have been reported mainly for *B. equi* and *B. caballi*, but rarely for other species such as *B. bigemina*, *B. divergens* (= *B. bovis*), *B. ovis*,[18,189] or *B. bovis* (= *B. berbera*).

Babesia equi appears to be the only species which is transmitted frequently to the fetus.[191,192] However, quantitative data are not available. Attempts to demonstrate intrauterine infection with other *Babesia* species have failed.[147,189]

C. Iatrogenic Infections

Infections with *B. equi* were demonstrated when the same hypodermic needle was used to inject a drug.[193] We know of at least three outbreaks of *B. equi* babesiosis in Germany and Switzerland caused by contaminated instruments,[50] but transmission of *B. equi* by instruments is probably uncommon and not of epidemiological significance.

VII. CONCLUDING REMARKS

It is nearly impossible to cover all of the pertinent literature. Thus, I should be grateful to anybody who draws my attention to a relevent publication not considered here. Negative statements, like negative results, are hardly to be confirmed; however, they usually provoke a response. The most pronounced negative statement concerns transmission of *Babesia* species of other mammals than domestic animals and of birds. The lack of knowledge on *Babesia* species not of veterinary or medical importance, however, is evident.

ACKNOWLEDGMENTS

I am grateful to Dr. G. Büscher and Dr. I. Kunstyr for translation of Russian literature. I dedicate this contribution to Dr. K. Enigk on occasion of his 80th birthday.

REFERENCES

1. **Friedhoff, K. T. and Smith, R. D.,** Transmission of babesia by ticks, in *Babesiosis*, Ristic, M. and Kreier, J. P., Eds., Academic Press, New York, 1981, 267.
2. **Friedhoff, K. T.,** Morphologic aspects of babesia in the tick, in *Babesiosis*, Ristic, M. and Kreier, J. P., Eds., Academic Press, New York, 1981, 143.
3. **Mehlhorn, H. and Schein, E.,** The piroplasms: life cycle and sexual stages, *Adv. Parasitol.*, 23, 37, 1984.

4. **Friedhoff, K. T.**, Development and transmission of *Babesia* spp. A review, in *Agriculture. Some Important Parasitic Infections in Bovines Considered from Economic and Social (Zoonosis) Points of View*, Parasitol. Symp., Euzeby, J. and Gevrey, J., Eds., Commission of the European Communities, Luxemburg, 1984, 3.
5. **MacLeod, J.**, Arthropod transmission of micro-organisms, *Nature (London)*, 191, 885, 1961.
6. **Smith, D. D. and Frenkel, J. K.**, Cockroaches as vectors of *Sarcocystis muris* and of other coccidia in the laboratory, *J. Parasitol.*, 64(2), 315, 1978.
7. **Levine, N. D.**, *Veterinary Protozoology*, Iowa State University Press, Ames, 1985.
8. **Rudzinska, M. A., Spielman, A., Riek, R. F., Lewengrub, S. J., and Piesman, J.**, Intraerythocytic "gametocytes" of *Babesia microti* and their maturation in ticks, *Can. J. Zool.*, 57, 424, 1979.
9. **Friedhoff, K. T. and Scholtyseck, E.**, Fine structural indentification of erythrocytic stages of *Babesia bigemina, B. divergens*, and *B. ovis, Protistologica*, 13, 195, 1977.
10. **Weyreter, H., Wattendorff, S., and Friedhoff, K. T.**, Initial stages of *Babesia ovis* and *B. bigemina* in the gut of engorged female ticks *(Rhipicephalus bursa, Boophilus microplus), Zentralbl. Bakteriol. Parasitenkd. Infektionskr. Hyg., Abt.1 Ref.*, 267, 300, 1980.
11. **Weber, G. and Walter, G.**, *Babesia microti* (Apicomplexa: Piroplasmida): electron microscope detection in salivary glands of the tick vector *Ixodes ricinus* (Ixodoidea: Ixodidae), *Z. Parasitenkd.*, 64, 113, 1980.
12. **Mehlhorn, H., Moltmann, U. G., Schein, E., and Voigt, W. P.**, Fine structure of supposed gametes and syngamy of *Babesia canis* (Piroplasmea) after in-vitro development, *Zentralbl. Bakteriol. Parasitenkd. Infektionskr. Hyg., Abt. 1 Orig. A*, 250, 248, 1981.
13. **Kähler, J.**, Untersuchung zur Immunogenität der Entwicklungsstadien von *Babesia ovis* aus dem Ovar von *Rhipicephalus bursa*, Dr. med. vet. thesis, Tierärztliche Hochschule, Hannover, West Germany, 1981.
14. **Friedhoff, K. T. and de Buhr, H.**, The effect of drugs on developmental stages of *Babesia bigemina* in *Boophilus microplus, 2nd Int. Conf. Malaria Babesiosis*, Marcel Merieux, Lyon, 1983, 170.
15. **Rudzinska, M. A., Spielman, A., Lewengrub, S. J., Trager, W., and Piesman, J.**, Sexuality in piroplasms as revealed by electron microscopy in *Babesia microti, Proc. Natl. Acad. Sci. U.S.A.*, 80, 2966, 1983.
16. **Agbede, R. I. S., Kemp, D. H., and Hoyte, H. M. D.**, *Babesia bovis* infection of secretory cells in the gut of the vector tick *Boophilus microplus, Int. J. Parasitol.*, 16, 109, 1986.
17. **Hoffmann, G.**, Infizierbarkeit verschiedener Boophilusstämme mit *Babesia bigemina* sowie Beeinflussung der Zecken durch Wirt oder Parasit, *Z. Tropenmed. Parasitol.*, 22(3), 270, 1971.
18. **Neitz, W. O.**, Classification, transmission, and biology of piroplasms of domestic animals, *Ann. N.Y. Acad. Sci.*, 64 (2), 56, 1956.
19. **Krylov, M. V.**, Some questions on the zoogeography and phylogeny of piroplasmids (in Russian), *Parazitologiya*, 5, 201, 1971.
20. **Büscher, G.**, The infection of various tick species with *Babesia bigemina*, its transmission and identification, *Parasitol. Res.*, in press.
21. **Purnell, R. E., Branagan, D., and Brown, C. G. D.**, Attempted transmission of some piroplasms by Rhipicephalid ticks, *Trop. Anim. Health Prod.*, 2, 146, 1970.
22. **Zeller, H. and Helm, R.**, Versuche zur Frage der Übertragbarkeit des Texasfiebers auf deutsche Rinder durch die bei uns vorkommenden Zecken *Ixodes ricinus* und *Haemaphysalis punctata cinnabarina, Berl. Tierärztl. Wochenschr.*, 39, 1, 1923.
23. **Friedhoff, K. T. and Liebisch, A.**, Piroplasmeninfektionen der Haustiere, *Tierärztl. Prax.*, 6, 125, 1978.
24. **Sergent, E., Donatien, A., Parrot, L., and Lestoquard, F.**, Transmission héreditaire de *Piroplasma bigeminum* chez *Rhipicephalus bursa*. Persistance du parasite chez des tiques nourries sur des chevaux, *Bull. Soc. Pathol. Exot.*, 24, 195, 1931.
25. **Pipano, E.**, Piroplasmosis — a review, *Refuah Vet.*, 28 (2), 51, 1965.
26. **Akinboade, O. A. and Dipeolou, O. O.**, Bovine babesiosis in Nigeria: the vectorial capability of *Boophilus geigyi* for *Babesia bigemina* and *Babesia bovis, Acarologia (Paris)*, 26, 235, 1985.
27. **Büscher, G., Friedhoff, K. T., and Yossef, L. S. E. A.**, Transmission of *Babesia bigemina* by *Boophilus* spp. and *Rhipicephalus evertsi, 2nd Int. Conf. Malaria Babesia*, Marcel Merieux, Lyon, 1983, 226.
28. **Yossef, L. S. E. A.**, Untersuchungen über die Möglichkeit einer vertikalen Infektion von *Boophilus microplus* mit *Babesia bigemina* (Piroplasmida), Dr. med. vet. thesis, Tierärztliche Hochschule, Hannover, West Germany, 1980.
29. **De Vos, A. J., Stewart, N. P., and Dalgliesh, R. J.**, The effect of chemotherapy on *Babesia bigemina* in the tick vector *Boophilus microplus, Int. J. Parasitol.*, 14, 249, 1984.
30. **Muangyai, M.**, Quantitative Untersuchungen zur transovarialen Infektion von *Boophilus microplus* (Ixodoidea) mit *Babesia bigemina* (Piroplasmea), Dr. med. vet. thesis, Tierärztliche Hochschule, Hannover, West Germany, 1974.
31. **Büscher, G.**, Untersuchungen zur Entwicklungsdynamik von *Babesia ovis (Piroplasmea) in Ovar und Haemolymphe weiblicher Zecken (Rhipicephalus bursa)* (Ixodoidea), Dr. med. vet. thesis, Tierärztliche Hochschule, Hannover, West Germany, 1975.

32. **Gray, J. S. and Potgieter, F. T.,** The retention of *Babesia bigemina* infection by *Boophilus decoloratus* exposed to imidocarb dipropionate during engorgement, *Onderstepoort J. Vet. Res.*, 48, 225, 1981.
33. **Friedhoff, K. T.,** Studies on the life cycle of *Babesia ovis*. II (Part 1), *J. Parasitol.*, 56 (4), 109, 1970.
34. **Callow, L. L.,** *Babesia bigemina* in ticks grown on non-bovine hosts and its transmission to these hosts, *Parasitology*, 55, 375, 1965.
35. **Akinboade, O. A.,** Experimental transmission of *Babesia bigemina* in sheep using infective larval tick of *Boophilus decoloratus, Rev. Elev. Med. Vet. Pays Trop.*, 34 (3), 271, 1981.
36. **Muratov, A. E. and Kheissin, E. M.,** Some data on the development of *Piroplasma bigeminum* in *Boophilus calcaratus* (in Russian), *Dokl. Acad. Nauk Tadzh. SSR.*, 1 (4), 47, 1958.
37. **Callow, L. L.,** The infection of *Boophilus microplus* with *Babesia bigemina, Parasitology*, 58, 663, 1968.
38. **Gray, J. S. and Potgieter, F. T.,** Studies on the infectivity of *Boophilus decoloratus* males and larvae infected with *Babesia bigemina, Onderstepoort J. Vet. Res.*, 49, 1, 1982.
39. **Dalgliesh, R. J. and Stewart, N. P.,** The use of tick transmission by *Boophilus microplus* to isolate pure strains of *Babesia bovis, Babesia bigemina* and *Anaplasma marginale* from cattle with mixed infections, *Vet. Parasitol.*, 13, 317, 1983.
40. **Morzaria, S. P., Young, A. S., and Hudson, E. B.,** *Babesia bigemina* in Kenya: experimental transmission by *Boophilus decoloratus* and the production of tick-derived stabilates, *Parasitology*, 74, 291, 1977.
41. **Hadani, A., Pipano, E., Tsafrir, N., Rauchbach, K., and Mayer, E.,** The transmission of *Babesia bigemina, Babesiella berbera* and *Anaplasma centrale* by *Boophilus annulatus* (Say), *Refuah Vet.*, 31, 149, 1974.
42. **Dalgliesh, R. J.,** Transmission of *Babesia bigemina* by transfer of adult male *Boophilus microplus, Aust. Vet. J.*, 54, 205, 1978.
43. **Potgieter, E. T. and Els, H. J.,** Light and electron microscopic observations on the development of *Babesia bigemina* in larvae, nymphae and non-replete females of *Boophilus decoloratus, Onderstepoort J. Vet. Res.*, 44, 213, 1977.
44. **de Buhr, H.,** Versuche zur Verbesserung von Babesicidprüfungen, Dr. med. vet. thesis, Tierärztliche Hochschule, Hannover, West Germany, 1981.
45. **Norval, R. A. I., Fivaz, B. H., Lawrence, J. A., and Daillecourt, T.,** Epidemiology of tick-borne diseases of cattle in Zimbabwe. I. Babesiosis, *Trop. Anim. Health Prod.*, 15, 87, 1983.
46. **Akinboade, O. A. and Dipeolu, O. O.,** Detection of *Babesia bovis* infections in *Boophilus geigyi* with egg crushing, larval smears, and haemolymph puncture., *Vet. Q.*, 3 (3), 143, 1981.
47. **Sergent, E., Donatien, A., Parrot, L., and Lestoguard, F.,** *Etudes sur les Piroplasmoses Bovines*, Institut Pasteur d'Algerie, Alger, 1945.
48. **Mahoney, D. F. and Mirre, G. B.,** A note on the transmission of *Babesia bovis* (Syn. B. argentina) by the one-host tick, *Boophilus microplus, Res. Vet. Sci.*, 26, 253, 1979.
49. **Potgieter, F. T. and Vuuren, A. S. van,** The transmission of *Babesia bovis* using frozen infective material obtained from *Boophilus microplus* larvae, *Onderstepoort J. Vet. Res.*, 41, 79, 1974.
50. **Friedhoff, K. T.,** unpublished data, 1986.
51. **Goussev, W. F., Rastegajeva, E. F., and Soussko, S. F.,** Die Zecken als Überträger der Babesiellosen und Françaiellosen der Rinder, *Arch. Wiss. Prakt. Tierheilkd.*, 71, 138, 1937.
52. **Markov, A.,** Parasitic diseases of the blood in domestic animals (piroplasmosis, babesiellosis, nuttalliosis, theileriosis, anaplasmosis) and the methods of their control in the U.S.S.R., *Bull. Off. Intern. Epizoot.*, 49, 96, 1958.
53. **Zolotarev, H. N.,** Hemosporidioses of cattle, in *Parasitology and Parasitic Diseases of Livestock* (in Russian), Ershov, V. S., Ed., State Publishing House for Agricultural Literature, Moscow, 1956, 372.
54. **Donnelly, J. and Peirce, M. A.,** Experiments on the transmission of *Babesia divergens* to cattle by the tick *Ixodes ricinus, Int. J. Parasitol.*, 5, 363, 1975.
55. **Garnham, P. C. C. and Bray, R. S.,** The susceptibility of the higher primates to piroplasms, *J. Protozool.*, 6, 352, 1959.
56. **Enigk, K. and Friedhoff, K. T.,** Zur Wirtsspezifität von *Babesia divergens, Z. Parasitenkd.*, 21, 238, 1962.
57. **Petrov, V. G.,** On the development of *Babesiella bovis* in *Ixodes ricinus* L. (in Russian), *Tr. Leningr. Piroplaz. Stn.*, 1, 58, 1939.
58. **Lewis, D. and Young, E. R.,** The transmission of a human strain of Babesia divergens by *Ixodes ricinus* ticks, *J. Parasitol.*, 66, 359, 1980.
59. **Adam, K. M. G., Blewett, D. A., Brocklesby, D. W., and Sharman, G. A. M.,** The isolation and characterization of a *Babesia* from red deer *(Cervus elaphus), Parasitology*, 73, 1, 1976.
60. **Nilsson, O., Nordkvist, M., and Ryden, L.,** Experimental *Babesia divergens* infection in reindeer *(Rangifer tarandus), Acta Vet. Scand.*, 6, 353, 1965.
61. **Tashkinov, N. I.,** Piroplasmidosen und Anaplasmose der Rentiere (in Russian), *Veterinariya (Moscow)*, 9, 48, 1976.

62. **Tashkinov, N. I.**, Piroplasmosis of reindeer and its control (in Russian), *Tr. Vses. Inst. Eksp. Vet.*, 56, 54, 1982.
63. **Petrov, V. G.**, The development of the piroplasm *Babesiella bovis* in the tick *Ixodes ricinus* L. and a method to examine the ticks for pathogen carriership (in Russian). III, *Soveshch. Parasitol. Probl.*, p. 29, 1941.
64. **Joyner, L. P. and Donnelly, J.**, The epidemiology of babesial infections, *Adv. Parasitol.*, 17, 115, 1979.
65. **Krylov, M. V.**, On the development of some Piroplasmidae in the vertebrate hosts (in Russian), *Acta Protozool.*, 2, 307, 1964.
66. **Knuth, P.**, Weitere Mitteilungen über *Haemaphysalis cinnabarina* und über umfangreiche Blutungen in die Muskulatur beim Rinde, *Berl. Tierärztl. Wochenschr.*, 52, 825, 1914.
67. **Brocklesby, D. W. and Barnett, S. F.**, Large Babesia species transmitted to splenectomized calves by field collections of British ticks *(Haemaphysalis punctata), Nature (London)*, 228, 1215, 1970.
68. **Vecherkin, S. S. and Esikov, V. I.**, The causative agents of bovine hemosporidioses in *Boophilus calcaratus (in Russian), Tr. Inst. Zool. Parazitol. Acad. Nauk Kirg. SSR*, 5, 129, 1956.
69. **Vecherkin, S. S. and Esikov, V. I.**, Bovine hemosporidioses in Kirgizia and control measures against them (in Russian), *Tr. 5. Konf. Prirod. Ochag. Bolez. Vop. Parazitol. Respub. Sred. Azii Kazakh.*, 4, 148, 1964.
70. **Uilenberg, G.**, Ruminant babesiosis in the Netherlands: parasites — epidemiology — clinical data, in *Agriculture. Some Important Parasitic Infections in Bovines Considered from Economic and Social (Zoonosis) Points of View*, Parasitol. Symp., Euzeby, J., Gevrey, J., Eds., Commission of the European Communities, Luxemburg, 1984, 17.
71. **Liebisch, A., Melfsen, J., and Rahman, M. S.**, Zum Vorkommen der Zecke *Haemaphysalis punctata* und von *Babesia major* beim Rind in Norddeutschland, *Berl. Münch. Tierärztl. Wochenschr.*, 89, 477, 1976.
72. **Cerny, V. and Liplich, L.**, Die erste Feststellung von Texasfieber des Rindes auf dem Gebiet der CSR (in Czech), *Vet. Cas.*, 7, 160, 1958.
73. **Morzaria, S. P., Brocklesby, D. W., and Harradine, D. L.**, Experimental transmission of *Babesia major* by *Haemaphysalis punctata, Res. Vet. Sci.*, 23, 261, 1977.
74. **Morzaria, S. P., Brocklesby, D. W., Harradine, D. L., and Barnett, S. F.**, *Babesia major* in Britain: inactivity of suspensions derived from ground-up *Haemaphysalis punctata* nymphs, *Int. J. Parasitol.*, 4, 437, 1974.
75. **Gray, J. S. and de Vos, A. J.**, Studies on a bovine *Babesia* transmitted by *Hyalomma marginatum rufipes* Koch, 1844, *Onderstepoort J. Vet. Res.*, 48, 215, 1981.
76. **Thomas, S. E. and Mason, T. E.**, Isolation and transmission of an unidentified *Babesia* sp. infective for cattle, *Onderstepoort J. Vet. Res.*, 48, 155, 1981.
77. **Dipeolu, O. O. and Amoo, A.**, The presence of kinetes of a *Babesia* species in the haemolymph smears of engorged *Hyalomma* ticks in Nigeria, *Vet. Parasitol.*, 17, 41, 1984/1985.
78. **Chaudhri, R. P., Gill, B. S., and Khan, M. H.**, Studies on transmission of *Babesia bigemina, Ann. Soc. Belge Med. Trop.*, 55(4), 327, 1975.
79. **Mamatkulov, S. S.**, On the transovarial transmission of blood parasites by the tick *Hyalomma anatolicum* (in Russian), *Veterinariya (Moscow)*, 47, 59, 1970.
80. **Mirzabekov, K. D.**, Infection rate and developmental forms of 3 *Babesia* and *Piroplasma* species (Piroplasmida) in *Ixodid* tick eggs (in Russian), *Parazitologiya*, 11 (3), 230, 1977.
81. **Minami, T. and Ishihara, T.**, *Babesia ovata* sp.n. isolated from cattle in Japan, *Natl. Inst. Anim. Health Q.*, 20, 101, 1980.
82. **Takahashi, K., Wada, A., Kawai, S., Yokota, H., Kurosowa, T., and Sonoda, M.**, Investigations of isolation, transmission and virulence of bovine *Babesia* sp. in Hokkaido (in Japanese), *J. Coll. Dairying*, 10, 171, 1983.
83. **Ishihara, T.**, Tick-Borne Diseases. I. Tick-Borne Diseases in Japan and Korea, Tech. Bull. No. 61, Food and Fertilizer Technology Center, Taiwan, 1982, 1.
84. **Hafez, M., El-Refail, A. H., and Michael, S. A.**, Experimental transmission of some blood parasites by *Rhipicephalus* species in Egypt. II. The role of *R. turanicus* in transmitting *Babesia ovis* in sheep, *J. Egypt. Soc. Parasitol.*, 12, 543, 1982.
85. **D'yakonov, L. P.**, Die Bedeutung von *Rhipicephalus turanicus* B. Pom. (1940) in der Epizootiologie der Haemosporidiosen der Schafe, *Veterinariya (Moscow)*, 36, 30, 1959.
86. **Rastegaieva, E. F. and Timofiev, P. S.**, Zur Frage der endoglobularen Parasiten der Schafe und Ziegen und ihrer Überträger, *Berl. Tierärztl. Wochenschr.*, 52, 584, 1936.
87. **Reid, J. F., Armour, J., Jennings, F. W., and Urquhart, G. M.**, *Babesia* in sheep — first isolation, *Vet. Rec.*, 99, 419, 1976.
88. **Büscher, G., Friedhoff, K. T., and El-Allawy, T. A. A.**, Quantitative description of the development of Babesia ovis in Rhipicephalus bursa (hemolymph, ovary, eggs), *Parasitol. Res.*, in press.

89. **Li, P. N.**, Spontaneous rate of infection with *Babesiella ovis* in *Rhipicephalus bursa* on the Crimea (in Russian), *Sb. Nauchn. Rab. Sarat. Nauchno Issled. Vet. Stn.*, 6, 163, 1963.
90. **Gasanov, A. A.**, The infection of *Rhipicephalus bursa* with *Babesia ovis* in the Nakhichevan region of the ASSR (in Russian), *Izv. Akad. Nauk Az. SSR Ser. Biol. Nauk.*, 4, 117, 1963.
91. **Markov, A. A. and Abramov, J. V.**, Results of a 20 years observation of repeated cycles of *Babesia ovis* in 44 generations of *Rhipicephalus bursa* (in Russian), *Tr. Vses. Inst. Eksp. Vet.*, 38, 5, 1970.
92. **Prokhorova, P. K.**, The duration of ovine piroplasmid infection in *Rhipicephalus bursa* fed on cattle (in Russia), *Nauchn. Tr. Stavrop. Skh. Inst.*, 5 (41), 25, 1978.
93. **Fine, P. E. M.**, Vectors and vertical transmission: an epidemiologic perspective, *Ann. N.Y. Acad. Sci.*, 266, 173, 1975.
94. **Abramov, I. V., Tsaprun, A. A., Stepanova, N. I., and Lebedev, E. M.**, The importance of the quantity of ticks for the infection of animals with hemosporidia (in Russian), *Tr. Vses. Inst. Eksp. Vet.*, 19(2), 44, 1952.
95. **Stepanowa, N. I.**, Zur Frage der Epizootologie der Babesiellose der Schafe (in Russian), *Tr. Vses. Inst. Eksp. Vet.*, 21, 123, 1957.
96. **Markov, A. A. and D'yakonov, L. P.**, Study of transmission of haemoparasites of sheep in the mammalian and arthropod host (in Russian), *Tr. Vses. Inst. Eksp. Vet.*, 38, 15, 1970.
97. **Pegreffi, G. and Mura, D.**, Sulla piroplasmosi delle capre, *Atti Soc. Ital. Sci. Vet.*, 2, 72, 1948.
98. **Uilenberg, G., Rombach, M. C., Perie, N. M., and Zwart, D.**, Blood parasites of sheep in the Netherlands. II. *Babesia motasi* (Sporozoa, Babesiidae), *Vet. Q.*, 2, 3, 1980.
99. **Enigk, K., Friedhoff, K., and Wirahadiredja, S.**, Die Piroplasmosen der Wiederkäuer in Deutschland, *Dtsch. Tierarztl. Wochenschr.*, 70, 422, 1963.
100. **Lewis, D. and Herbert, I.**, A large babesia of sheep from North Wales, *Vet. Rec.*, 107, 352, 1980.
101. **Christensson, D. and Thunegard, E.**, *Babesia motasi* in sheep on the island of Gotland in Sweden, *Vet. Parasitol.*, 9, 99, 1981.
102. **Enigk, K.**, Die Schafpiroplasmose in Deutschland, *Dtsch. Tierarztl. Wochenschr.*, 63, 161, 1956.
103. **Rastegaieva, E. F.**, Zur Frage der Überträger der Schafpiroplasmosen in Azerbaidschan (transl.), *Arch. Wiss. Prakt. Tierheilkd.*, 67, 176, 1933.
104. **Motas, C.**, La piroplasmose ovine, *C. R. Seances Soc. Biol. Paris.*, 54/55, 1522, 1903.
105. **Hashemi-Fesharki, R. and Uilenberg, G.**, *Babesia crassa* n. sp. (Sporozoa, Babesiidae) of domestic sheep in Iran, *Vet. Q.*, 3(1), 1, 1981.
106. **Hoogstraal, H.**, African Ixodoidea, I. Ticks of the Sudan, Res. Rep. NM 005050.29.07, U.S. Naval Medical Research Unit No. 3, Cairo, Egypt, 1956,1.
107. **Enigk, K.**, Die Überträger der Pferdepiroplasmose, ihre Verbreitung und Biologie, *Arch. Wiss. Prakt. Tierheilkd.*, 78 (3), 209, 1943.
108. **Enigk, K.**, Zur Epidemiologie der Pferdepiroplasmose, *Dtsch. Tropenmed. Z.*, 47, 333, 1943.
109. **Enigk, K.**, Weitere Untersuchungen zur Überträgerfrage der Pferdepiroplasmose, *Arch. Wiss. Prakt. Tierheilkd.*, 79 (1), 58, 1944.
110. **Holbrook, A. A., Anthony, D. W., and Johnson, A. J.**, Observations on the development of *Babesia caballi* (Nuttall) in the tropical horse tick (Dermacentor nitens Neumann, *J. Protozool.*, 15(2), 391, 1968.
111. **Abramov, I. V.**, The duration of preservation of the agent of equine piroplasmosis (*Piroplasma caballi*) in the tick *Hyalomma plumbeum* Panzer, 1795 (in Russian), *J. Agric. Sci. (Moscow)*, 32(3), 42, 1955.
112. **Abramov, I. V.**, The importance of *Hyalomma plumbeum* Panzer, 1795, as reservoir of *Piroplasma caballi* Nuttall and Strickland, 1910 (in Russian), Tr. *Vses. Inst. Eksp. Vet.*, 27, 241, 1957.
113. **Dzasokhov, G. S. and Tsaprun, A. A.**, The duration of the preservation of the infection of equine piroplasmosis in *Dermacentor silvarum* (in Russian) Cited by 115., *Sov. Vet.*, 5, 43, 1939.
114. **Budnik, V. S.**, The duration of preservation and transovarial transmission of *Piroplasma caballi* in *Dermacentor silvarum* (in Russian), *Veterinariya (Moscow)*, 20 (2), 15, 1941.
115. **Nikol'sky, S. N. and Prokhorova, P. K.**, Ixodid ticks — reservoir of babesiids (in Russian), *Veterinariya (Moscow)*, 49 (4), 56, 1972.
116. **Marchenko, G. F.**, On some conditions of preservation and transmission of *Piroplasma caballi* by *Dermacentor silvarum* (in Russian), *Sb. Nauchn. Tr. Stavropol. Inst. Epidemiol. Mikrobiol.*, 1, 67, 1947.
117. **Kartashev, M. V.**, Studies on the infection rate of *Dermacentor* with *Piroplasma caballi* under different agricultural conditions in the central region of the U.S.S.R. (in Russian), *Tr. Vses. Inst. Eksp. Vet.*, 27, 210, 1957.
118. **Klinckmann, G.**, Sporozoitenstabilate von *Babesia equi* aus *Hyalomma anatolicum* und *Rhipicephalus turanicus*, Dr. med. vet. thesis, Tierärztliche Hochschule, Hannover, West Germany, 1981.
119. **Abramov, J. V., Tsaprun, A. A., and Lebdev, E. M.**, The importance of a single tick in the transmission of the pathogen of equine piroplasmosis (in Russian), *Veterinariya (Moscow)*, 27(3), 12, 1950.
120. **Toit, P. J. du**, Experimentelle Studien über die Pferdepiroplasmose, *Arch. Schiffs Trop. Hyg.*, 23, 141 and 359, 1919.

121. **Stiller, D. and Frerichs, W. M.**, Experimental transmission of *Babesia caballi* to equines by different stages of the tropical horse tick, *Rec. Adv. Acarol.*, 2, 263, 1979.
122. **Salyaev, N. A.**, Haemosporidioses of horses, in *Parasitology and Parasitic Diseases of Livestock*, Ershov, V. S., Ed., State Publishing House for Agricultural Literature, Moscow, 1956, 353.
123. **Anastos, G.**, The Ticks, or Ixodides, of the U.S.S.R. — A Review of the Literature, Public Health Service, National Institutes of Health, U.S. Department of Health, Education, and Welfare. Washington, D.C., 1957.
124. **Theiler, A.**, Transmission of equine Piroplasmosis by ticks in South Africa, *J. Comp. Pathol. Ther.*, 19, 283, 1906.
125. **Young, A. S. and Purnell, R. E.**, Observations on *Babesia equi* in the salivary glands of *Rhipicephalus evertsi*, *Bull. Epizoot. Dis. Afr.*, 21, 377, 1973.
126. **Enigk, K.**, personal communication, 1985.
127. **Tenter, A. M. and Friedhoff, K. T.**, Serodiagnosis of experimental and natural *Babesia equi and B. caballi* infections, *Vet. Parasitol.*, 20, 49, 1986.
128. **Friedhoff, K. T., Klinckmann, G., and Sigrist, B.**, Transmission of *Babesia equi* by *Hyalomma a. anatolicum* and *Rhipicephalus turanicus*, *Abstr. World Assoc. Adv. Vet. Parasitol.*, No. 209, 47, 1985.
129. **Sigrist, B.**, Übertragung von *Babesia equi* durch *Hyalomma anatolicum anatolicum und Rhipicephalus turanicus*, Dr. med. vet. thesis, Tierärztliche Hochschule, Hannover, West Germany, 1983.
130. **Budnik, V. S.**, New data on the mechanism of transmitting the organism of nutalliosis in horses by the pasture tick (*Dermacentor marginatus* Sulz) (in Russian), *Veterinariya (Moscow)*, 32(8), 36, 1955.
131. **Abramov, I. V.**, A new type of transmission of *Nuttallia equi* by tick vectors (in Russian), *Veterinariya (Moscow)*, 32(8), 43, 1955.
132. **Abramov, J. V.**, The transmission of *Nuttallia equi* by *Dermacentor marginatus* within one generation (in Russian), *Tr. Vses. Inst. Eksp. Vet.*, 19(2), 42, 1952.
133. **Brumpt, E. and Larrousse, F.**, Transmission de la Piroplasmose canine francaise par le *Dermacentor venustus*, *Bull. Soc. Pathol. Exot.*, 15, 540, 1922.
134. **Enigk, K.**, Das Vorkommen der Hundepiroplasmose in den besetzten Ostgebieten, *Dtsch. Tropenmed. A.*, 48 (3-6), 88, 1944.
135. **Brumpt, E.**, Cycle evolutive de *Piroplasma canis* chez les Ixodines, *C. R. Seances Soc. Biol. Paris*, 124, 928, 1937.
136. **Liebisch, A. and Gillani, S.**, Experimentelle Übertragung der Hundebabesiose (*Babesia canis*) durch einheimische deutsche Zeckenarten. I. Die braune Hundezecke (*Rhipicephalus sanguineus*), *Dtsch. Tierärztl. Wochenschr.*, 86, 149, 1979.
137. **Reichenow, E.**, Übertragungsweise und Entwicklung der Piroplasmen, *Zentralbl. Bakteriol.*, *Parasitenkd. Infektionskr. Hyg. Abt. 1 Orig.*, 135, (Beiheft) 108, 1935.
138. **Shortt, H. E.**, *Babesia canis*: the life cycle and laboratory maintenance in its arthropod and mammalian hosts, *Int. J. Parasitol.*, 3, 119, 1972.
139. **Regendanz, P. and Reichenow, E.**, Beitrag zur Übertragungsweise von *Babesia canis* durch Zecken, *Zentralbl. Bakteriol. Parasitenkd. Infektionskr. Hyg. Abt. 1 Orig.*, 124, 471—478, 1932.
140. **Belitzer, A. and Markov, A.**, L'agent vecteur de la piroplasmose canine dans les regions centrales de l'U.R.S.S. (Russie), *Ann. Parasitol. Hum. Comp.*, 8, 598, 1930.
141. **Uilenberg, G., Verdiesen, P. A. H. M., and Zwart, D.**, Imidocarb: a chemoprophylactic experiment with *Babesia canis*, *Vet. Q.*, 3, 118, 1981.
142. **Brumpt, E.**, Transmission de la piroplasmose canine française par le *Dermacentor reticulatus*. Embolies parasitaires dans les capillaires de l'encephale, *Bull. Soc. Pathol. Exot.*, 12, 651, 1919.
143. **Lounsbury, C. P.**, Ticks and malignant jaundice of the dog. *J. Comp. Pathol. Ther.*, 17, 113, 1904.
144. **Sen, S. K.**, The vector of canine piroplasmosis due to Piroplasma gibsoni, *Indian J. Vet. Sci.*, 3, 356, 1933.
145. **Walter, G.**, The transmission and course of parasitaemia of *Babesia microti* (Hannover I strain) in *Clethrionomys glareolus* and *Microtus agrestis*, *Acta Trop.*, 41, 259, 1984.
146. **Walter, G. and Weber, G.**, Untersuchungen zur Übertragung (transstadial, transovarial) von *Babesia microti*, Stamm "Hannover I", in *Ixodes ricinus*, *Ann. Trop. Med. Parasitol.*, 32, 209, 1981.
147. **Young, A. S.**, Studies on the Blood Parasites of Small Animals with Special Reference to Piroplasms, Ph.D. thesis, University of London, 1970.
148. **Hussein, H. S.**, *Ixodes trianguliceps*: seasonal abundance and role in the epidemiology of *Babesia microti* infection in north-western England, *Ann. Tropenmed. Parasitol.*, 74, 531, 1980.
149. **Piesman, J. and Spielman, A.**, Human Babesiosis on Nantucket Island: prevalence of *Babesia microti* in ticks, *Am. J. Trop. Med. Hyg.*, 29, 742, 1980.
150. **Spielman, A., Clifford, C. M., Piesman, J., and Corwin, M. D.**, Human Babesiosis on Nantucket Island, USA: description of the vector *Ixodes* (Ixodes) *dammini*, n.sp. (Acarina: Ixodidae), *J. Med. Entomol.*, 15, 218, 1979.

151. **Spielman, A.,** Human Babesiosis on Nantucket Island: transmission by nymphal *Ixodes* ticks, *Am. J. Trop. Med. Hyg.*, 25, 784, 1976.
152. **Healy, G. R., Spielman, A., and Gleason, N.,** Human Babesiosis: reservoir of infection on Nantucket Island, *Science*, 192, 479, 1976.
153. **Oliveira, M. R. and Kreier, J. P.,** Transmission of *Babesia microti* using various species of ticks as vectors, *J. Parasitol.*, 65, 816, 1979.
154. **Gengu, U. E. and Kreier, J. P.,** Transmission experiments with *Babesia microti* (Gray strain) using *Dermacentor andersoni* tiles as a vector, *Ohio J. Sci.*, 76(4), 188, 1976.
155. **Kreier, J. P.,** personal communication, 1979.
156. **Shiota, T., Kurimoto, H., Haguma, N., and Yoshida, Y.,** Studies on *Babesia* first found in Japan. I. Epidemiology and morphology (in Japanese), *Jpn. J. Parasitol.*, 32, 165, 1983.
157. **Adler, S. and Feldman-Muhsam, B.,** Transmission of *Nuttallia* of a Gerbil by *Rhipicephalus sanguineus*, *Nature (London)*, 169, 552, 1952.
158. **Feldman-Muhsam, B.,** A note on the transmission of a *Nuttallia* of *Meriones shawi* by the tick *Rhipicephalus secundus*, *J. Protozool.*, 5, 225, 1958.
159. **Hadani, A. and Theodor, O.,** Studies on the transmission of Nuttallia danii (Tsur, Hadani and Pipano, 1960) by ticks to the gerbil (Meriones tristrami), *J. Protozool.*, 18 (Abstr.), 34, 1971.
160. **Krylov, M. V.,** The development of Nuttallia tadzhikistanica Krylov et Zanina, 1962 in the tick *Hyalomma anatolicum* (in Russian), *Acta Protozool.*, 3, 369, 1965.
161. **Hadani, A. and Theodor, O.,** Factors affecting the rate of infection of *Hyalomma excavatum* nymphs with *Nuttallia danii* (Tsur, Hadani und Pipano, 1960), *J. Protozool.*, 20 (Abstr.), 532, 1973.
162. **Feldman-Muhsam, B.,** On *Nuttallia adleri* of *Meriones tristrami*, *11th Int. Kongr. Entomol., Wien*, 1962, 3.
163. **Fay, F. H. and Rausch, R. L.,** Parasitic organisms in the blood of arvicoline rodents in Alaska, *J. Parasitol.*, 55, 1258, 1969.
164. **Gunders, A. E. and Hadani, A.,** An argasid tick, *Ornithodoros erraticus* (Lucas) a natural vector of *Nuttallia meri* Gunders, *Z. Trop. med. Parasitol.*, 24, 536, 1973.
165. **Gunders, A. E.,** Transmission of mammalian piroplasm by an argasid tick, *Nature (London)*, 247, 225, 1973.
166. **Gunders, A. E.,** Piroplasmal sporozoites in the argasid *Ornithodoros erraticus*, *Experientia*, 33, 892, 1977.
167. **Killick-Kendrick, R.,** Parasitic protozoa of the blood of the rodents. II, *Acta Trop.*, 31, 1, 1974.
168. **Levine, D.,** Taxonomy of the piroplasms, *Trans. Am. Microsc. Soc.*, 90, 2, 1971.
169. **Kurchatov, V. I. and Markov, A. A.,** (in Russian), *Veterinariya Moscow*, 2, 63, 1940; as cited by **Neitz, W. O.,** *Ann. N.Y. Acad. Sci.*, 64(2), 56, 1956.
170. **Vercruysse, J. and Parent, R.,** Observation d'une epizootie de babesiose porcine a *Babesia perroncitoi* Cerruti 1939 au Senegal, *Afr. Med.*, 20, 435, 1981.
171. **Pegreffi, G. and Mura, D.,** Sulla piroplasmosi delle capre, *Atti Soc. Ital. Sci. Vet.*, 2, 72, 1948.
172. **Nikol'sky, S. M. and Pozov, S. A.,** *Ixodes ricinus* ticks as carriers of *Babesia capreoli* in the roe deer (in Russian), *Veterinariya (Moscow)*, 49 (4), 62, 1972.
173. **Yakimov, W. L. and Kolmakov, W. D.,** Zur Frage über die Piroplasmose des Rentieres, *Arch. Protistenk.*, 66, 322, 1929.
174. **Chambers, F.,** Piroplasmosis of the reindeer, *Vet. J.*, 77, 415, 1921.
175. **Perry, B. D., Nichols, D. K., and Cullom, E. S.,** *Babesia odocoilei* Emerson and Wright, 1970 in white-tailed deer, *Odocoileus virginianus* (Zimmermann), in Virginia, *J. Wildl. Dis.*, 21, 149, 1985.
176. **Dalgliesh, R. J. and Stewart, N. P.,** Some effects of time, temperature and feeding on infection rates with *Babesia bovis* and *Babesia bigemina* in *Boophilus microplus* larvae, *Int. J. Parasitol.*, 12, 323, 1982.
177. **Dalgliesh, R. J. and Stewart, N. P.,** The extraction of infective *Babesia bovis* and *Babesia bigemina* from tick eggs and *B. bigemina* from unfed larval ticks, *Aust. Vet. J.*, 54, 453, 1978.
178. **Ronald, N. C. and Cruz, D.,** Transmission of *Babesia bovis*, using undifferentiated embryonic cells from *Boophilus microplus* tick eggs, *Am. J. Vet. Res.*, 43, 544, 1981.
179. **O'Sullivan, P. J. and Callow, L. L.,** Loss of infectivity of a vaccine strain of *Babesia argentina* for *Boophilus microplus*, *Aust. Vet. J.*, 42, 252, 1966.
180. **Dalgliesh, R. J., Stewart, N. P., and Duncalfe, F.,** Reduction in pathogenicity of *Babesia bovis* for its tick vector, *Boophilus microplus*, after rapid blood passage in splenectomized calves, *Z. Parasitenkd.*, 64, 347, 1981.
181. **Dalgliesh, R. J. and Stewart, N. P.,** Failure of vaccine strains of *Babesia bovis* to regain infectivity for ticks during long-standing infections in cattle, *Aust. Vet. J.*, 53, 429, 1977.
182. **Kahl, L. P., Mitchell, G. F., Dalgliesh, R. J., Stewart, N. P., Rodwell, B. J., Mellors, L. T., Timms, P., and Callow, L. L.,** *Babesia bovis*: proteins of virulent and avirulent parasites passaged through ticks and splenectomized or intact calves, *Exp. Parasitol.*, 56, 222, 1983.

183. **Stewart, N. P., Dalgliesh, R. J., and De Vos, A. J.,** Effect of different methods of maintenance on the development and morphology of *Babesia bigemina* in the gut of *Boophilus microplus, Res. Vet. Sci.*, 40, 94, 1986.
184. **Wright, I. G., Mirre, G. B., Mahoney, D. F., and Goodger, B. V.,** Failure of *Boophilus microplus* to transmit irradiated *Babesia bovis, Res. Vet. Sci.*, 34, 124, 1983.
185. **Gray, J. S.,** The effects of the piroplasm *Babesia bigemina* on the survival and reproduction of the blue tick, *Boophilus decoloratus, J. Invertebr. Pathol.*, 39, 413, 1982.
186. **Davey, R. B.,** Effects of *Babesia bovis* on the ovipositional success of the southern cattle tick, *Boophilus microplus 2.3, Ann. Entomol. Soc. Am.*, 74, 331, 1981.
187. **Friedhoff, K. T.,** Microorganisms in engorged females of the tick *Rhipicephalus bursa. I, J. Parasitol.*, 56 (4), 111, 1970.
188. **Ristic, M. and Healy, G. R.,** Babesiosis, in *CRC Handbook Series in Zoonoses. Section C, Vol. 3, Steele, J. H., Ed.*, CRC Press, Boca Raton, Fla., 1982, 1.
189. **Enigk, K.,** Die intrauterine Übertragung von Blutprotozoen bei den Haustieren, *Dtsch. Tropenmed. Z.*, 46, 153, 1942.
190. **Aragon, R. S.,** Bovine babesiosis: a review, *Vet. Bull.*, 46, 903, 1976.
191. **Correa, W. M., Correa, C. N. M., and Fanton, E. B.,** Equine abortion by foetal babesiosis, *Arch. Escola Vet. Univ. Fed. Minas Gerais*, 30, 303, 1978.
192. **Friedhoff, K. T.,** Die Piroplasmen der Equiden — Bedeutung für den internationalen Pferdeverkehr, *Berl. Münch. Tierärztl. Wochenschr.*, 95, 368, 1982.
193. **Knowles, R. C.,** Equine piroplasmosis (babesiasis) of the *Babesia caballi* type, *Equine Pract.*, 5, 18, 1983.
194. **Muratov, E. A. and Kheissin, E. M.,** Die Entwicklung von Piroplasma bigeminum in der Zecke *Boophilus calcaratus* (in Russian), Zool. J., 38 (7), 970, 1959.
195. **Kossel, H., Weber, A., Schütz, W., and Miessner, H.,** Über die Haemoglobinurie der Rinder in Deutschland, *Arb. Gesund. Berlin*, 20, 1, 1904.
196. **Tsaprun, A. A.,** Ixodid ticks as reservoir hosts of *Haemosporidia* and the contemporary development of different stages of these parasites with metamorphosis of the tick hosts (in Russian), *Sb. Nauchn. Rab. Sib. Nauchno Issled. Vet. Inst.*, 5, 275, 1954.
197. **Enigk, K.,** Zur Frage der Einschleppung der Pferdepiroplasmosen nach Deutschland, *Dtsch. Tierärztl. Wochenschr.*, 54, (1/2), 1, 1947.

Chapter 3

HOST CELL-PARASITE INTERACTIONS IN BABESIOSIS

Ikuo Igarashi, Masamichi Aikawa, and Julius P. Kreier

TABLE OF CONTENTS

I. INTRODUCTION

Babesia is a tick-transmitted, intraerythrocytic protozoan parasite of animals and humans. *Babesia* parasites are introduced into the host by the bites of infected ticks and multiply within the erythrocytes of the host. They cannot survive away from the host cells, since feeding, development, and reproduction depend on environmental and nutritional factors provided by host cells. The interaction between *Babesia* and host cells is complex and has many aspects. Electron microscopy has contributed greatly to the understanding of host-parasite interactions. There are still, however, many gaps in our knowledge of the specific mechanisms that are involved in the interaction between parasite and host. It is very important to understand the relationship between *Babesia* and the host cell clearly, because the understanding of host-parasite interaction may lead to the development of new anti-*Babesia* drugs or of vaccines against *Babesia*. In this chapter, we will discuss the following four features involved in the interaction of *Babesia* and the host cell: entry into erythrocytes, growth and development of the parasite, alteration of the host cell, and development of the sexual stages in ticks.

II. FINE STRUCTURE OF THE MEROZOITE

Merozoites of *Babesia* are small (1 to 5 μm) and elongated in shape (Figure 1). The body of the merozoite is covered by a single plasma membrane with interrupted double membranes beneath it. In the anterior region of the merozoite are various organelles composing the so-called "apical complex". These include polar rings, micronemes, and rhoptries. The polar ring is formed by the double membranes under the plasma membrane at the anterior end. The rhoptries are electron-dense, teardrop-shaped structures located in the anterior region of the cell. The anterior neck portion of the organelles is narrow and extends into the polar ring area. The number of rhoptries varies among *Babesia* spp.; in *B. microti*, the merozoite contains one large rhoptry,[1] while merozoites of *B. canis* contain several rhoptries.[2,3] The micronemes are small, osmophilic structures present around the rhoptries in the anterior region of the merozoite and appear to be connected with the rhoptries.[4] The function of rhoptries and micronemes is not known exactly. There are, however, some indications that these organelles contain proteolytic enzymes and that they may play an important role in the invasion of the merozoite into the erythrocyte.[5] Two large electron-dense spherical bodies are often seen anterior to the nucleus in the merozoite of *B. bovis*.[6] The function is unknown. The cytoplasm of the merozoites contains many free ribosomes as well as endoplasmic reticulum and mitochondria-like organelles. The nucleus is usually located at the midpoint of the merozoite and is surrounded by two membranes. The contents of the nucleus are homogeneous and of low density.

The free, extracellular merozoite is covered by a surface coat (Figure 2a). This coat is composed of fibrils that adhere to the plasma membrane of the free merozoites. The surface coat disappears when the merozoites enter erythrocytes. The characteristics of the surface coat are not yet fully known. In *B. canis*, surface coat antigens are water soluble and can be separated from the merozoites by centrifugation.[7] Incubation of merozoites of *B. bovis* with immune serum causes thickening of the surface coat, aggregation with other merozoites, and finally, lysis of the merozoites (Figures 2b and 3).[8] A surface coat antigen derived from cultures of *B. bovis* is a prime candidate of the preparation of a vaccine.[9] A similar surface coat was found on merozoites of several species of *Plasmodium*.[10-12]

III. ENTRY OF THE MEROZOITE INTO THE ERYTHROCYTE

There are two possible ways for extracellular parasites to invade host cells.[5] First, as in

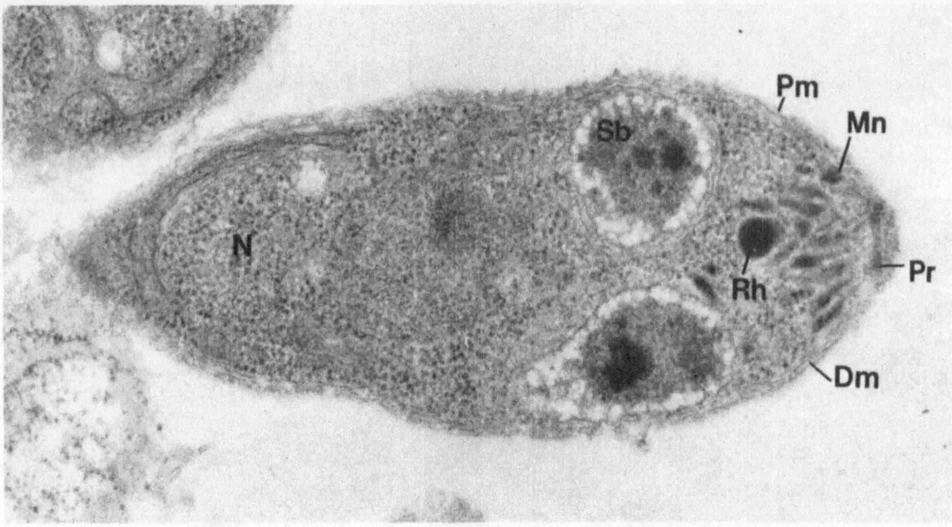

FIGURE 1. Free merozoite of *B. bovis* showing rhoptries (Rh), micronemes (Mn), plasma membrane (Pm), double membrane (Dm), spherical bodies (Sb), polar rings (Pr), and nucleus (N). Preparation from an in vitro culture. (Magnification × 35,000.)

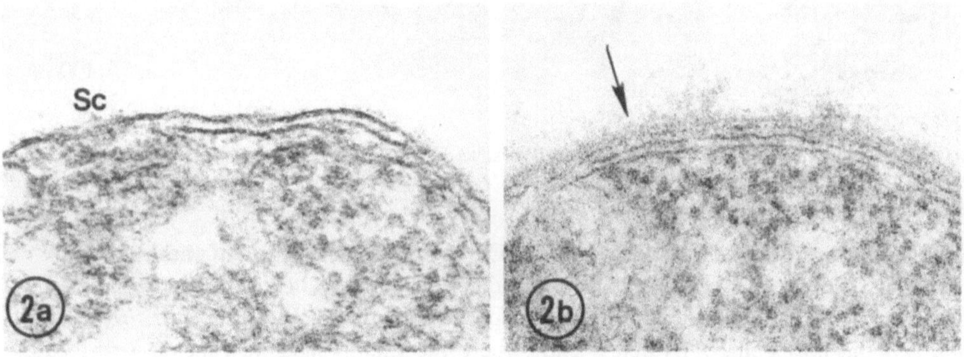

FIGURE 2. Effect of immune serum on *B. bovis* merozoites. (a) A merozoite incubated with control serum. The surface coat (Sc) is normal. (Magnification × 84,000.) (b) A merozoite incubated with immune serum showing the thickened coat (arrow). (Magnification × 78,000.) (From Smith, R. D., James, M. A., Ristic, M., Aikawa, M., and Murguia, C. A. V., *Science*, 212, 335, 1981. With permission.)

Plasmodium and *Babesia*, they may enter the host cell by endocytosis. In this process they are engulfed by the host cell.[1,12] Second, as in *Nosema*, they may invade the host cell by boring through its membrane.[13]

Entry of host cells by endocytosis has been studied extensively since 1969 when Ladda et al.[10] reported on the invasion of erythrocytes by merozoites of *P. berghei* and *P. gallinaceum*. Entry of the parasite into the erythrocyte is very complex and can be divided into three stages: (1) recognition of and attachment to the erythrocyte membrane of the merozoite, (2) invagination of the erythrocyte around the merozoite to form a parasitophorous vacuole, and (3) sealing of the erythrocyte membrane after completion of merozoite invasion.[14] Electron microscopic studies of *B. microti*[1] have shown that following attachment to the erythrocyte, the rhoptries discharge, the merozoite becomes interiorized, and parasitophorous vacuole forms in a way very similar to the process which occurs with *Plasmodium*.

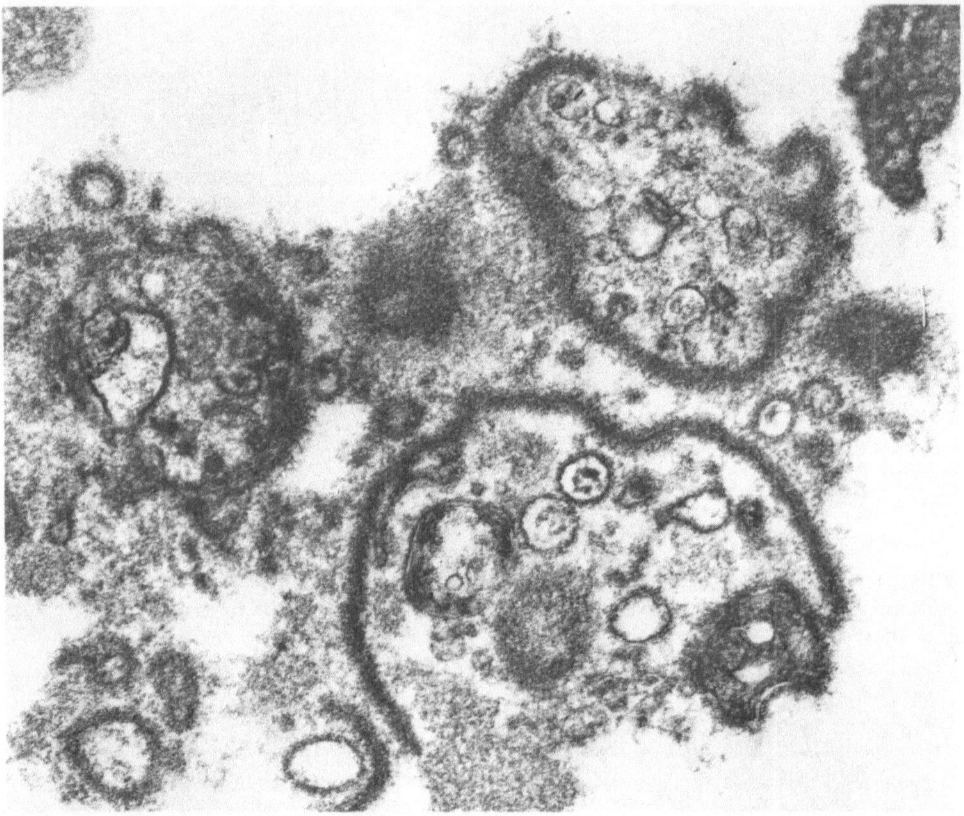

FIGURE 3. Effect of immune serum on *B. bovis* merozoites, a strong lytic effect is apparent. (Magnification × 25,000.)

A. Recognition of and Attachment to the Erythrocyte Membrane of the Merozoite

The first step in merozoite invasion is initiated by the attachment to and orientation of the merozoite to the erythrocyte surface. Before penetration can occur, the merozoite must attach to the host cell by its apical end, where the rhoptry and micronemes are located (Figure 4a). This specific orientation leads to strong adherence and invagination of erythrocyte membrane (Figure 4b). An initial loose attachment mediated by the surface coat may precede orientation and apical end attachment. It is not known exactly what determinants are involved in the specific interaction. However, an important role for complement has been suggested since the in vivo depletion of complement by injection of rats with C3 inactivator isolated from cobra venom inhibited the development of *B. rodhaini* parasitemia.[15] Treatment of rats with trypan blue, a drug that blocks C3b receptors on cell membranes, also impairs development of parasitemia.[16]

Additional evidence of a role for complement in the penetration of merozoites into erythrocytes was found in in vitro experiments.[17] Serum deficient in C3 did not support the entry of *B. rodhaini* into human red cells. When serum was reconstituted with purified human C3, the penetration rate of merozoites recovered to normal levels. Furthermore, evidence has been accumulated to indicate that both rat erythrocytes and *B. rodhaini* contain C3b receptors.[17,18] These results suggested strongly that C3b receptors are involved in the penetration of merozoites into red cells. However, they also provided conflicting evidence to explain the mechanism by which complement facilitates penetration of red cell by *Babesia*. Ward and Jack[19] suggested two possibilities for the role of C3b receptors in the entry of

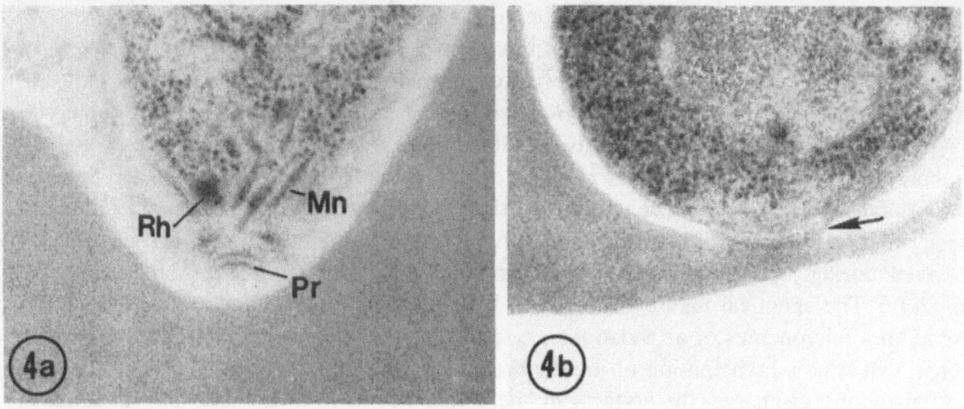

FIGURE 4. The first step of merozoite entry into the erythrocyte. (a) *Babesia* merozoite showing the initial contact with an erythrocyte at the appical end. Rhotry (Rh); micronemes (Mn), and polar ring (Pr). (Magnification × 31,000.) (b) Strong attachment of *Babesia* merozoite with the erythrocyte membrane (arrow). (Magnification × 40,000.)

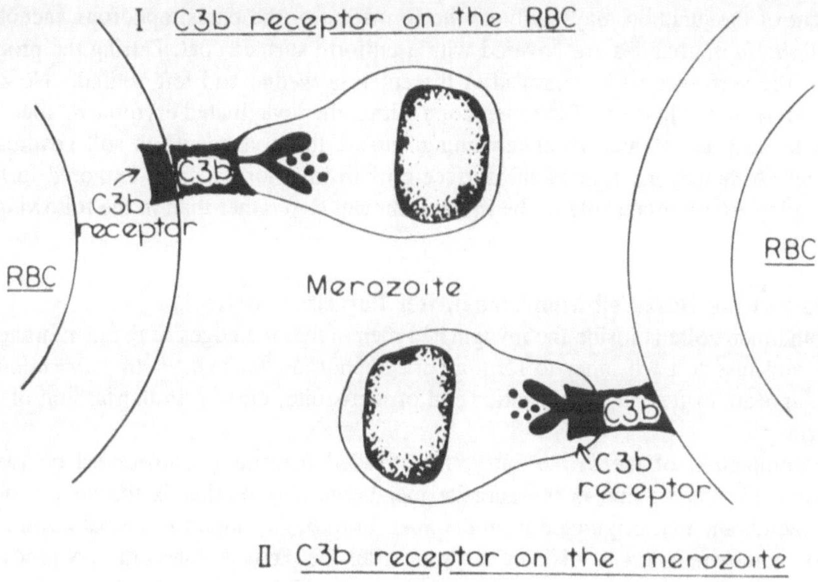

FIGURE 5. Hypothetical mechanisms on the role of C3b receptor in penetration of merozoite to red blood cells. (1) Merozoites which have an acquired C3b are attached to red cells by a C3b receptor on their surfaces. (2) C3b-bearing red bells attached to a merozoite by means of a C3b receptor on the parasites. (From Ward, P. A. and Jack, R. M., *Am. J. Pathol.*, 102, 109, 1981. With permission.)

Babesia parasites into erythrocytes (Figure 5). First, contact of merozoites with plasma activates the alternative complement pathway, and merozoites acquire C3b on their surfaces and then make the attachment to the red cell by the C3b receptor. Second, merozoites attach to C3b-containing red cells by means of the C3b receptor on the parasite. At the present time there is no way to distinguish clearly which of the two mechanisms accounts for the role of complement in the penetration of *B. rodhaini* merozoites into erythrocytes. The C3b receptor is also needed for the development of *B. bigemima*, but not for development of *B. bovis*.

B. Invagination of the Host Cell Membrane

After the merozoite attaches to the erythrocyte membrane, rhoptry discharge seems to be required to initiate invagination of the erythrocyte membrane and formation of a parasitophorous vacuole. When merozoites of *B. microti* attached by their apical ends to the erythrocyte membrane were studied, a dense substance was found between the rhoptry and the erythrocyte.[1] This substance appeared to be continuous with a common duct of the rhoptry. During invasion, the electron density of the rhoptry decreases, which suggests a release of rhoptry contents. The participation in invasion of secretory products form rhoptries was also observed during studies of invasion by merozoites of *Plasmodium*[12] and sporozoites of *Eimeria*.[20] The chemical nature of rhoptries is largely unknown. Kilejian[21] suggested that rhoptry and micronemes of an avian malaria parasite. *P. lophurae*, contain a histidine-rich protein that causes invagination of the erythrocyte membrane.

After rhoptry discharge, the erythrocyte membrane begins to invaginate. The invaginated membrane broadens and deepens, forming an ovoid vacuole, which accommodates the entering merozoite (Figure 6). The edges of the invaginating membrane have tight contact with the entering merozoites during invasion. At the point of attachment to a merozoite, however, the erythrocyte membrane does not become thickened as in *Plasmodium*.[12] The junction between the thickened erythrocyte membrane and the malaria merozoite moves along the entering parasite and maintains its position at the orifice of invagination. The movement of the junction may facilitate the formation of the parasitophorous vacuole.

Free *Babesia* merozoites are covered with a uniform surface coat. During the process of invasion, the surface coat is removed at the entering wedge and left behind.[1] No surface coat is visible on the portion of the merozoite within the invaginated erythrocyte membrane, whereas the portion of merozoite remaining outside the erythrocyte is still covered by a fuzzy coat. Although the role of the surface coat in invasion must be explored further, it seems to play an important role in the first attachment step rather than in the following steps of invasion.

C. Sealing of the Host Cell Membrane After Parasite Penetration

Once the merozoite is inside the invaginated membrane, the edges of the membrane come together and fuse to each other to form a parasitophorous vacuole.[1,23] In *Plasmodium*, the junction appears to fuse at the posterior end of merozoite, closing in the fashion of an iris diaphragm.[12]

After completion of merozoite entry, the invaded parasite is surrounded by two unit membranes (Figure 7): one is the parasite membrane and another is the parasitophorous vacuole membrane that originated from the host. In *Babesia*, unlike *Plasmodium*, however, the outer membrane starts to disintegrate soon after the parasite enters the erythrocyte. As a result, in all subsequent stages, the parasite is covered by a single plasma membrane, and the parasite is in direct contact with the erythrocyte cytoplasm (Figure 8). Similarly, the vacuole membrane disappears after the entry of *Eimeria* sporozoites or *Toxoplasma* merozoites. However, with *Eimeria* and *Toxoplasma*, unlike *Babesia*, a new membrane is soon formed around the parasite.[23,24] The absence of a parasitophorous vacuole membrane may give *Babesia* flexibility during development, resulting in a variety of parasite shapes such as pyriform, round oval, elongate (Figure 9), twisted, or folded.

IV. GROWTH AND DEVELOPMENT OF ASEXUAL STAGES OF THE PARASITE

Once a merozoite enters the erythrocyte, it transforms to a trophozoite and grows and reproduces by budding; reproduction results in two or four daughter cells. Although the generation time varies with species,[25,26] it takes about 8 hr for one binary fussion. It is of

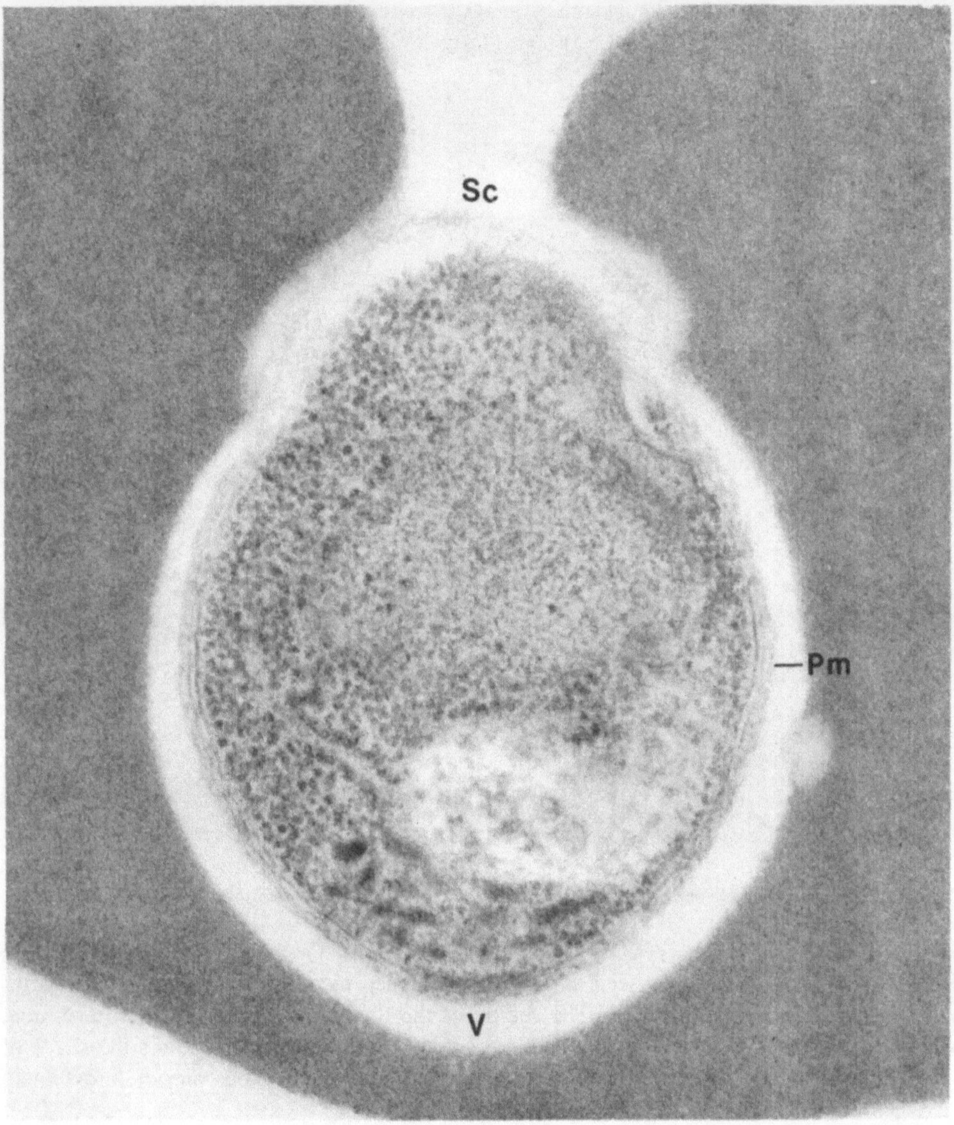

FIGURE 6. A merozoite entering the erythrocyte. The invaginated erythrocyte membrane broadens and deepens. Parasitophorous vacuole (V); plasma membrane (Pm), and surface coat (Sc) of merozoite. (Magnification × 60,000.)

interest to note that a fast-growing clone of *B. bovis* was less pathogenic than the original isolate and that the cattle receiving cloned parasites were totally immune to challenge with the original virulent parasite.[27,28] The cloned, fast-growing parasite appears to be a promising candidate for a live-type vaccine.

During the process of merozoite invasion, the rhoptry becomes less dense, and it cannot be identified after completion of invasion. In the young trophozoite, the cytoplasm of the parasite contains many free ribosomes, the smooth and rough-surfaced vesicles of the endoplasmic reticulum, and mitochondria-like structures composed of two concentric membranes (Figure 10). In the growing trophozoite, differentiation of the parasite begins with the appearance of micronemes.[4,29] These structures originate most probably from the surface vesicles of a primitive Golgi apparatus. At the next step of differentiation, segments of

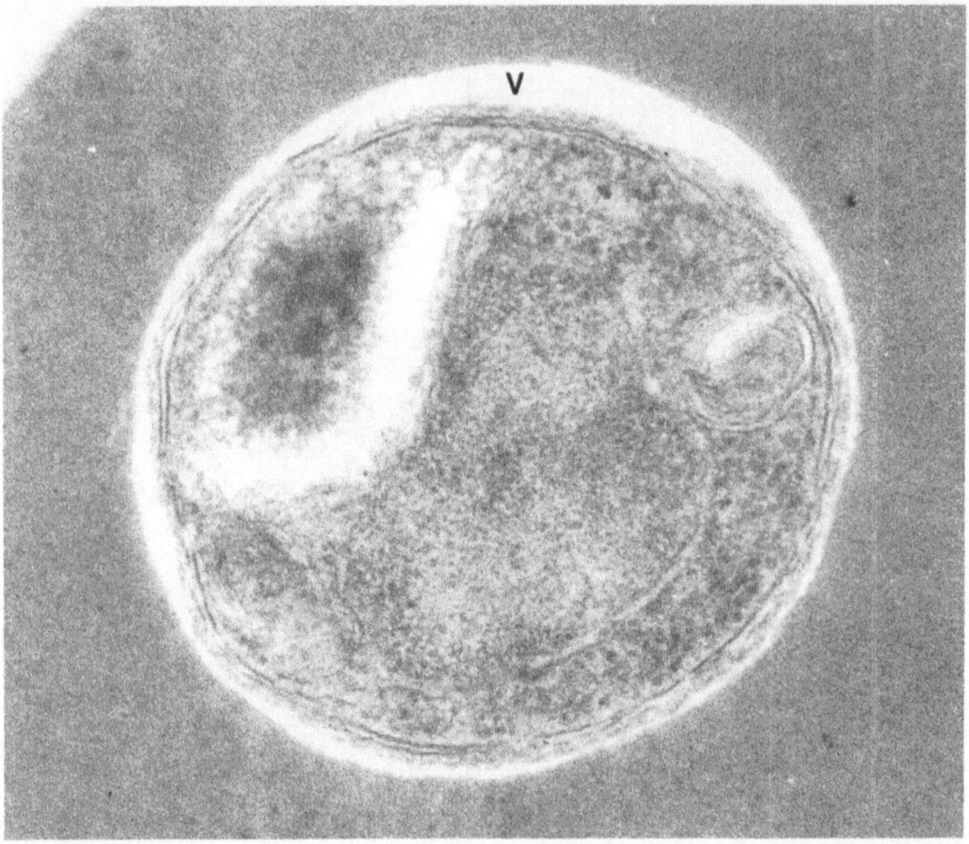

FIGURE 7. A merozoite after invasion into erythrocyte. The parasite is enclosed in a parasitophorous vacuole (V). (Magnification × 72,000.)

double membrane, sometimes seen as a single thickened membrane, appear with rhoptries beneath the plasma membrane.[4,29] The region of the double membrane starts to protrude, becoming the anterior end of the forming merozoite. The bulge soon acquires the shape of a bud. In an advanced bud, the newly formed merozoites appear to be connected to a small cytoplasmic mass or bud residue (Figure 11).[23]

During differentiation of the cytoplasm, the nucleus increases in size and becomes loboid. During division in *Babesia microti*,[4] however, neither microtubules nor fine filaments, such as those which usually appear in the nucleus of dividing *Plasmodium* parasites,[30] are observed. The lobes of the nucleus enter the newly forming merozoites and remain connected with the main body of the nucleus, even in advanced stages of division. At the early stage of differentiation, the mode of division in *Babesia* resembles schizogony in *Plasmodium*, but at later stages differences between *Babesia* and *Plasmodium* become apparent. In *Plasomodium*, nuclear division occurs first and is followed by cytoplasmic divisions. At the final stage of division, the newly formed merozoites separate from each other completely and from a residual body. In *Babesia*, the division of the cytoplasm starts prior to the development of nuclear lobes. The nucleus and cytoplasm of new merozoites remain connected with the main body, even up to the final stage of division. Finally, *Babesia* merozoites separate from one another and escape from the erythrocyte and invade new erythrocytes.

Intracellular parasites depend on the host cell as for most nutrients during development. In *Plasmodium*, the parasites ingest erythrocyte cytoplasm through a cytostome[31] and digest hemoglobin with proteolytic enzymes in digestive vesicles.[32,33] Electron-dense pigment,

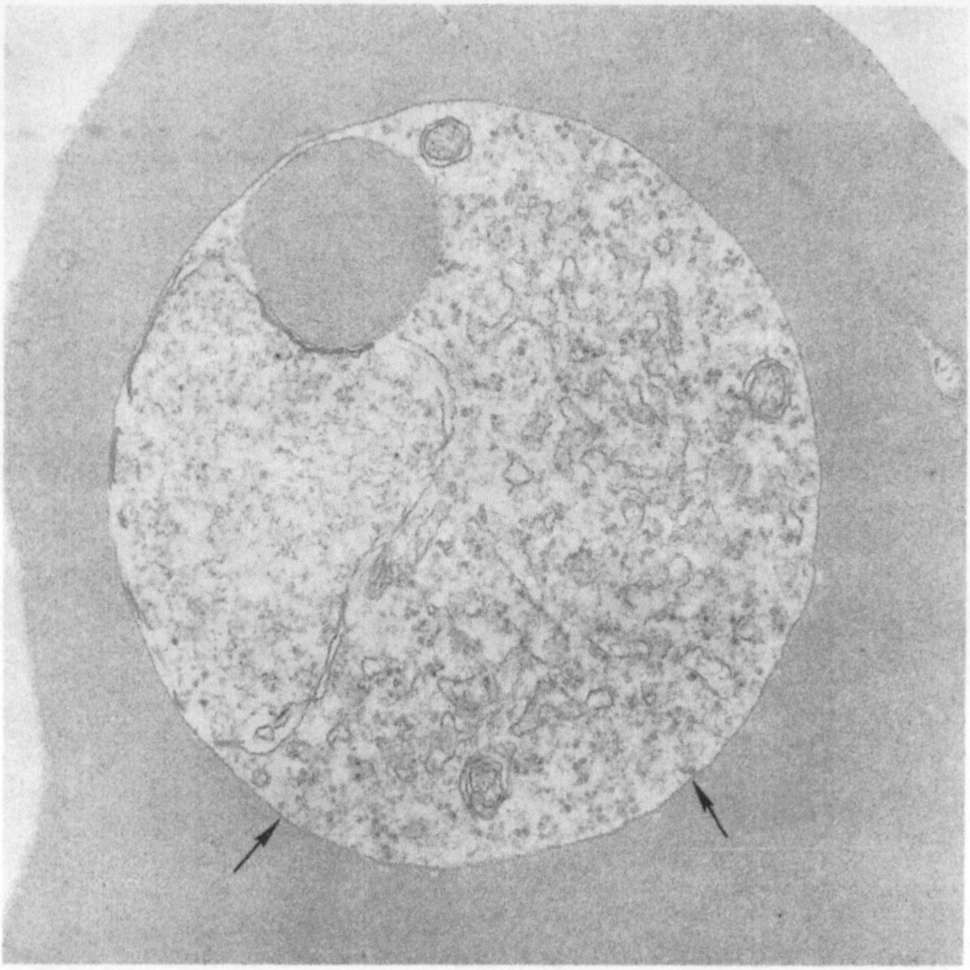

FIGURE 8. A trophozoite within an erythrocyte. The plasma membrane is in direct contact with the erythrocyte membrane (arrow). (Magnification × 48,000.)

hemozoin, is formed in food vacuoles as the final product of hemoglobin digestion. The feeding mechanism of *Babesia* is not as fully understood as that of *Plasmodium*. In *Babesia*, the existence of the cytostome, a specific organelle responsible for formation of food vacuoles, is controversial. Some workers have found cytostomes in *B. microti*,[5] in *B. equi*,[34] and in *B. canis* (Figure 10),[7] while others have failed to observe this organelle.[35,37] The formation of a food vacuole by the phagocytosis of hemoglobin has been observed by electron microscopic studies,[36,37] but the food vacuole disappeared when the host cell was removed by continuous-flow ultrasonication,[38] immune lysis,[39] and sponin treatment.[40] Serial sections of infected erythrocytes revealed that the dense bodies were not incorporated into the parasite, but were partially surrounded by the parasite due to parasite pleomorphism.[40] These results indicate that food vacuoles of *Babesia* are not true food vacuoles formed by phagocytosis and have been called ''pseudo-food vacuoles''.[40] These results suggest that the feeding mechanisms of *Babesia* differ from those of *Plasmodium*. *Babesia* parasites may obtain their nutrients by diffusion through the parasite membrane. If this is true, then the single plasma membrane surrounding the parasite and the pleomorphic shape of the parasite may seem to increase contact with the cytoplasm of the erythrocyte. In *B. equi*, the intake of nutrients seems to be carried out by a tubular structure.[36] The structure extends from the parasite to

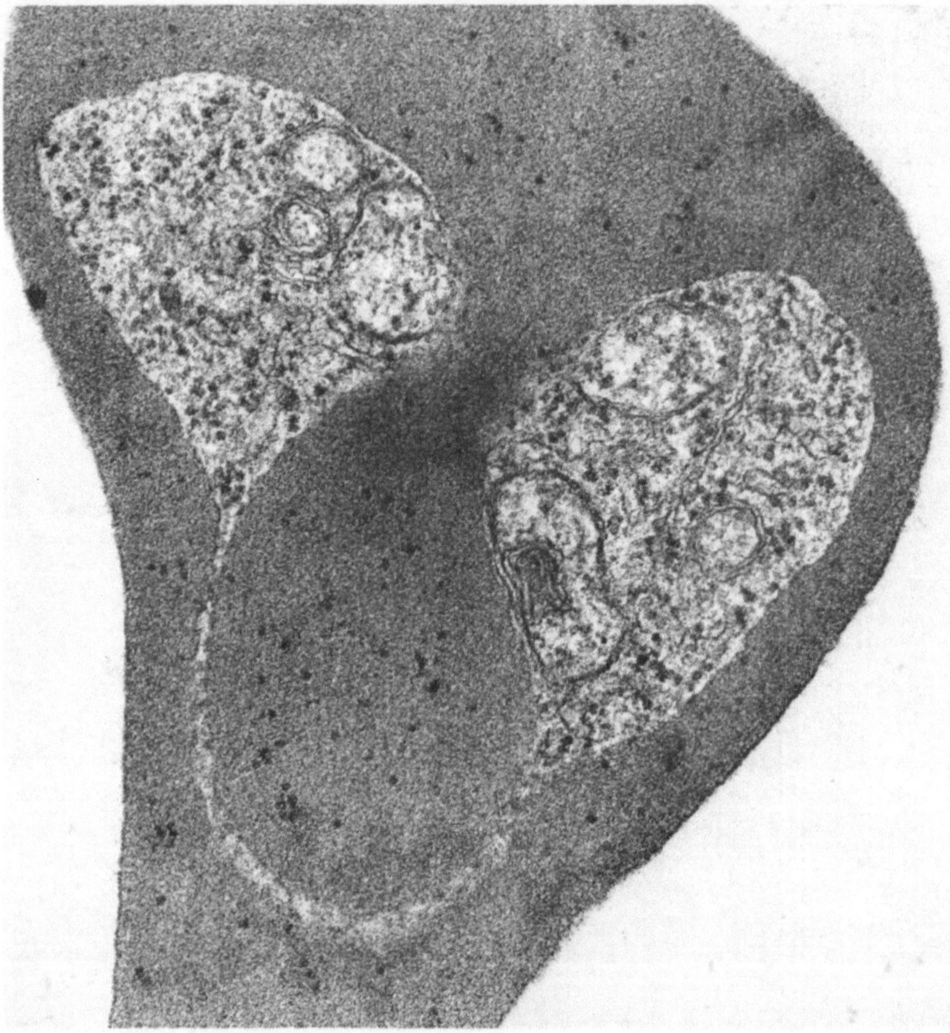

FIGURE 9. An elongated trophozoite surrounded by a single membrane. (Magnification × 60,000.)

the periphery of the erythrocyte where the lumen of the tubular structure often has direct contact with the blood plasma.[34,36,41] The tubule appears to ingest blood plasma.[34]

Rudzinska[40] also suggested that *Babesia* may feed by absorption following extracorporeal digestion. The coiled organelle encircles the cytoplasm of the erythrocyte and seems to digest it with enzymes. After digestion of the enclosed cytoplasm, the digestion products may be incorporated into the parasite. *Babesia* were shown to be capable of pinocytosis by labeling of the multimembranous structure with ferritin.[39] This result may support the suggestion that extracellular digestion occurs. Further studies are required to determine how *Babesia* ingests nutrients.

It has been reported that *Babesia* parasites lack the hemozoin pigments observed in *Plasmodium*.[37] *Babesia* may not have hemozoin because the parasites may digest hemogloblin more effectively than *Plasmodium*. Alternatively, they may not utilize hemoglobin as a source of nutrients.[22,40] Wright and Goodger[43] have demonstrated that extracts of *B. argentina* and *B. bigemina* have proteolytic enzymes for both denatured and native hemoglobin. This is the first biochemical indication that an enzyme might be involved in the utilization of

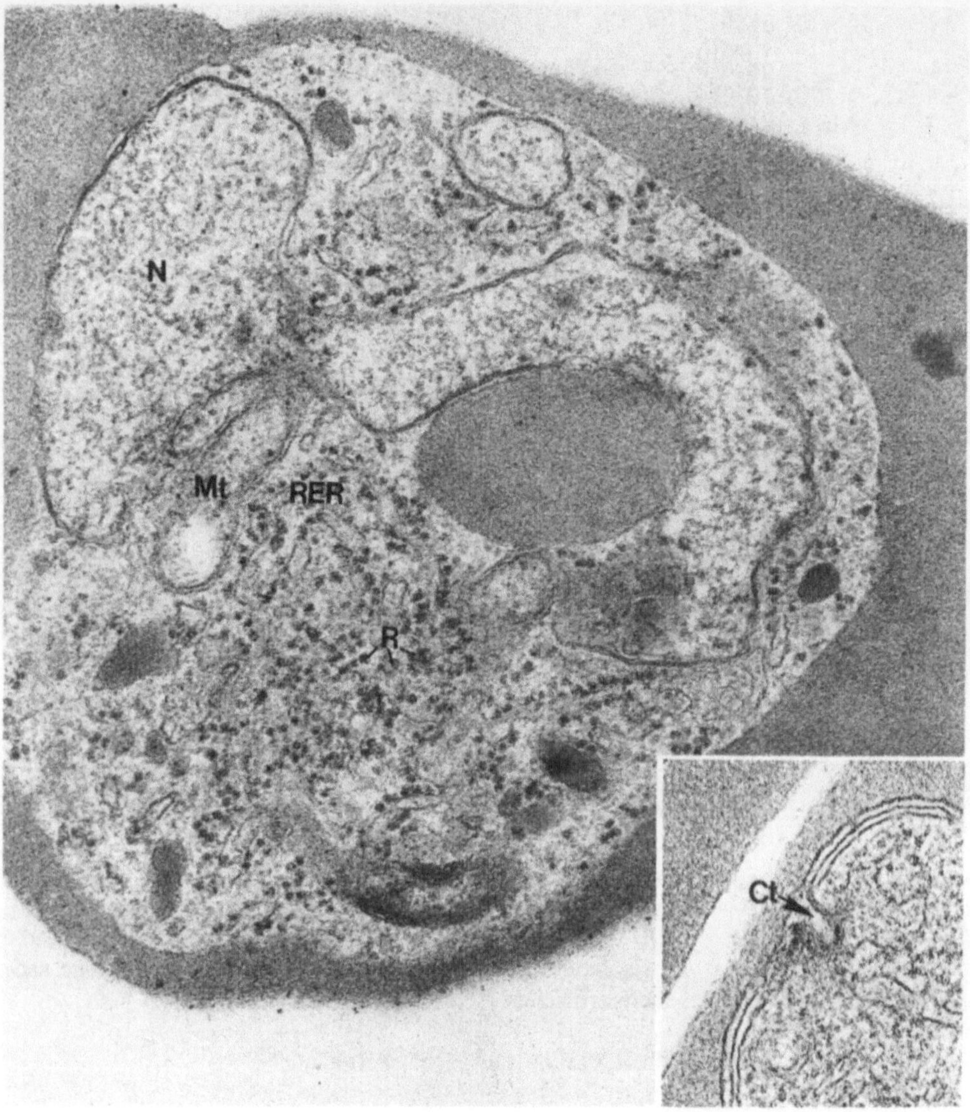

FIGURE 10. A growing trophozoite showing free ribosomes (R), rough endoplasmic recticulum (RER), mitrochondria-like structure (Mt), and nucleus (N). Insert shows a cytostome (Ct). (Magnification × 58,000.)

hemoglobin by living *Babesia* parasites. Recently, it was demonstrated that *B. hylomysci* has an aminopeptidase[43] with properties very similar to those of *Plasmodium*. Differences in the microenvironment or the active site of the proteolytic enzyme of *Babesia* and *Plasmodium* may explain the absence of hemozoin in *Babesia*. By ultrastructural enzyme cytochemistry,[33] aminopeptidase was detected throughout the cytoplasm of *B. hylomysci*, with heavier labeling in growing trophozoites than in younger forms. In *P. chabaudi*, aminopeptidase and endocrylamidase are first detected on the ribosomes, and then in the membranes of the pinocytic vesicles and within the contents of vesicles. In later stages of development, the enzyme activity decreased as malarial pigment became more and more prominent. Activity was not detected in schizonts. These observations indicate that there are differences in the methods of digestion of hemoglobin by *Babesia* and *Plasmodium*. It is assumed that *Plasmodium* incorporates large volumes of hemoglobin through a cytostome and digests it

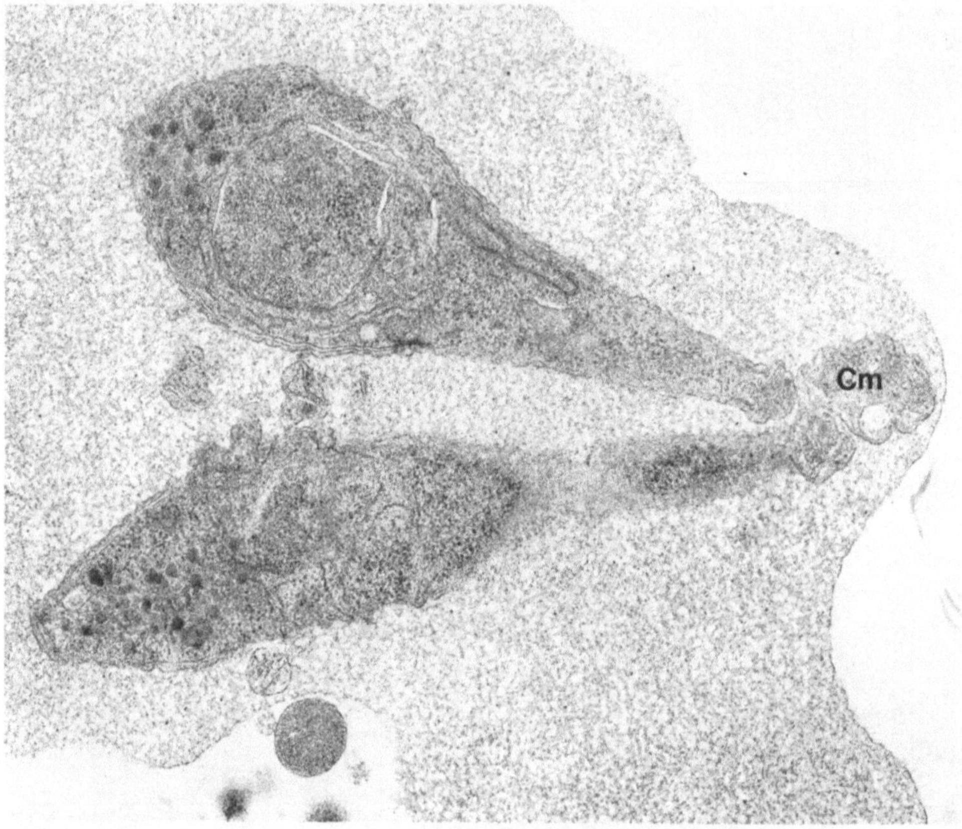

FIGURE 11. Newly formed merozoites of *B. canis* in an erythrocyte. The merozoites appear to be connected to a small cytoplasmic mass (Cm). (Magnification × 24,000.)

inefficiently, while *Babesia* takes its nutrients from the interface between the parasite and the host cell and assimilates them efficiently.[39]

V. ALTERATION OF ERYTHROCYTES

Once parasites have successfully entered host cells, they develop and multiply, and in the process, affect the host cell. Morphologic alterations of host erythrocytes are observed. Erythrocytes infected with *B. argentina* decrease in mean corpuscular diameter (MCD) and acquire spherocytic shapes, while erythrocytes parasitized with *B. bigemina* are bigger than uninfected erythrocytes. As infection develops, reticulocytosis and polychromasia appear.[44] The parasite might affect the osmoregulation of the infected erythrocyte by changing the concentrations of erythrophilic lipid.[45] In erythrocytes of humans infected with *B. microti,*[29] scanning electron microscopic studies demonstrated surface membrane changes. Deformed cells had one to four protrusions, and by transmission electron microscopy it was revealed that these protrusions were caused by parasites located beneath the erythrocyte membrane. Sometimes perforations were also seen in the center or between protrusions, but the knobs that are characteristic of erythrocyte membranes of *P. falciparum* parasitized erythrocytes were not observed. Stellate protrusions were observed on the membranes of *B. bovis*-infected erythrocytes (Figures 12 and 13).[46] Although the fine structure of the stellate protrusions on the membranes of erythrocytes infected with *B. bovis* is distinct from the electron-dense, cone-shaped knobs of erythrocytes infected with *P. falciparum*,[47] the roles of these two

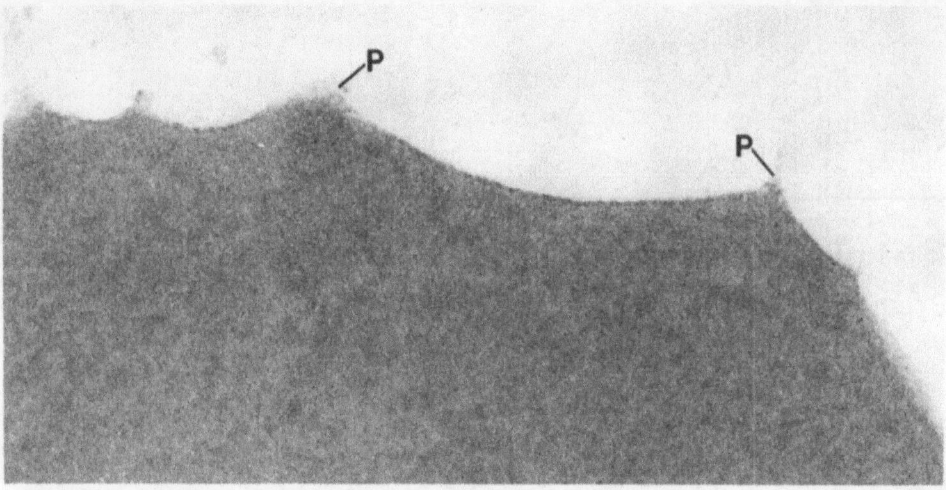

FIGURE 12. Stellate protrusions (P) on a *B. bovis*-infected erythrocyte membrane. (Magnification × 70,000.) (From Aikawa, M., Rabbege, J., Uni, S., Ristic, M., and Miller, L. M., *Am. J. Trop. Med. Hyg.*, 34, 45, 1985. With permission.)

protrusions appear to be similar. The stellate protrusions connect to endothelial cells of capillaries and cause cerebral babesiosis[48] in the same way that knobs attach to endothelial cells of brain capillaries and cause cerebral malaria.[49]

Some strains of *B. bovis* do not bind to kidney capillaries by the stellate protrusions, while others do.[50] This result suggests that different strains of *B. bovis* induce different effects in the erythrocyte membrane. In vitro studies confirmed this speculation: virulent strains of *B. bovis* induced more stellate protrusions on the erythrocyte membrane than did avirulent strains of *B. bovis*.[46] This finding indicates that the number of protrusions on infected erythrocyte membranes, as well as the growth rate of the parasite,[28] could be criteria for estimating the virulence of *B. bovis*. Development of criteria for virulence is very important if we are to screen parasites for vaccine development.

Intramembrane particles (IMP) were also affected by infection (Figure 13).[46] The number of IMP on the E face of the erythrocyte membrane increases when the erythrocyte is parasitized by *B. bovis*. Although the significance of IMP is not fully understood, they may alter the shape of infected erythrocytes by the specific orientation of the IMP on the erythrocyte membrane.

The parasite also induces changes in the cytoplasm of host cells. In host cells infected with *B. microti*, tight, concentric, double membranes and aggregates of small vesicles were observed.[40] The tight, concentric, double membranes encircled the cytoplasm of the erythrocyte and was sometimes continuous with the coiled organelle of the parasite. The aggregates of small vesicles were grouped into clusters, and their content varied from low density to a density similar to that of the host cytoplasm. Changes in density may be related to the digestion of host cytoplasm.[40] Other structures similar to Maurer's clefts and dense accumulations of ribosomes are occasionally observed in *Babesia*-infected erythrocytes. Although the exact mechanisms for the morphological alternation of host cells is not known, mechanical or biochemical factors induced by the parasite may be responsible for alteration in host cell structure.

VI. DEVELOPMENT OF SEXUAL STAGES OF THE PARASITES IN TICKS

Babesia are transmitted to mammalian hosts by female ticks. The life cycle of the tick

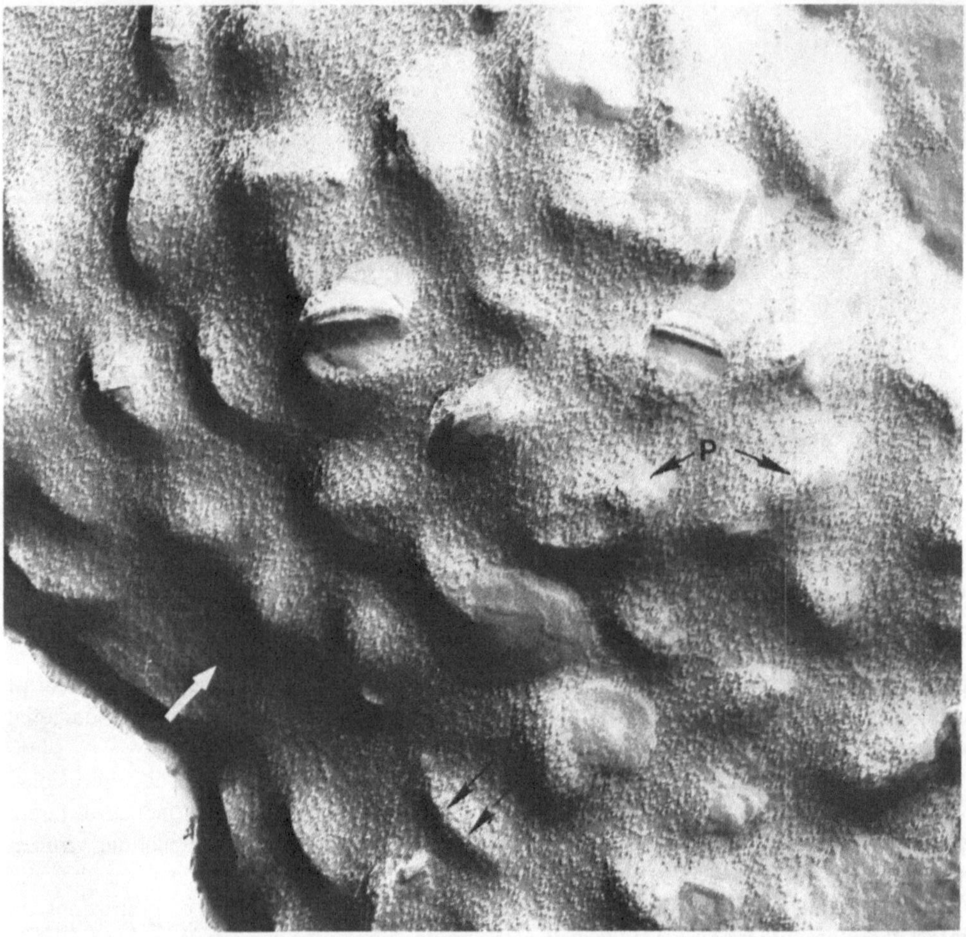

FIGURE 13. Freeze-fracture electron microscopy showing elliptical protrusions (P) of an erythrocyte infected with *B. bovis*. Aligned IMP (arrow) are present along the axis of protrusions. (Magnification × 78,000.)

stage is not known completely. Whether sexual reproduction occurs in *Babesia* as it does in other apicomplexa has been controversial for a long time.

In the early 1900s, Koch[51] described Strahlenkörper, an unusual form with a long ray or spike in ticks. He suggested that the Strahlenkörper might be a sexual form of *Babesia*. However, these stages have been considered to be unimportant or doubtful for many years. Recently, Strahlenkörper were again described in *Babesia* of several species,[52-55] and hypothetical life cycles were proposed which included a sexual stage.[53,55] Most *B. microti*[56] parasites develop into trophozoites and produce new merozoites. A small percentage of trophozoites, however, do not follow this path, but become larger and acquire an odd shape by folding, coiling, or twisting. These parasites are assumed to be gametocytes. In the gut of the tick, these forms are believed to develop into gametes by acquiring new organelles such as cytosomes, microtubules, an arrowhead-like structure, and a long, thin tail. Both the arrowhead-like structure and the long, thin tail are characteristic of gametes. The arrowhead-like structure is electron-dense and located at the anterior end of the parasite. The tail is long, very narrow, and located at the posterior end of the gamete. One to three very thin, short arms are also observed. Gametes equipped with these organelles are considered to be the Strahlenkörper which were first found by Koch.

The close aposition of two gametes and the fusion of their cytoplasm has been described.[56,57] Furthermore, it was clearly demonstrated that parasites with the arrowhead structure fused with parasites without an arrowhead.[58] The two cells touched and fused their cytoplasm at the point of close contact. This is the first description of a sexual stage in piroplasms since Koch suggested its existence in 1906. There are still unsolved questions such as how microgamete or macrogamete differentiates and how nuclear fission occurs. Further studies are required to fully understand the sexual stage of *Babesia* spp.

After fertilization, fused gametes develop into zygotes in the gut of the ticks. The zygotes still have arrowheads at their anterior end. The zygotes penetrate and pass through the peritrophic membrane of the gut with the aid of the arrowhead organelles and enter the epithelial cells of the gut.[59] As the zygote passes through the peritrophic membrane, the arrowhead structure loses its density and finally disappears.[60] This suggests that the arrowhead structure contains enzymes which dissolve the peritrophic membrane to allow the zygotes to pass through the membrane.

In the intestine, the zygotes develop into kinetes and then leave the intestine and enter the various organs of the tick via the hemolymph.[57,60] They produce new kinetes in a variety of organs including hemocytes, muscle fibers Malpighian tube cells, and in the ovarian cells of female ticks.

Finally, the parasites reach the salivary glands of the tick and develop into sporozoites. When the tick feeds on the blood of the vertebrate, the sporozoites are passed with the saliva into the circulation of the host. In the blood, sporozoites of the majority of *Babesia* species invade erythrocytes directly. Some of the small *Babesia*, e.g., *B. equi* and *B. microti*, proliferate in lymphocytes before invading erythrocytes by a process similar to that which occurs among *Theileria*.[57,61] Because of these observations, the classification of the babesiosis, particularly *Babesia equi* and *Babesia microti*, may require modification.

REFERENCES

1. **Rudzinska, M. A., Trager, W., Lewengrub, S. J., and Gubert, E.,** An electron microscopic study of *Babesia microti* invading erythrocytes, *Cell Tissue Res.*, 169, 323, 1976.
2. **Büttner, D. W.,** Vergleichende Untersuchung der Feinstruktur von *Babesia gibsoni* und *Babesia canis*, *Tropenmed. Parasitol.*, 19, 330, 1968.
3. **Smith, R. D. and Ristic, M.,** Immunization against bovine babesiosis with culture-derived antigens, in *Babesiosis*, Ristic, M. and Kreier, J. P., Eds., Academic Press, New York, 1981, 485.
4. **Rudzinska, M. A. and Trager, W.,** Formation of merozoites in intraerythrocytic *Babesia microti*: an ultrastructural study, *Can. J. Zool.*, 55, 928, 1977.
5. **Aikawa, M. and Sterling, C. R.,** *Intracellular Parasitic Protozoa*, Academic Press, New York, 1974, 5.
6. **Potgieter, F. T. and Els, H. J.,** An electron microscopic study of intra-erythrocytic stages of *Babesia bovis* in the brain capillaries of infected splenectomized calves, *Onderstepoort J. Vet. Res.*, 46, 41, 1979.
7. **Ristic, M., Smith, R. D., and Kakoma, I.,** Characterization of babesia antigens derived from cell cultures and ticks, in *Babesiosis*, Ristic, M. and Kreier, J. P., Eds., Academic Press, New York, 1981, 337.
8. **Smith, R. D., James, M. A., Ristic, M., Aikawa, M., and Murguia, C. A. V.,** Bovine babesiosis: protection of cattle with culture-derived soluble *Babesia bovis* antigen, *Science*, 212, 335, 1981.
9. **Kuttler, K. L., Levey, M. G., and Ristic, M.,** Cell culture-derived *Babesia bovis* vaccine: sequential challenge exposure of protective immunity during a 6-month postvaccination period, *Am. J. Vet. Res.*, 44, 1456, 1983.
10. **Ladda, R., Aikawa, M., and Sprinz, H.,** Penetration of erythrocytes by merozoites of mammalian and avian malarial parasites, *J. Parasitol.*, 55, 633, 1969.
11. **Miller, L. H., Aikawa, M., and Dvorak, J. A.,** Malaria (*Plasmodium knowlesi*) merozoites: immunity and the surface coat, *J. Immunol.*, 114, 1237, 1975.
12. **Aikawa, M., Miller, L. H., Johnson, J., and Rabbege, J.,** Erythrocyte entry by malarial parasites: a moving junction between erythrocyte and parasite, *J. Cell Biol.*, 77, 72, 1978.
13. **Weidner, E.,** Ultrastructural study of microsporidian invasion into cells, *Z. Parasitenkd.*, 40, 227, 1972.

14. **Aikawa, M.,** Host-parasite interaction: electron-microscopic study, in *Molecular Biology of Parasites*, Guardiola, J., Luzzatto, L., and Trager, W., Eds., Raven Press, New York, 1983, 1.
15. **Annable, C. R.,** Role of complement in intraerythrocytic protozoan (*Babesia*) parasitemias, *Fed. Proc. Fed. Am. Soc. Exp. Biol.*, 31(Abstr.), 655, 1972.
16. **Jack, R. M. and Ward, P. A.,** The role *in vivo* of C3 and the C3b receptor in babesial infection in the rat, *J. Immunol.*, 124, 1574, 1980.
17. **Chapman, W. E. and Ward, P. A.,** *Babesia rodhaini:* requirement of complement for penetration of human erythrocytes, *Science*, 196, 67, 1977.
18. **Jack, R. M. and Ward, P. A.,** *Babesia rodhaini* interactions with complement: relationship to parasitic entry into red cells, *J. Immunol.*, 124, 1566, 1980.
19. **Ward, P. A. and Jack, R. M.,** The entry process of *Babesia* merozoites into red cells, *Am. J. Pathol.*, 102, 109, 1981.
20. **Jensen, J. B. and Edgar, S. A.,** Possible secretory function of rhoptries of *Eimeria magna* during penetration of cultured cells, *J. Parasitol.*, 62, 988, 1976.
21. **Kilejian, A.,** Does a histidine-rich protein from *Plasmodium lophurae* have a function in merozoite penetration?, *J. Protozool.*, 23, 272, 1976.
22. **Aikawa, M. and Sterling, C.,** *Intracellular Parasitic Protozoa*, Academic Press, New York, 1974, 54.
23. **Jensen, J. B. and Hammond, D. M.,** Ultrastructure of the invasion of *Eimeria magna* sporozoites into cultured cells, *J. Protozool.*, 22, 411, 1975.
24. **Lycke, E., Carlberg, K., and Norrby, R.,** Interactions between *Toxoplasma gondii* and its host cells: function of penetration-enhancing factor of *Toxoplasma*, *Infect. Immun.*, 11, 853, 1975.
25. **Roberts, J. A., Kerr, J. D., and Tracey-Patte, P.,** Function of the spleen in controlling infections of *Babesia rodhaini* in mice, *Int. J. Parasitol.*, 2, 217, 1972.
26. **Mahoney, D. F., Wright, I. G., and Ketterer, P. J.,** *Babesia argentina*: The infectivity and immunogenicity of irradiated blood parasites for splenectomized calves, *Int. J. Parasitol.*, 3, 209, 1973.
27. **Rodriguez, S. D., Buening, G. M., and Green, T. J., and Carson, C. A.,** Cloning of *Babesia bovis* by *in vitro* cultivation, *Infect. Immun.*, 42, 15, 1983.
28. **Buening, G. M., Kuttler, K. L., and Rodriguez, S. D.,** Evaluation of a clonded *Babesia bovis* organism as a live immunogen, *Vet. Parasitol.*, 22, 242, 1986.
29. **Sun, T., Tenenbaum, M. J., Greenspan, J., Teichberg, S., Wang, R.-T., Degnan, T., and Kaplan, M. H.,** Morphologic and clinical observations in human infection with *Babesia microti*, *J. Infect. Dis.*, 148, 239, 1983.
30. **Aikawa, M.,** Ultrastructure of the pellicular complex of *Plasmodium fallax*, *J. Cell Biol.*, 35, 103, 1967.
31. **Aikawa, M., Hepler, P. K., Huff, C. G., and Sprintz, H.,** The feeding mechanism of avian malarial parasites, *J. Cell. Biol.*, 28, 355, 1966.
32. **Aissi, E., Charet, P., Bouquelet, S., and Biguet, J.,** Endoprotease in *Plasmodium yoelii nigeriensis*, *Comp. Biochem. Physiol. B*, 74, 559, 1983.
33. **Slomianny, C., Charet, P., and Prensier, G.,** Ultrastructural localization of enzymes involved in feeding process in *Plasmodium chabaudi* and *Babesia hylomysci*, *J. Protozool.*, 30, 376, 1983.
34. **Frerichs, W. M., and Holbrook, A. A.,** Feeding mechanisms of *Babesia equi*, *J. Protozool.*, 21, 707, 1974.
35. **Rudzinska, M. A. and Trager, W.,** The fine structure of intraerythrocytic merozoites in *Babesia microti*, *Fed. Proc. Fed. Am. Soc. Exp. Biol.*, 33(Abstr.), 445, 1974.
36. **Simpson, C. F., Kirkham, W. W., and Kling, J. M.,** Comparative morphologic features of *Babesia caballi and Babesia equi*, *Am. J. Vet. Res.*, 28, 1693, 1967.
37. **Rudzinska, M. A. and Trager, W.,** Intracellular phagotrophy in *Babesia rodhaini* as revealed by electron microscopy, *J. Protozool.*, 9, 279, 1962.
38. **Kreier, J. P., Gravely, S. M., Seed, T. M., Smucker, R., and Pfister, R. M.,** *Babesia* spp.: the relationship of stage of development to structure of intra- and extracellular parasites, *Tropenmed. Parasitol.*, 26, 9, 1975.
39. **Langreth, S.,** Feeding mechanisms in extracellular *Babesia microti* and *Plasmodium lophurae*, *J. Protozool.*, 23, 215, 1976.
40. **Rudzinska, M. A.,** Ultrastructure of intraerythrocytic *Babesia microti* with emphasis on the feeding mechanism, *J. Protozool.*, 23, 224, 1976.
41. **Simpson, C. F. and Neal, F. C.,** Ultrastructure of *Babesia equi* in ponies treated with imidocarb, *Am. J. Vet. Res.*, 141, 267, 1980.
42. **Wright, I. G. and Goodger, B. V.,** Proteolytic enzyme activity in the intra-erythrocytic parasites *Babesia argentina* and *Babesia bigemina*, *Z. Parasitenkd.*, 42, 213, 1973.
43. **Aissi, E. and Charet, P.,** Proteolytic system in *Babesia hylomysci*, *Comp. Biochem. Physiol. B*, 70, 133, 1981.
44. **Wright, I. G.,** Osmotic fragility of erythrocytes in acute *Babesia argentina* and *Babesia bigemina* infections in splenectomized *Bos taurus* calves, *Res. Vet. Sci.*, 15, 299, 1973.

45. **Cooper, R. A. and Jandl, J. H.,** The role of membrane lipids in the survival of red cells in hereditary spherocytosis, *J. Clin. Invest.*, 48, 736, 1969.
46. **Aikawa, M., Rabbege, J., Uni, S., Ristic, M., and Miller, L. H.,** Structural alteration of the membrane of erythrocytes infected with *Babesia bovis*, *Am. J. Trop. Med. Hyg.*, 34, 45, 1985.
47. **Aikawa, M., Rabbege, J., Udeinya, I., and Miller, L. H.,** Electron microscopy of knobs in *P. falciparum*-infected erythrocytes, *J. Parasitol.*, 62, 435, 1983.
48. **Wright, I. G.,** An electron microscopic study of intravascular agglutination in the cerebral cortex due to *Babesia argentina* infection, *Int. J. Parasitol.*, 2, 209, 1972.
49. **Oo, M. M., Aikawa, M., Than, T., Aye, T. M., Myiant, P. T., Igarashi, I., and Schoene, W. C.,** Human cerebral malaria: a pathological study, *J. Neuropathol. Exp. Neurol.*, 46, 223, 1987.
50. **Wright, I. G.,** Ultrastructural changes in *Babesia argentina*-infected erythrocytes in kidney capillaries, *J. Parasitol.*, 59, 735, 1973.
51. **Koch, R.,** Beiträge zur Entwicklungsgeschichte der Piroplasmen, *Z. Hyg. Infectionskr.*, 54, 1, 1906.
52. **Friedhoff, K. T. and Büscher, G.,** Rediscovery of Koch's "Strahlenkörper" of *Babesia bigemina*, *Z. Parasitenkd.*, 50, 345, 1976.
53. **Weber, G. and Friedhoff, K. T.,** Preliminary observations of the ultrastructure of supposed sexual stages of *Babesia bigemina* (Piroplasmea), *Z. Parasitenkd.*, 53, 83, 1977.
54. **Mehlhorn, H., Schein, E., and Voigt, W. P.,** Light and electron microscopic study on developmental stages of *Babesia canis* within the gut of the tick *Dermacentor reticulatus*, *J. Parasitol.*, 66, 220, 1980.
55. **Droleskey, R. E., Holman, P. J., Craig, T. M., Wagner, G. G., and Hollenhauer, H. H.,** Ultrastructure of *Babesia bovis* sexual stages as observed in *Boophilus microplus* cell cultures, *Res. Vet. Sci.*, 34, 249, 1983.
56. **Rudzinska, M. A., Spielman, A., Riek, R. F., Lewengrub, S. J., and Piesman, J.,** Intraerythrocytic "gametocytes" of *Babesia microti* and their maturation in ticks, *Can. J. Zool.*, 57, 424, 1979.
57. **Mehlhorn, H. and Schein, E.,** The piroplasms: life cycle and sexual stages, in *Adv. Parasitol.*, 23, 37, 1984.
58. **Rudzinska, M. A., Spielman, A., Lewengrub, S., Trager, W., and Piesman, J.,** Sexuality in piroplasms as revealed by electron microscopy in *Babesia microti*, *Proc. Natl. Acad. Sci. U.S.A.*, 80, 2966, 1983.
59. **Rudzinska, M. A., Lewengrub, S., Spielman, A., and Piesman, J.,** Invasion of *Babesia microti* into epithelial cells of the tick gut, *J. Protozool.*, 30, 338, 1983.
60. **Friedhoff, K. T.,** Morphologic aspects of *Babesia* in the tick, in *Babesiosis*, Ristic, M. and Kreier, J. P., Eds., Academic Press, New York, 1981, 143.
61. **Moltmann, U. G., Mehlhorn, H., Schein, E., Rehbein, G., Voigt, W. P., and Zweygarth, E.,** Fine structure of *Babesia equi* Laveran, 1901 within lymphocytes and erythrocytes of horses: an *in vivo* and *in vitro* study, *J. Parasitol.*, 69, 111, 1983.

Chapter 4

ADVANCES IN THE IN VITRO CULTIVATION OF *BABESIA* SPECIES

Gottfried Kellermann, K. Roger Tsang, and Ibulaimu Kakoma

TABLE OF CONTENTS

I. INTRODUCTION

The development of a continuous culture system for *Babesia bovis*[1] represented a major landmark in the history of babesiology since the discovery of the organism,[2] and the demonstration of the definitive role of an arthropod vector[3]; in the transmission of the disease. This technique facilitated research on morphological characteristics,[4] role of serum components,[5] chemotherapeutic efficacy of drugs,[6] and offered unprecedented potential for immunoprophylactic control of bovine babesiosis.[7] The most practical application of the in vitro culture system has been the production of immunogens in various parts of the world including Venezuela,[8] Mexico,[9] Israel,[10] and Australia.[11] (Table 1 summarizes the most important practical applications of the in vitro culture systems for *Babesia* species.) Short-term culture systems have been developed for *B. canis*[12] and other babesia species,[13] and recently these research efforts culminated in the manufacture of the first protozoan vaccine against canine babesiosis named "Pirodog®" — now commercially available (Rhone Merieux, Lyon, France).[14] In spite of the major strides made in the in vitro cultivation of *Babesia* species, there is need and scope for the introduction of more innovative techniques to meet increasing requirements for laboratory research and field immunization programs.

This review will highlight the major developments in the in vitro cultivation of *Babesia* species elaborating particularly on *B. bovis* for which optimal in vitro conditions have been well documented. In addition, a brief account will be presented describing salient innovative modifications of the conventional microaerophilous stationery phase (MASP) technique.[1]

II. CULTURE SYSTEMS FOR *BABESIA*

A. In Vitro Culture for *Babesia bovis*

The successful continuous cultivation of *Plasmodium falciparum* by Trager and Jensen[15] gave hope for in vitro cultivation of other protozoa. However, many *Babesia* species were not readily adaptable to the conditions defined for malaria parasties, and only *B. microti* could be propagated under the same conditions as *P. falciparum*. Consequently, Erp et al.[16] resorted to suspension cultures for short cultivation of *B. bovis*, and later the technique was successfully utilized for *B. canis*.[12] Minor modifications of the Erp technique resulted in continuous in vitro growth of the important bovine parasite *B. bovis*.[16] Although spinner flask (SF) cultures supported the growth of *B. bovis*, the large volumes and extensive manipulations required by these cultures and the relatively slow growth rate achieved limited utlization of this method. These limitations have been circumvented by the development of the MASP system.[1] High parasitemia adaptability to microtiter culture plates and growth characteristics similar to those observed in the bovine host make this an ideal system for experimental investigation.

The MASP system also overcame other practical drawbacks of the SF method, such as the requirement for daily centrifugation, the tendency for excessive accumulation of extracellular parasites with poor invasive ability for new cells, and the overall slow growth rate. With the MASP system, cultures are initiated using defibrinated infected bovine blood when the parasitemia is between 1.0 and 2%. Similar to the SF cultures, the use of standard anticoagulants, e.g., heparin, ethylenediaminetetraacetic acid (EDTA), or acid-citrate dextrose (ACD) in the collection of infected or normal blood is avoided. Infected erythrocytes are suspended to a final packed cell volume of 5 to 10% in a medium consisting of 60% Medium 199 and 40% normal bovine serum supplemented with 15 m*M* HEPES (see Section II.A.1. for definition), 100 μg streptomycin, and 100 IU penicillin G per milliliter. Serum obtained from mature *Bos taurus* cattle is preferable, as commercially obtained fetal calf, newborn calf, and bobby calf serum will not support growth. Additionally, the sera of calves (mixed dairy breeds under 3 months of age) obtained locally vary in their ability to support growth, with the sera of many failing to support parasite growth.

Table 1
PRACTICAL APPLICATIONS OF THE *IN VITRO* CULTURE SYSTEMS FOR *BABESIA* SPECIES

Application	Major species	Key refs.
Immunoprophylaxis	*B. bovis*	7—9
	B. canis	12, 14
Immunodiagnosis	*B. bovis*	28, 29
Assessment of immunologic effector mechanisms	*B. bovis*	6, 27
	Rodent babesias	25, 26
Ultrastructural studies	*B. bovis*	31
	B. canis	33
Cloning	*B. bigemina*	18
Identification of protective immunogens	*B. bovis*	32
Metabolic pathways	*B. bigemina*	22—24
	B. bovis	30
	B. divergens	22—24
	Rodent babesias	22—24
Role of serum factors	*B. bovis*	5
	B. rhodhaini	25
	B. microti	26
Drug sensitivity assays	*B. bovis*	6
	Rodent babesia	22

The critical factors affecting *B. bovis* in MASP culture are pH, depth of culture, CO_2, relative humidity, and frequency of change of medium. In order to ensure the ideal growth conditions, Levy and Ristic[1] have recommended that following titration to pH 7.0, culture suspensions should be aliquoted to a ratio of 0.62 mℓ/cm^2 and incubated at 37 to 38°C under an atmosphere of 5% CO_2 air. At approximately 24-hr intervals, the overlying medium is replaced with fresh babesia tissue culture medium #4 (BTCM4) consisting of 40% serum and 60% M199, supplemented with HEPES and antibiotics and titrated to pH 7. Following 48 to 72 hr, subcultures are accomplished by diluting old culture two- to threefold with new culture (NC) containing freshly collected uninfected bovine erythrocytes. Parasite multiplication during each subculture does not persist beyond 3 to 4 days, and maximum rates of growth are achieved when percent parasitized erythrocyte (PPE) values are reduced to 0.5 to 1% at each pass. Infected cultures may be stored for up to 3 weeks at 4°C and should be diluted with freshly prepared NC in order to reinitiate active growth. Satisfactory results are obtained with NC stored for up to 7 days at 4°C, providing the pH is readjusted to 7 before use.

For routine maintenance, flat-bottomed microtiter culture plates are utilized. Each well has a diameter of 6.4 mm and receives 0.2 mℓ of culture suspension. Cultures can be grown in a variety of containers, and as long as this ratio of culture volume to area is maintained, the volume of a single culture is limitless.

Under these conditions, the layer of settled erythrocytes becomes dark red to black, indicating low oxygen tension (deoxyhemoglobin). Resuspension of the erythrocytes in fresh oxygenated medium or lowering the temperature to 4°C restores the bright red appearance of oxygenated erythrocytes. The darkening of the erythrocytes cannot be explained by acidic conditions as the culture pH increases to approximately 7.2 during incubation. Generally, cultures containing less than 1% infected erythrocytes fail to darken. These observations indicate that the depletion of oxygen from the settled cell layer is a result of parasite

Table 2
KINETICS OF *BABESIA BOVIS* IN MASP CULTURES AND THE EFFECT OF VARYING SERUM CONCENTRATIONS

		Experiment 1	
% Serum	Initial PPE[a]	Max PPE	Cumulative increase (CI)
20	0.5	0.3	41
30	0.5	9.9	1451
40	0.5	22.8	3958
40 (Frozen)	0.5	26.4	4092

Note: Freshly collected erythrocytes and serum were used each time cultures were initiated. Each figure represents results on three sequential subcultures. C.I. = $(FP/_{IP})$ CD cumulative dilution of the factor by which the erythrocytes present at the start of experiments are diluted with fresh erythrocytes during subcultures.

[a] PPE = Percent parasitized erythrocytes.

metabolism. The column of medium above the parasitized cells acts as a barrier to oxygen exchange, and its depth is crucial to successful cultivation. Reducing the depth to 0.16 cm maintains the cultures in an oxygenated state, and such cultures cannot be continuously maintained. Cultures which fail to achieve high enough PPE to darken the cells within 4 days produce aberrant and degenerate parasite forms (large rings with diffuse, poorly staining cytoplasm) and die out, indicating the requirement for a minimum seed density. The culture conditions allow the parasites to establish, as a result of their metabolic activity, a reduced oxygen environment, thereby preventing the toxic effect of high concentrations of this gas. It is for this reason that the term microaerophilous stationary phase (MASP) was used to describe this culture system. Numerous attempts to decrease or eliminate the seeding density requirement by placing cultures in candle jars similar to those used for the cultivation of *Plasmodium falciparum*[15] have so far been unsuccessful. *Babesia bovis* could not be propagated even under 5% CO_2, 5% O_2, and 90% nitrogen. When optimal conditions are used, the kinetics of growth during primary culture is indistinguishable form those of later subcultures indicating that there is no period of adaptation and/or selection. Table 2 summarizes the typical kinetics of growth and the unusually high serum requirements of the same isolates. Of special interest is that rates of parasite growth in vitro are comparable to those observed in vivo.[34] These organisms can be maintained in continuous cultivation for up to 3 months with subcultures every 2 to 3 days with no significant alteration in morphology or virulence of *B. bovis*. Statistical data on long-term cultures are summarized in Table 3.

In SF cultures, extracellular *Babesia* merozoites are frequently observed. In MASP cultures, extracellular forms are rare, particularly in the exponential growth phase. This reflects a high degree of efficiency of *B. bovis* for invading susceptible erythrocytes. However, transferring MASP cultures to an incubator containing ambient air (CO_2 deprived) results in an accumulation of merozoites in the medium. Isolation of infectious *B. bovis* merozoites is accomplished by depriving MASP cultures of CO_2 for 4 to 6 hr. The cultures are then gently centrifuged (400 × *g* for 5 min) to sediment erythrocytes. The top two thirds of the supernatant is sedimented twice at 750 × *g* for 10 min to bring down the remaining erythrocytes. The resulting supernatant is centrifuged at 1000 × *g* for 20 min in order to

Table 3
STATISTICAL ANALYSIS OF PROLONGED CULTURE AND REPEATED SUBCULTURE OF *B. BOVIS* IN MASP SYSTEM

Days in culture	82.6
Number of subcultures	33
PPE of 2-day subculture (44.5 hr)	19.9 ± 7.3%
PPE of 3-day subculture (68.6 hr)	26.1 ± 7.1%
Maximum PPE	38.1 ± 3.0%
Cumulative dilution	2.49×10^{26} fold
Cumulative increase	1.74×10^{27}-fold
1 log increase (routine)	73.2 hr[a]

[a] When cultures were subjected to cumulative dilution of 3.15×10^{17}-fold over 31 days, an average of 42.3 hr was required for a tenfold increase. In the exponential growth phase, this would represent a generation (doubling) time of 7 to 10 hr.

pellet the merozoites. All processing is at 4°C as spontaneous agglutination of parasites occurs at higher temperatures. Additionally, cultures should be prepared utilizing erythrocytes from cattle which have not been repeatedly bled, since reticulocytes are difficult to separate from merozoites by centrifugation.

Merozoites stored at 4°C for as long as 16 hr are still viable and can be used to initiate new cultures.

1. Optimization and Modification of the MASP System

The first attempt to modify and optimize the Levy-Ristic MASP system was by Palmer et al.[17] These authors incorporated Zwitter ionic buffers TES (2,2-hydroxy-1-1-bis hydroxyonethyl-ethyl-) aminoethane sulfonic acid) or HEPES (4-2 hydroxyethyl-1-piperazine ethan sulfonic acid). The pH was adjusted to 7.0 and cultures in 96-well microtiter plates were incubated at 37°C lin a humidified 5% CO_2 incubator. Under these conditions, cultures buffered with TES reached a maximum PPE of 30% and healthy parasites at different stages of development were observed. Subsequently, Rodriguez et al.[18] were able to synchronize *B. bovis* cultures and develop a cloning system. In the latter modification, the authors reported that by starting at low PPE, fluid levels were optimal at 4.0 to 5.0 mm in contrast to the original MASP system which required a fluid level of 6.2 mm for maximal growth. Further, by lowering the oxygen tension to less than 2%, wide ranges of fluid levels could be used.

2. Recent Commercial Innovations in the In Vitro Cultivation of Babesia bovis

Although much progress has been made identifying culture systems and conditions for the continuous propagation of *Babesia*, present methods are not very suitable for industrial-scale production. The ideal system would require a cell line in which the organism can be propagated indefinitely. The fact that the organism is adapted only to erythrocytes is a major constraint. Consequently, efforts to overcome some of the difficulties have concentrated on two areas:

1. Optimization and automation of present culture methods.
2. Search for alternate host cell systems.

Table 4
CULTURE PARAMETERS FOR IN VITRO REPLICATION OF ERYTHROCYTIC STAGES OF *BABESIA* SPECIES

Parameter	Culture devices	Evaluation criteria
pH	T-flasks	PPE[a]
pO_2,pCO_2	Roller bottles	Parsite stages and morphology
Temperature	Spinner flasks	Parasite stages and morphology
Glucose	Hollow fiber bioreactors[b]	Efficacy of production
Lactate and other essential nutrients or metabolites		Growth kinetics

[a] PPE = percent parasitized erythrocytes.
[b] Hollow fiber bioreactors = Devices for effective exchange of macromolecules.

The success of in vitro cultivation of many hemoparasites hinges on the understanding and the ability to provide the correct combination of physiological parameters which are crucial for survival and continuous replication of the parasites. These parasites are obligate intracellular parasites and, therefore, rely heavily on the well-being of their host cells. Thus, it is equally important to fully realize the physiological needs of the host cell and those of the parasite to survive as an inseparable unit. The next step is to design the correct culture hardware and establish the optimal culture conditions (medium, gas, temperature, etc.) which permit a dynamic culture and mimic an environment that is similar to that in vivo (Table 4). Most of the above listed culture devices are inadequate by themselves for industrial scale-up. However, the combination of several of these devices proved to be useful and advantageous. Combining new hollow fiber technology with spinner culture devices allowed the successful propagation of *B. bovis* on a large scale where gas tensions and culture media conditions were tightly controlled. Various criteria for evaluating different culture systems are shown in Table 4.

B. Search for Alternate Host Cell Systems

Erythrocytes cannot be maintained in continuous culture for over 2 weeks without losing a significant portion of their functional properties. Thus, any erythrocyte culture system is self-limiting. From a strategic point of view, a technique would be preferred that uses a nonlimiting, immortal host cell system for support of parasitic growth. It is obvious that such a system has to meet the needs for certain physical, chemical, and metabolic parameters of any particular red blood cell culture.

Preliminary results showed some promise in this area. Using invertebrate cell cultures as host cell systems, it was possible to cultivate certain *Babesia* species for a period of 12 months. Although the morphology of the parasites has changed under these conditions, the antigenic properties as judged by the indirect immunofluorescent antibody (IFA) test were retained for the period of the experiment. Thus, it is anticipated that protozoal culture systems can be upgraded to a scale comparable to continuous monolayer cultures.[19]

C. In Vitro Cultivation of *Babesia bigemina*

A modification of Trager and Jensen's technique was applied to the in vitro cultivation of *B. bigemina* by Timms.[13] The resulting short-term cultures were adequate for the multiplication of the parasite, and the infected cells could be maintained for up to 96 hr by regular changes of medium. The culture medium incorporated 5.14 IU/mℓ of heparin, polymycin and neomycin at 20 IU/mℓ and 50 μg/mℓ, respectively, in medium RPMI 1640 pH 7.4 containing 2 g $NaHCO_3$ and 5.73 g TES to a final concentration of 0.2% and 25 m*M*, respectively, and supplemented with 15% fresh bovine plasma. The rest of the procedure is essentially a combination of Trager and Jensen[15] incorporating some concepts from Erp's procedure.[16]

Subsequently, Vega et al.[20] modified Timms technique into a contiuous system. These authors utilized the Vega Y Martinez phosphate buffered saline for washing the infected cells and set up cultures using 5 to 10% erythrocyte suspensions in Medium 199 containing 20 to 50% fresh normal bovine serum. The depth of the cultures was maintained at 4 mm in a 5% CO_2, 2% O_2, and 93% N_2 atmosphere. They were able to propagate the organism for 36 days in vitro and carried out nine subcultures. The cultivated strain was reported as having maintained normal morphology and pathogenicity for susceptible cattle. They reported also that the original isolate had been maintained in vitro for more than 99 days.

In general, however, the in vitro cultivation of *B. bigemina* has not advanced as that of *B. bovis*, particularly with regard to the maximum PPE attainable.

D. In Vitro Cultivation of *Babesia divergens* and *B. major*

The MASP technique[1] has recently been adapted to the continuous culture system of *B. divergens* isolated from Finland.[21] The organism had been previously passaged through splenectomized cattle and intermittently stored at −70°C between passes. The organism was cultured in multichambered containers at 38°C in a humidifed atmosphere by the candle jar method of Trager and Jensen.[15] The *B. divergen* cultures could be maintained through 27 subcultures during a period of 85 days, and the organisms continued to multiply. An average PPE of 5 to 10% was attained, and the cultured organism remained virulent for splenectomized cattle. The authors also noted that extracellular merozoites wer frequent and often appearing in clumps with a number of lysed, agglutinated forms.

For *B. major*, an in vitro technique was developed by Irvin and Young.[22] The authors reported good success and subsequently used the method for metabolic studies.[23,24]

E. In Vitro Cultivation of *Babesia canis*

Molinar et al.[12] adapted *B. canis* for growth in short-term cultures. The methodology utilized by these investigators was similar to that reported by Erp et al.[16] Briefly, blood was collected when the parasitemia in infected splenectomized dogs reached 3 to 4%. Following defibrination by shaking with glass beads, the blood was centrifuged at 400 × *g* for 20 min at 4°C. After removal of the buffy coat, the erythrocytes were washed three times in tissue culture medium. Washed erythrocytes were resuspended in four volumes of medium that consisted of RPMI 1640 pH 7.4, supplemented with 25 m*M* HEPES and 20% normal canine serum. Cultures were incubated at 38°C under an atmosphere at 5% CO_2/air with gentle stirring. After 48 hr of in vitro growth, the cultures were sedimented at 400 × *g* for 10 min at 4°C, and the supernatant was replaced with fresh medium.

Under these conditions, the PPE increased from an initial 3.9 to 7.6% after 24 hr. By 48 hr, the PPE decreased to 3.3%, and the cultures were discontinued at 96 hr, at which time few parasitized cells were observed. Similar to the report of Erp et al.,[16] large numbers of extraerythrocytic forms were observed.

F. Cultivation of Rodent Babesias

Short-term cultivation of *B. rodhaini* was accomplished by Jack and Ward.[25] Their method consisted of collecting infected rat erythrocytes and diluting these cells with uninfected erythrocytes to yield a final parasitemia of 0.3 to 1.0%. The medium consisted of equal volumes of RPMI 1640 and serum (human or rat) with the final packed cell volume not reported. Then, 200 $\mu\ell$ of this material was added to each well of a microtiter plate and incubated under an atmosphere of 5% CO_2 in humidified air. Medium was changed at 24-hr intervals and cultures were maintained for 3 days. A threefold increase in parasitemia occurred during the first 48 hr of in vitro cultivation.

Babesia microti has been cultivated for up to 4 days using a modification[26] of the Trager-Jensen method for the cultivation of *P. falciparum*. The methodology differed from Trager-Jensen in that fetal bovine serum was substituted for human serum and hamster erythrocytes were the host cells. A two- to threefold increase in parasitemia occurred during the first 48 hr of cultivation. These investigators were able to adapt *B. microti* to growth in culture by alternate passages in culture (72 hr) and hamsters. Following ten such passages, a strain of *B. microti* was obtained which grew better in short-term cultures than the original strain, i.e., it produced almost twice the maximum parasitemia as compared to the original isolate. The presence of immune hamster macrophages or the supernatants from such macrophage cultures significantly suppressed growth of *B. microti* in vitro, demonstrating nonantibody-mediated inhibition of *B. microti* growth.

III. PRACTICAL APPLICATIONS OF IN VITRO CULTURE SYSTEMS FOR *BABESIA*

Four major categories of the practical applications of in vitro cultures of *Babesia* species can be identified:

1. Immunoprophylaxis, diagnosis, and assessment of immune status
2. Identification of potentially protective immunogens
3. Study of metabolic pathways and optimal growth requirements for various *Babesia* species
4. Drug sensitivity assays

Since the development of continuous culture systems of *B. bovis*,[1] much innovative research has been undertaken by scientists world-wide exploiting the in vitro system for various applications. Most of the emphasis has been the development of vaccines for field and laboratory research against bovine and canine babesiosis, but many studies have been directed to the in vitro measurement of functional immunity, serodiagnosis, and epidemiology.[25-29] In addition, the in vitro cultures have facilitated investigation of the growth requirements and metabolism[23,24,30] and detailed ultrastructural studies.[31] Finally, significant research has been carried out on the in vitro assays of drug sensitivity.[6,22] This development has obviated reliance on animal models for drug assays. The latter technique has too many in vivo variables and is not cost effective. The various applications of cell cultures of babesia are summarized in Table 2. It must be appreciated that, given the current pace of biotechnology, the future potential of in vitro cultures can only be limited by individual, institutional, and commercial insight, and innovativeness to explore and exploit the existing protocols.

REFERENCES

1. **Levy, M. G. and Ristic, M.,** *Babesia bovis:* continuous *in vitro* cultivation in microaerophilous stationary phase culture, *Science*, 207, 1218, 1980.
2. **Babes, V.,** Sur l'hemoglobinuria bacterine Du boeuf, *C. R. Acad. Sci.*, 107, 692, 1888.
3. **Smith, T. and Kilbourne, F. L.,** Investigations into the nature, causation and prevention of Texas or southern cattle fever, *U.S. Dep. Agric. Bur. Anim. Ind. Bull.*, 1, 1, 1893.

4. **Aikawa, M., Rabegge, J., Uni, S., Ristic, M., and Miller, L. H.,** Structural alterations of the membrane of erythrocytes infected with *Babesia bovis, Am. J. Trop. Med. Hyg.,* 35, 45, 1985.
5. **Levy, M. G., Kakoma, I., Clabaugh, G. W., and Ristic, M.,** Complement does not facilitate *in vitro* invasion of bovine erythrocytes by *Babesia bovis, Ann. Trop. Med. Parasitol.,* 80, 377, 1986.
6. **Ristic, M. and Levy, M. G.,** unpublished observations, 1984.
7. **Ristic, M.,** Research on babesiosis vaccines, in *Malaria and Babesiosis,* Ristic, M., Ambroise-Thomas, P., and Kreier, J. P., Eds., Martinus Nijhoff, Dordrecht, Netherlands, 1984, 103.
8. **Montenegro-James, S., Toro Benitez, M., Leon, E., Lopez, R., and Ristic, M.,** Heterologous strain immunity in bovine babesiosis using a culture-derived soluble *Babesia bovis* immunogen, *Vet. Parasitol.,* 18, 321, 1985.
9. **Smith, R. D., James, M. A., Ristic, M., Aikawa, M., and Vega, C. A.,** Bovine-babesiosis: protection of cattle with culture-derived soluble *Babesia bovis* antigen, *Science,* 212, 335, 1981.
10. **Pipano, E.,** personal communication, 1986.
11. **Timms, P., Dalgliesh, R. J., Barry, D. N., Dimmock, C. K., and Rodwell, B. J.,** *Babesia bovis*: comparison on culture-derived parasites, non-living antigen and conventional vaccine in the protection of cattle against heterologous challenge, Aust. Vet. J., 60, 75, 1983.
12. **Molinar, E., James, M. A., Kakoma, I., Holland, C. J., and Ristic, M.,** Antigenic and immunogenic studies on cell culture-derived *Babesia canis, Vet. Parasitol.,* 10, 29, 1982.
13. **Timms, P.,** Short-term cultivation of *Babesia* species, *Res. Vet. Sci.,* 29, 102, 1980.
14. **Moreau, Y. and Laurent, N.,** Antibabesial vaccination using antigens from cell cultures: industrial requirements, in *Malaria and Babesiosis,* Ristic, M., Ambroise-Thomas, P., and Kreier, J. P., Eds. Martinus Nijhoff, Dordrecht, Netherlands, 1984, 129.
15. **Trager, W. and Jensen, B.,** Human malaria parasites in continuous culture, *Science,* 193, 673, 1976.
16. **Erp, E. E., Smith, R. D., Ristic, M., and Osorno, B. M. O.,** Continuous *in vitro* cultivation of *Babesia bovis, Am. J. Vet. Res.,* 41, 1141, 1980.
17. **Palmer, D. A., Buening, G. M., and Carson, C. A.,** Cryopreservation of *Babesia bovis* for *in vitro* cultivation, *Parasitology,* 62, 221, 1982.
18. **Rodriguez, S. D., Buening, G. M., Green, T. J., and Carson, C. A.,** Cloning of *Babesia bovis* by *in vitro* cultivation, *Infect. Immun.,* 42, 15, 1983.
19. **Aranthoon, W. R. and Birch, J. R.,** Large-scale cell culture in biotechnology, *Science,* 232, 1390, 1986.
20. **Vega, C. A., Buening, G. M., Green, T. J., and Carson, C. A.,** *In vitro* cultivation of *Babesia bigemina, Am. J. Vet. Res.,* 46, 416, 1985.
21. **Vaayrynen, R. and Tuomi, J.,** Continuous *in vitro* cultivation of *Babesia divergens, Acta Vet. Scand.,* 23, 471, 1982.
22. **Irvin, A. D. and Young, E. R.,** Possible *in vitro* test for screening drugs for activity against *Babesia* and other bloood protozoa, *Nature (London),* 169, 407, 1977.
23. **Irvin, A. D., Young, E. R., and Purnell, R. E.,** The *in vitro* uptake of tritiated nucleic acid precursors by *Babesia* species of cattle and mice, *Int. J. Parasitol.,* 8, 19, 1978.
24. **Irvin, A. D. and Young, E. R.,** Further studies on the uptake of tritiated nucleic acid precursors by *Babesia* species of cattle and mice, *Int. J. Parasitol.,* 9, 109, 1979.
25. **Jack, R. M. and Ward, P.,** *Babesia rodhaini* interaction with complement: relationship to parasitic entry into red cells, J. Immunol., 124, 1566, 1980.
26. **Bautista, C. R. and Kreier, J. P.,** Effect of immune serum on the growth of *Babesia microti* in hamster erythrocytes in short-term culture, *Infect. Immun.,* 25, 470, 1979.
27. **Montealegre, F., Levy, M. G., Ristic, M., and James, M. A.,** Growth inhibition of *Babesia bovis* by secretions from bovine mononuclear phageocytes, *Infect. Immun.,* 50, 523, 1985.
28. **Montenegro-James, S., James, M. A., Levy, M. G., Preston, M. D., Exparza, H., and Ristic, M.,** Utilization of culture-derived soluble antigen in the latex agglutination test for bovine babesiosis and anaplasmosis, *Vet. Parasitol.,* 8, 291, 1981.
29. **Carson, C. A. and Phillips, R. A.,** Immunologic response of the vertebrate host to *Babesia* , in *Babesiosis,* Ristic, M. and Kreier, J. P., Eds., Academic Press, New York, 1981, 411.
30. **Conrad, P. A.,** Uptake of tritiated nucleic acid precursors by *Babesia bovis in vitro, Int. J. Parasitol.,* 16, 263, 1986.
31. **Aikawa, M., Rabbege, J., Uni, S., Ristic, M., and Miller, L. H.,** Structural alteration of the membrane of erythrocytes infected with *Babesia bovis, Am. J. Trop. Med. Hyg.,* 34, 45, 1985.
32. **Montenegro-James, S., James, M. A., and Ristic, M.,** Localization of culture-derived soluble *Babesia bovis antigens in the infected erythrocyte, Vet. Parasitol.,* 13, 311, 1983.
33. **Molinar, E.,** *In vitro* Propagation of *Babesia canis*: morphological Ultrastructural and Immunologic Studies on Antigens Derived from Primary Cell Cultures, M.S. thesis, University of Illinois, Urbana, 1979.
34. **Ristic, M.,** personal communication.

Chapter 5

EPIDEMIOLOGY OF BABESIOSIS

Alan S. Young

TABLE OF CONTENTS

I. INTRODUCTION

The epidemiology of babesiosis is a difficult subject since as a vector-borne disease, babesiosis has a very complex interrelationship between the vector and the host.[1] In diseases caused by *Babesia* and the related genus *Theileria*, the most complex epidemiology is attributable to the intricate feeding patterns of the invertebrate vector. The tick remains attached to the mammalian host for a substantial period before it transmits *Babesia* sporozoites. During this time, the host can respond by either rejecting the tick or by impairing tick feeding capacity, thus reducing both transmission to the host and infection of ticks feeding on the host.[2]

In many epidemiologic studies of babesiosis, there has been a disregard for the population dynamics of the teck vectors. As a result, two distinct disciplines of tick ecology and babesia parasitology have emerged.

Recent reports on the epidemiology of babesiosis are sparse, but Mahoney[3] has reviewed the epidemiology of cattle babesiosis in Australia, Joyner and Donnelly,[4] the epidemiology of babesiosis in general, and Friedhoff and Smith[5] concentrated on the epidemiology aspects of babesiosis in the tick vector. In addition, Dallwitz et al.[6] have reviewed recent approaches to the epidemiology of babesiosis and other tick-borne diseases.

A. Veterinary Epidemiology

There appear to be some misconceptions in the veterinary world concerning the definition of epidemiology. Schwabe et al.[7] defined epidemiology as the study of disease in populations, or alternatively, the study of the health status of populations. McDonald[8] defined epidemiology as the study of reasons for the prevalence of disease and the nature and causes of variation in it. In this review, I will discuss epidemiology as the study of the persistence of organisms under field conditions and the situations by which these organisms cause disease. Schwabe et al.[7] have divided studies in epidemiology into three areas, which I have generally followed in this review.

1. **Descriptive epidemiology** is the collection of field observations in order to determine how the epidemiology of the parasite exists in the field.[9,10] These epidemiological observations should allow the formulation of an epidemiological hypothesis.
2. This leads to **experimental epidemiology** whereby epidemiological hypotheses are tested. Common epidemiological experiments include vaccination and drug trials as well as clinical trials involving susceptible cattle populations. Useful experimental epidemiological studies are those in which cattle ticks and parasites are manipulated in paddocks to study various components of the epidemiology.[11,15]
3. **Analytical epidemiology** is a mathematical approach using simple, descriptive, and analytical mathematical models.
4. **Simulation epidemiology** is a new field of study dependent on the development of relatively sophisticated computers in which the epidemiological components are simulated as accurately as possible.[6]

B. History of the Epidemiology of Babesiosis

The epidemiology of babesiosis was first studied in the U.S. A veterinarian working for the U.S. government, D.E. Salmon, had noted by 1885 that a disease of cattle known as Texas fever showed a pattern of occurrence which could not be explained by any known mechanism of disease transmission. This disease did not occur north of a geographic line referred to as Salmon's Texas fever line. As is the case in many epidemiological studies, cattle ranchers associated the cause of Texas fever with ticks before scientists discovered this correlation. The American veterinarian, Frederick Kilborne, wanted to test the American cattlemen's hypothesis, since he believed that Salmon's Texas fever line coincided with the cattle tick's natural range. He persuaded Salmon that he should undertake epidemiological experiments in paddocks to test this hypothesis. Salmon approved of the study and brought in C. Curtice to work out the life history of the cattle tick and T. Smith to study the pathology and possible etiology of the disease. An experiment station of the Bureau of Animal Industries was divided into six well-separated, fenced paddocks.

Susceptible cattle from north of the Texas fever line were distributed as control cattle in the paddocks with two different possibilities for exposure to ticks. In addition, North Carolina cattle from south of the Texas fever line, some of which were tick free, were utilized. The results showed that only the susceptible cattle exposed to ticks contracted Texas fever. These experiments were repeated in the summers of 1890, 1891, and 1892. Meanwhile, Smith showed that the agent responsible for Texas fever was a protozoa, now called *Babesia bigemina.* Kilborne and Curtice then postulated that Smith's agent was being transmitted from ticks to cattle and that transmission could occur from one generation to another. This was later proven and was the first time that an etiologic agent was shown to be transmitted by a vector. By 1906, Curtice was able to use arsenic solution in North Carolina to effectively control cattle ticks. Eventually the tick vectors, *Boophilus microplus* and *B. annulatus*, and the parasite, *Babesia bigemina*, were eliminated from the U.S. by control and elimination of the tick vector, illustrating the importance of the experimental epidemiological approach.

C. Modern Approaches to the Epidemiology of Babesiosis

Since the studies of Kilborne and colleagues, much progress has been made on the understanding of the biology of *Babesia*, previously reviewed by Young and Morzaria.[16] Given the current understanding of the life cycle of the *Babesia* parasites and pertinent immunologic data, it should now be possible to improve and utilize analytical and simulation models of the epidemiology of babesiosis which have been developed recently.[3,6,17] Such models should contribute to the introduction of new control measures into the field such as vaccination using living and dead preparations,[18,19] optimal use of chemotherapy,[20] improved methods of tick control,[21] and selection of tick-resistant breeds of cattle.[22]

II. TECHNIQUES USED IN MONITORING THE EPIDEMIOLOGY OF BABESIOSIS

The main epidemiological components of babesiosis requiring quantitation are the population dynamics of the tick vector, the exposure of mammalian hosts to *Babesia*, and the capacity of the mammalian population to infect ticks. In addition to these major components, a full understanding of the biology of ticks and *Babesia* parasites is necessary.

The most important requirement in any geographic area is to determine the major tick vectors of babesiosis and the host range for both the tick vectors and *Babesia* parasites. The quantitation of various aspects of the epidemiology of babesiosis is considered in this section.

A. Monitoring the Tick Vector Populations

The tick vector is an extremely important component of the epidemiology of babesiosis. An accurate identification of tick vectors is essential. In Africa, for example, it has only been recently realized that *Boophilus* species are the major vectors of cattle babesias. This was because some early studies wrongly incriminated tick species as vectors, presumably due to extraneous infections.[23,24] It is now recognized that *Babesia bigemina* is transmitted in most of Africa by both *Boophilus decoloratus* and *B. microplus*, and *Babesia bovis* transmitted only by *Boophilus microplus.* The distribution of the parasites is determined by the distribution of the tick species. Once the tick vector for a *Babesia* species has been identified, the distribution of the vector can be mapped out. After further critical work on the biology and ecology of the tick vector, the potential ecological range of the vector can be established.[25]

The population dynamics of the tick vector need to be determined. This entails the investigation of the seasonality of tick infection of hosts, the potential generation time of the tick, viability of its unfed instars, and an accurate assessment of the climate within the distribution of the tick (particularly the microclimate in which the tick instars live). The relationship between the tick vector and host then must be studied (e.g., feeding patterns of the tick instars and the proportion of ticks that can develop through their life cycle infesting animals of different breeds with varying degrees of resistance to ticks). Most of these techniques have been described by Sutherst et al.[26] Once the required data are obtained from field and laboratory studies, such as the establishment of tick-infested paddocks and exposure of different instars to natural developmental conditions, it is possible to construct simulation models of tick population dynamics.[6,27]

B. Monitoring *Babesia* Infections in Ticks

Two components can be studied: first, the capacity of infected animals to subsequently infect ticks upon exposure; second, the tick infection rates in the field can be determined. Methods used to analyze the infection rates of bovine babesias in ticks have been described by Mahoney and Mirre[28] and Friedhoff and Smith.[5] An important epidemiological component is the inoculation rate (h) which is dependent on the numbers of ticks feeding successfully on a host and the infection rates in the tick population. The only tick that effectively transmits *Babesia* spp. is the one that can feed and successfully inoculate sporozoites. In *Boophilus* transmission of bovine babesias, the infection rate within tick salivary glands has not been regularly quantitated, but it is relatively easy to detect parasites in the hemolymph of the engorged female (the sporokinetes) or in the feeding larvae. In *Babesia bovis* infections of *Boophilus microplus,* approximately 50% of the infections will be missed as low numbers of sporokinetes are produced. It is easier to detect *Babesia bigemina* infections in the hemolymph of ticks because comparatively higher numbers of sporokinetes are present. Mahoney et al.[19] attempted to determine the infecton rate of *B. bovis* in *Boophilus microplus* larvae using the methods described by Mahoney and Mirre.[28] Several thousand larvae were examined microscopically, and they found that the infection rates in larvae varied from 0.037 to 0.05% from year to year in a paddock near Brisbane.

C. Monitoring *Babesia* in Mammalian Hosts

The determination of the incidence and prevalence of *Babesia* infections in mammalian hosts is absolutely essential. A simple method used effectively by Mahoney[29] is the examination of Giemsa-stained thick and thin blood smears. He found that cattle infected with *B. bovis*, and thereafter kept under tick-free conditions, showed subclinical relapses every 3 to 8 weeks, and that the numbers of parasites fell so low in the intervening periods that even with subinoculation of 500 to 1000 mℓ into susceptible, splenectomized calves, infections could not be detected. Subinoculation of blood from donor animal to a splenectomized recipient is the most sensitive method of detecting infective *Babesia*, particularly from carrier blood. Another relevant method is application of uninfected ticks to experimentally infested animals to determine the infection rates in the resultant tick instars.[28]

After acute or primary infections, recovered animals may develop subclinical infection, which can last for a considerable period of time. This is accompanied by the development of immunity.[3] The occurrence of disease is avoided upon subsequent challenge with the same species. This does not prevent entry and limited multiplication of the parasite introduced by the challenge infection, particularly if different strains are involved. Apparently, immunity does not necessarily depend on the continuing presence of infection. The duration of immunity after a single infection is important and has been measured using several methods for *B.*

bigemina and *B. bovis*.[30] The carrier state in recovered animals is probably facilitated by antigenic variation, demonstrated in *B. bovis* by Curnow[31,32] who studied the specificity of an agglutinogen on the surface of *B. bovis*-infected erythrocytes and found significant differences in successive recrudescences of parasitemia during the chronic phase of infection. The antigens appeared to be altered by the immune system of the host, since the antigen remained stable during rapid passage of the organism through splenectomized calves. A striking feature of these variable agglutinogens was their high degree of antigenic specificity. In each infected animal, all variant populations reverted to one antigenic type after infecting the vector, but this basic strain-specific antigen was also variable. Antigenic variation may also be one of the factors involved in differences in the duration of subclinical infection observed between various *Babesia* species.

A number of serological tests have been used to detect antibodies to *Babesia* parasites as a useful measure of prevalence and incidence. The main tests have been the complement fixation test (CFT), indirect fluorescent antibody tests (IFAT), indirect hemagglutination test (IHA), gel diffusion, capillary agglutination test (CA), bentonite agglutination test, latex particle agglutination test (LA), parasite agglutination, indirect radio immunoassay (RIA), and the micro-ELISA test. The CFT shows a high degree of specificity, giving only 1 to 2% false positives in noninfected cattle, but its sensitivity is low as antibodies remain at detectable levels for only a few months.[33] Therefore, it is not suitable for epidemiological studies in endemic areas, but can be used to detect foci of infections and residual cases after outbreaks in areas where the disease has been eradicated.[34] IFAT has been widely used.[35,36] Johnston et al.[36] found that 3 to 4% of noninfected cattle gave false positive reactions to *B. bovis* antigens, but the sensitized IFAT was 97 to 98% effective in detecting subclinical cases. Cross reactions between *B. bovis* and *B. bigemina* were minimal.

The IHA test has been evaluated by Goodger and Mahoney[37] who found 0.5% false positives and a high degree of reliability in detecting subclinical infections (97.5%). However, it was not as reliable for detecting *B. bigemina* infection in the field.[3]

Rapid agglutination tests, though not particularly accurate, are useful for evaluation of the epidemiological status of a herd so that control decisions can be made immediately.[3] New techniques such as the use of monoclonal antibodies against *Babesia* species and DNA probes may have applications in the epidemiology of babesiosis, at least as population markers.[19,38,39]

III. DESCRIPTIVE EPIDEMIOLOGY OF BABESIOSIS

A. Cattle Babesiosis in Australia

In Australia there are two species of *Babesia*, *B. bovis* and *B. bigemina*, infecting cattle. Both parasites are transmitted by the one-hose tick *Boophilus microplus*. Transovarian transmission occurs with both species of *Babesia*: *B. bigemina* is transmitted by nymphs and adults of the next generation after the adult female is infected, and *B. bovis* is transmitted by larvae.[16]

Basically, the situation in Australia is relatively simple and probably reflects the more complex tick-borne disease situation in Africa.[21] Both *Babesia* species were introduced along with their tick vector into the northern part of the country in the 1890s via Indonesia. The ticks and the parasites moved south until they occupied the present range. Humidity appears to be the major determinant for *Boophilus microplus* distribution. The ticks are found in savannah woodlands and tend to be scarce or absent from open savannah or deserts, but can penetrate via riverine areas. Western Australia, south of 20° latitude, has been shielded from *B. microplus* penetration by low rainfall, desert barriers, and effective quarantine measures applied to cattle migrating from the north. In the southern part of the range, winter temperatures control the spread of the tick. According to MacKerras et al.,[40] *B. microplus* could not survive a winter of more than 5 months between an autumn mean temperature of 20°C and a spring mean of 15°C. The subject of climate and *B. microplus* on a world-wide basis has been discussed by Sutherst and Maywald.[25]

Both *Babesia* species and *Boophilus microplus* do not have alternative hosts in Australia except for the feral water-buffalo in the Northern Territory. The present distribution of the tick and *Babesia* spp. includes a northeastern corner of New South Wales, a large section of Queensland, the northern portion of the Northern Territory, and a northern part of Western Australia. In many of the ranching areas where *Boophilus microplus* occurs, there is minimal contact between the rancher and cattle due to low stocking density, huge ranches, and an economic situation where most cattle are raised for export. Because of the rapid completion of the life cycle, up to six generations of *B. microplus* can occur within a year. It has been estimated in Australia that cattle lose 1 kg of live weight for each 1300 larvae maturing into engorged females.[41] Even given the relatively simple situation in Australia, successive crises of acaricide resistance caused by excessive reliance on chemical control have occurred,[42] Alternative approaches to tick control have been pasture spelling, i.e., keeping the pasture free until the larvae have died off,[43] and the use of tick resistant cattle.[22,44-46] The importance of this problem can be illustrated by the cost in 1975 of over A$40 million per annum due to ticks and babesiosis (control, A$14 million, production losses, A$27 million).[47] Pasture spelling has not proved acceptable to the cattle producer who finds the strategies difficult to apply.[46]

The rapid development of *B. microplus* resistance to acaricides in northern Australia due to poor strategies and dip management,[27,46] leaves the increased use of tick-resistant cattle as the only answer to tick infestation.[22,27] Phenotypic and genotypic variation of tick resistance exists between and within breeds. Thus, there is opportunity for improving cattle resistance to tick feeding by selection.[48-51] Therefore, beef cattle breeders can have several choices for improving tick resistance in their herds, e.g., by changing to more tick-resistant breeds, Hereford to Brahman or Brahman × Hereford, and or increasing resistance by selection within the breed or cross breed of choice.[22]

It is recognized that *Babesia bovis* causes more acute disease than *B. bigemina*.[3] Several concepts in the epidemiology of tick-borne diseases, in general, have emerged from the studies by Mahoney and colleagues.[6] The endemic stability of *Babesia* infections is an important concept developed by Mahoney.[3] It appears applicable to most tick-borne diseases.[9,10,52] In an endemic situation, calves infected before approximately 6 months of age show very mild infections. This protective factor is claimed to be due to the passive transfer of colostral antibodies from the dam,[3] which could be sporozoite neutralizing.[53] It also could be due to other factors such as the prevention of merozoite entry into erythrocytes,[54] transfer of immune cells from the dam to the calf, and an age immunity as demonstrated in *Babesia* infection of rats. Barnett[55] noted that Zebu calves born from dams in *Theileria parva*-endemic areas and deprived of colostrum had a prolonged resistance to *T. parva* infection, which could be broken down only if a massive challenge was given. For whatever reason, endemic stability for *Babesia* occurs and is dependent on the level of challenge. If the number of infected ticks feeding on calves is too low or too high, an enzootically unstable situation is likely to develop, resulting in clincial babesiosis.[3]

Another important concept developed by Mahoney et al.[30] was the persistence of infection and the carrier state (that is the ability of the infected animal to infect ticks feeding on them) of *B. bovis* and *B. bigemina* infections. This also relates to the immunity of the population. By thick blood smear examination, Mahoney[29] found that both *B. bigemina* and *B. bovis* showed regular recrudesences from subpatent to patent infections, which persisted up to less than 2 years from a single infection in the case of *B. bigemina* and up to 4 years in the case of *B. bovis*.[29,30] However, cattle at 4 years remained immune to both *B. bigemina* and *B. bovis* challenge. The characteristics of these studies were the subclinical relapses of *B. bovis*, and to a lesser extent *B. bigemina*, which may be explained by antigenic variation.[31] Therefore, the longevity of infection might be a function of the total number of different antigenic types that any population of parasites is able to produce. With limited variation, the host may be able to eliminate the parasite.[30]

The next important observation on the epidemiology of babesiosis in Australia was to determine the infection of the ticks from experimentally infected carrier cattle and from infected ticks in the field.[28] They found that examination of hemolymph from female ticks 5 to 7 days after engorgement failed to reveal a considerable proportion of those which transmitted infection to their progeny. In the case of *B. bovis*, the method probably detected less than 50% of infected ticks, but with *B. bigemina*, the hemolymph smears detected a considerably higher proportion of infected ticks. As a result of these intensive epidemiological studies, Mahoney and colleagues were able to construct mathematical models for the epidemiology of babesiosis in Australia,[3] which have been subsequently followed by mathematical models for tick population dynamics.[27]

The Australian strategy for the control of ticks and *Babesia* infection has already been described. Callow[18] developed a piroplasm-infected blood vaccine for both *B. bovis* and *B. bigemina*, which relied on attenuation during rapid passage through splenectomized calves. This vaccination method has had great success in Australia and, to a lesser extent, elsewhere.

Because it consists of living material, there are still potential dangers, and it is hoped that these will be overcome eventually by the use of killed antigen vaccine.[19,39] Treatment in Australia of *Babesia* infections is important both for the treatment of natural infection and also if the cattle react to live *Babesia* vaccination. Finally, another exciting development in Australia was the development of a vaccine against *Boophilus microplus* using antigens derived form the mid-gut of the tick, which can in immunized cattle stop the feeding of adult females.[55]

B. Cattle Babesiosis in Britain

There are two vectors for cattle babesiosis in Britain, *Ixodes ricinus*, which transmits *Babesia divergens*, and *Haemaphysalis punctata*, transmitting *B. major*. The *B. major* problem has only been recognized recently in Britain, but the epidemiology of *B. divergens* has been extensively studied.[4]

Ixodes ricinus is a three-host tick that feeds virtually on any warm-blooded host it encounters. Its distribution in Britain and elsewhere is dependent on environmental conditions.[56,57] The tick can occur in a variety of vegetational types, its range including forest, woodland, scrubland, moorland, and pastures. Joyner and Donnelly[4] indicated that ticks are limited in their movement, thus their frequency within a suitable ecology niche often depends on the availability of wild mammals. Permanent pasture, if well maintained, seldom provides adequate humidity for long-term survival of *I. ricinus* within a given locality. Copses, hedges, headlands, and windbreakers with an associated wildlife population may be sufficient to provide a focus of babesial infection threatening to susceptible cattle.[58] McLeod[59] indicated that temperature controls the activity of *I. ricinus*, whereas its local distribution is largely determined by humidity as the tick can survive only for a few days at relative humidity below 80%. Therefore, the tick needs to find suitable conditions all year long at the base of permanent vegetation.

Donnelly and McKellar[60] investigated the seasonal occurrence of over 7500 cases of babesiosis in a restricted area of Devonshire and found that the monthly incidence of cases fitted a biomodal curve with peaks from April to June and August to October. This approximated the seasonal activity pattern of *I. ricinus*.[57,61] Donnelly and McKellar[60] attempted to correlate the disease incidence of babesiosis to the climatic conditions (e.g., air temperature and rainfall). They found that rainfall did not have an obvious affect, but with multiple regression analysis, they could relate seasonal variation of disease to air temperature at a

90% significance level. There was better correlation of disease incidence to temperature in the period from January to June than from July to December. Joyner and Donnelly[4] reported that disease incidence due to *B. divergens* was highly correlated with tick activity.

From these studies, it is obvious that the epidemiology of babesiosis due to *B. divergens* is integrally associated with the activity of *I. ricinus*. An important feature is the bimodal nature of the seasonal distribution curve of *I. ricinus* infestation on hosts. Since the activities of the tick are temperature related, several suggestions for the biphasic seasonal activity have been put forward. In some areas of Britain, a single peak of activity is seen in late spring and early summer,[56,62] which led Campbell[63] to postulate that there were two separate populations, one exclusively active in the spring and the other active in the autumn. McLeod[59] previously showed the different ecological requirements for engorged or developing ticks and those unfed actively seeking attachment to hosts. He found that only habitats providing relative humidity of 80% or more throughout the year could support an *I. ricinus* population, and in these areas tick activity was limited to periods in the summer and autumn with temperatures between 14 and 24°C. Therefore, lack of summer activity in summer was due to the adverse effects of higher temperature. Additional effects were discovered by Chmela[64] who placed all instars in vials of *I. ricinus* at the base of vegetation. Whether he placed the vials in spring or summer, the larvae and nymphs moulted during the autumn. The spring population of active ticks, therefore, must have been the residue from the previous autumn. From the data of Lees and Milne,[65] all ticks from the previous year should have died by June. Therefore, the autumn population of active ticks must arise from the spring and early summer feeding instars. Their activity declined with the lower temperatures of winter and resumed with the temperature increases in the spring.[4] Babesiosis cases that occur in the spring are produced by ticks which have fed 12 or 18 months earlier, and those cases of *Babesia* occurring in the autumn must have fed approximately 6 or 12 months previously.

Donnelly et al.[66] studied an outbreak of babesiosis in a dairy herd on a farm that was originally tick-free, but on which infected ticks had been introduced in the previous year. Of 100 dairy cattle exposed in the newly infected paddocks, 54 became serologically positive with only 13 showing clinical disease. They calculated that 36% of the cattle were becoming infected each month. Donnelly[67] extended this work to 650 cattle diagnosed with babesiosis. He calculated the inoculation rate (h) in the early part of the year was 1.54, and during the latter part it was 0.69, giving a ratio of 2.25:1.

In Britain, babesiosis is associated with small numbers of cattle which become infected in small, defined foci within farms. Endemic stability can build up within the herd, but babesiosis can occur within susceptible imported cattle. Epidemic babesiosis can occur due to cattle movement allowing the temporary introduction of ticks. Joyner and Donnelly[4] indicated that of 100,000 cattle imported from Ireland via Glasgow, 30% were serologically positive for *B. divergens*.

In Britain, other than ecological control of ticks by improving pasture, the only control measure used against cattle babesiosis is treatment of clinical red water fever when diagnosed.

Considerable work on vaccination against *B. divergens* was undertaken by Purnell,[68] but the problem was not considered sufficiently important to warrant continued support. In Britain, because of the ecology of *I. ricinus*, babesiosis is generally considered to be a disease of marginally productive areas.

C. Cattle Babesiosis in Africa

Two species of *Babesia* of cattle have been reported in Africa, *B. bigemina* and *B. bovis*. *Boophilus microplus* transmits *Babesia bigemina* and *B. bovis*, while *Boophlus decoloratus*, a native of Africa, transmits *Babesia bigemina*. Hence, the distribution of the *Babesia* parasites is totally dependent on the respective distribution of these *Boophilus* species. Mason and Norval[69] have shown that *B. microplus* and *B. decoloratus* cannot coexist since *B.*

microplus will predominate if the environment is suitable. Therefore, there are dangers of *B. microplus* increasing its distribution with the concomitant introduction of *Babesia bovis*. The epidemiology of *Babesia* infections in Africa has not been studied in detail, except in South Africa. This is because tick-borne diseases have been dominated by theileriosis. In Kenya, it is considered that tick-borne diseases in order of their importance, are theileriosis, anaplasmosis, heartwater, and babesiosis.[10] The use of acaracides in many areas of Kenya is intensive enough to control *Boophilus decoloratus* populations, as long as the ticks do not become resistant to the acaricides used.

In relation to the problem in Kenya, studies in South Africa may be useful. Babesiosis in Natal, South Africa caused 8000 deaths in 1 year.[70] *Babesia bovis* and *B. bigemina* have very different distributions in S. Africa,[71] due to the distribution of the tick vectors for the two *Babesia* species,[72] with *Boophilus decoloratus* being more widespread.[73] In Natal, *Babesia bigemina* is the most common cause of babesiosis. It causes a problem because enzootic stability does not occur on many farms, probably due to the seasonal occurrence of ticks, since South Africa has more marked winters than East Africa. Many of the large outbreaks in South Africa are attributed to the introduction of susceptible cattle into endemic areas.

Boophilus microplus, and therefore, *Babesia bovis* infections are more sensitive to cold climates and require higher temperatures than *Boophilus decoloratus*.[74] De Vos and Every[74] suggested that the eradication of *B. decoloratus* in Natal would be the best answer, but it was impractical.[72,75]

An important factor in the epidemiology of babesiosis is the potential role of wildlife for both the maintenance of the tick vector and, in some cases, the *Babesia* parasite itself. Although wild game may be important for the maintenance of *Boophilus* species, it is considered that maintenance of *Babesia* infections in wild game is negligible.[72] Adequate control measures will require more definitive studies on the epidemiology of babesiosis in wild animals, particularly the role of the host tick in maintaining infection in the wild ruminants.

D. Cattle Babesiosis in South and Central America

The vector for cattle babesias in South and Central American is *Boophilus microplus,* which has a very wide distribution, and *B. annulatus*, which has a narrow distribution. *Boophilus microplus* has the potential for an even wider distribution, according to the climate-matching model, Climex.[6,25] The cattle babesias in South and Central America are *Babesia bigemina* and *B. bovis* (previously called *B. argentina*). This situation would appear to be the same as that in Australia, including similarities in the epidemiologic picture.

The distribution of *Boophilus microplus* in Central and South America appears to be restricted by: (1) dryness, (2) rain forest and lack of suitable hosts, and (3) cold climate in high-altitude areas such as the Andes and the southern part of South America.

Otherwise, the climate in South and Central American would appear to be ideal for *B. microplus* to thrive. Nari et al.[76] found in studies in the extreme south of Uruguay that climatic conditions from May to August were too cold for egg production by the engorged female *B. microplus*, as opposed to the optimal conditions occurring between December and March. The survival of *B. microplus* during winter months depends on the egg production of engorged female ticks dropping from cattle in February and March. They estimated in this area that *B. microplus* underwent two and a half to three generations per annum compared with six in the most favorable areas.

Endemic stability of both *Babesia bigemina* and *B. bovis* exists in the cattle populations in many areas of Central/South America and the Caribbean, reflecting the regular life cycles of *Boophilus microplus*. There have been many seroepidemiological studies in these regions using a variety of serological tests, particularly the IFAT and CFT.[76-83]

In many areas, the prevalence of antibody titer is very high, indicating endemic stability. Few disease problems occur since calves become infected rapidly after birth when they show resistance to *Babesia* infection. In all studies, the prevalence of antibodies to *B. bigemina* were higher than those to *B. bovis*.[76-83] This may indicate a more efficient transmission of *B. bigemina* than *B. bovis* by *Boophilus microplus*. The most detailed study available is by Corrier et al.[84] in Colombia. They found that the epidemiology of bovine babesiosis was influenced greatly by zones of tropical, subtropical, and cooler mountain climate. *Boophilus microplus* is widely distributed in the tropical and subtropical zones. Colombia's estimated 19 million head of beef cattle, consisting of various crosses between native cattle of Spanish origin (*Bos taurus*) with Zebu (*Bos indicus*), are located in the tropical and subtropical zones where they satisfactorily withstand the climatic conditions and tick infestation. The estimated three million dairy cattle of the country, mainly Holstein-Friesian and Brown Swiss breeds, are most frequently located in tick-free, mountainous areas, and to a lesser extent in the subtropical intermediate zone. *Babesia* infection was reported to be absent in altitudes above 2200 m.

Corrier et al.[84] carried out field studies in three areas of Colombia, corresponding to tropical grassland and forested savannah, the tropical Caribbean coast, and in a subtropical area. They found there was a generally higher prevalence of *B. bigemina* antibodies in cattle than those of *B. bovis*. The highest serological prevalence was on the Caribbean coast followed by the eastern plains and the subtropical area. The age prevalence of *B. bigemina* antibodies in calves in the three areas was slightly different, since on the Caribbean coast prevalence declined as calves became older, but in the east plain and the subtropical area, it increased, Corrier et al.[84] indicated that *B. bovis* was more pathogenic than *B. bigemina*, but endemic stability existed throughout the tropical and subtropical zones. Except in the Caribbean coast, epidemic situations did occur, due to either interrupted or low levels of transmission since the prevalence of antibodies failed to reach 100%. The use of an Australian-type immunization program for *B. bigemina* and *B. bovis* has been investigated in the subtropical zone of Colombia, calves being immunized at 3 to 6 months of age. Management practices on diary ranches in the subtropical zone required that the calves are removed from their dams and reared in sheds until 7 months of age, by which time the maternal antibodies have waned. At this stage, when they are exposed to tick infestation, they are likely to suffer from epidemic babesiosis with serious losses. Immunization is likely to overcome this problem, which appears widespread in dairy cattle in South America.[78-83]

Immunization of cattle against babesiosis using an Australian-type system has been attempted in Argentina, Brazil, Bolivia, and Colombia, generally where susceptible cattle are being introduced into endemic and epidemic situations or where transmission is likely to be interrupted.

E. Babesiosis of Small Mammals in Britain

There has been a considerable amount of work on the small animal *Babesia* of rodents and insectivores in Britain. One of the main purposes was to study the effect of *Babesia* infections on small animal health and to investigate the possibility that these small animals babesias may provide a laboratory model for studies on domestic animal babesias. In the U.S., these small animal babesias received more attention recently because it was evident that they could infect human beings either clinically or subclinically.[85]

The parasite in Britain was identified as *B. microti* and was detected in six species of small rodents and insectivores.[86] Additional hosts were reported by Shortt and Blackie.[87] It was possible to isolate and maintain *B. microti* from four species of wild animals in laboratory rats and mice. The main vector of *B. microti* in Britain was identified as *Ixodes trianguliceps*, a three-host tick. It transmitted *B. microti* transtadially between the larval and nymphal stages, which is quite unusual for *Babesia* species which are normally transmitted transovarially. Young[86] found that *B. microti* was distributed widely in England, Wales, Northern Ireland, and the Republic of Ireland. Similar studies were carried out by Cole,[88] Baker et al.,[89] Healing,[90] and Kirner et al.[91] in the U.S. It was believed that *B. microti* occur through the Britich Isles associated with the widespread distribution of the tick vector *I. trianaguliceps*. Young[86] sampled 1446 small mammals, 401 once and 1045 twice or more. This sample consisted of 610 *A. sylvaticus* (long-tailed wood mouse), 395 *C. glareolus* (bank vole), 210 *M. agrestis* (field vole), 119 *Sorex araneus* (common shrew), 44 *Sorex minutus* (pigmy shrew), 39 *Mus musculus* (House mouse), 6 *Rattus norvegicus* (Norway rat), 4 *Erinaceus europaeus* (hedgehog), and one *Micromys minutus* (harvest mouse).

No infections were detected in hedgehogs, house mice, or Norwegian rats, probably due to the fact that the tick vector, *I. trianguliceps*, did not feed on these hosts due to ecological reasons. Other hosts had patent piroplasm infections of 1.0 to 16% with a total prevalence of 9.8% in this sample. After splenectomy of a few animals, it was shown that a much higher prevalence occurred. Healing[90] took a slightly different approach as he studied the prevalence of *Babesia* infections in three populations of small British rodents by capture-recapture methods over a period of 3 years, a fourth population for 1 year, and fifth group sampled annually for 4 years. He found *B. microti* infections in all populations of species (*A. sylvaticus, C. glareolus*, and *M. agrestis*), with patent infections varying from 2.1 to 22.3%. One population which was not infected with *B. microti* resided on the Isle of Skomer off the Welsh coast, indicating that *I. trianguliceps* probably does not exist on this isolated island.

Both Young[86] and Healing[90] generally found a higher prevalence of infection in male rodents than females. Young[86] has suggested that the male rodent, known to have a larger territorial range, could come in contact with more ticks as he observed a higher level of *I. trianguliceps* infestation of male rodents. Young[86] showed a seasonal incidence of both *Babesia* parasitemia and *I. trianguliceps* infestation; however, this was not confirmed in the Healing study.[90]

Young[86] produced a scheme on how *B. microti* was maintained in the field. It was difficult to determine what direct effect *Babesia* infection had on small animal health, although with experimental infections a high level of mortality could be produced. Anemia and splenomegaly were noted in feral populations infected with *B. microti* parasites. *Babesia microti* maintained by blood passage proved to be a useful tool for study of *Babesia* immunology; however, the tick was too difficult to maintain in the laboratory as a good model for tick-transmitted *Babesia* parasites.

IV. ANALYTICAL EPIDEMIOLOGY OF BABESIOSIS

The analytical approach to veterinary epidemiology is well described.[7] After basic work in descriptive and experimental epidemiology is completed, proper analysis of the work is required. For example, Blewett et al.[92] in the study of babesiosis in deer in the U.K. used an arcsine transformation, which allowed for the decay of antibodies in the host population. Presuming an end-point titer of antibodies was occurring at different times in individuals, and given this is normally distributed, the decline of antibodies would describe a sine curve. The proporation of the animals (p) which have lost detectable antibodies is given by the expression $p = (\text{Sin } 0)^2$.

A. Mathematical Modeling of the Epidemiology of Babesiosis

The use of mathematical models in the epidemiology of parasitic diseases has received much recent attention. There appear to be several schools of thought, all of which have respective merits. One approach emphasizes parasite population ecology where epidemiology is concerned, following parasites through vector and host populations.[93-95] These studies have been hindered by the lack of long-term investigations of the dynamics of parasite populations. A large collection of theoretical studies has been undertaken on a vast array of parasites. The aim of epidemiological models is to make predictions of disease incidence or prevalence, to have an understanding of the biological mechanisms involved, and to test hypotheses concerning these mechanisms. The requirements of the models are that they are mathematically and biologically reasonable and should mimic real epidemiological situations.[7] Unfortunately, these criteria have not always been forthcoming. Mahoney et al.[12] have attempted to test a *Babesia* model[3] under experimental field situations. This mathematical model is derived from that developed by McDonald[96] for malarial parasites, which had the simple assumption that in a situation of endemic disease the risk of infection is constant over a limited time period.

An important basis of the Mahoney[3] model is the inoculation rate of *Babesia* spp., which was defined as the probability of the infection of any member of the population during 1 day. This determines the incidence of the parasite and the proportion and age structure of the noninfected and infected populations. The inoculation rate (h) is the number of infected bites received from the vector by each host per day (m), the proporation of the vector population carrying the infective forms of *Babesia* (a), and the proportion of bites which successfully infect the host (b). Hence:

$$\text{Inoculation rate (h)} = mab \qquad (1)$$

Critical levels for these measurements are the number of ticks on the host and the infection rates of the ticks, but this cannot be established routinely. The rate at which the young animals acquire infection is also related to the inoculation rate. Hence I = proportion of animals infected, can be defined as

$$I = 1 - e^{-ht} \qquad (2)$$

The inoculation rate can be calculated from the measured levels of I and t (mean age of the group studied). Mahoney states that a serological diagnostic test is required for the measurement of I in practical circumstances.[3]

The ranges of the inoculation rate which can develop a stable state depend on the length of time after birth for which the protection by passively acquired immunity or some other factor is effective. If this period is known, the critical value of the inoculation rate for enzootic stability can be calculated by substituting that value for t in Equation 2. It is the value of h calculated for specific time and for I approaching 1.0 (i.e., 100% infection). Values of the inoculation rate equal to or above this critical level indicate a stable situation, whereas those below the critical level would indicate some risk of developing disease. The minimum level of tick infectation (m in Equation 1) associated with stability may also be calculated if data on babesial infection rates of ticks are available. Mahoney[3] found that at least 12 *Boophlus microplus* larvae were required to attach to each cow daily to produce a critical inoculation rate of 0.005 for *Babesia bovis*.

The value b in Equation 1 represents the ability of the tick to transmit *Babesia*, and the host resistance to *Babesia* influences this parameter. In some situations where all bites are infective to the host (b = 1), but in more resistant cattle, such as *Bos indicus*, b will be less than 1. The inoculation rate can be predictive since it can be used to predict the disease before it occurs. As the inoculation rate falls below the critical level for stability, there is a range of values which produce a high incidence of clinical cases because a low proportion of animals may contract infection before passive immunity wanes. In spite of this, the inoculation rate may be high enough to expose the rest of the population within 2 or 3 years.

If the inoculation rate is further reduced, the incidence of babesiosis may fall so low that the occurrence of disease may not be recognized and a large proportion of the population might not be exposed.

Another concept is the basic reproduction rate. This is a hypothetical number representing the number of secondary cases of a disease disseminated by a single primary case in a nonimmune individual in an environment where neither host nor vector populations were previously infected. All the important factors and relationships concerning transmission should be known. If the basic reproduction rate equals or is greater than 1, the disease should persist, while if it is below 1, eradication should occur. Mahoney[3] has identified factors concerned with the basic reproduction rate in *Babesia* transmitted ovarially as (1) the duration of infectivity for ticks of a primary case in days, (2) the number of adult female ticks dropped per day, (3) the number of larvae produced per tick, (4) the proportion of larvae that bite a new host, and (5) the proportion of larvae infected with *Babesia*. An approximate basic reproduction rate is given by

$$2\,d\,n\,a \tag{3}$$

where d = duration of infectivity in days, n = number of female ticks dropped on the pasture per day, and a = the average babesial infection rate of all the progeny. The critical situation is represented by the expression:

$$2\,d\,n\,a = 1$$

The critical level of tick infestation (n) required for maintenance of *Babesia* spp. in the environment by a one-host tick can be determined.[3]

Many aspects of the Mahoney model were tested.[3] The key to the assessment of the epidemiological status of the herd was the value of the inoculation rate (h) which is determined by measuring the proportion of young animals infected in a defined age group by an appropriate serological test. This gives values to substitute for I and t given in Equation 2, and h may then be calculated. If the inoculation rate exceeds the critical level for stability, there is no disease risk because all the animals are immunized by tick-transmitted infecton with the period of protection covered by maternal antibodies. In "high-risk" situations, cattle should be vaccinated while passive immunity remains in order to prevent economic losses. Even in a low-risk situation, vaccination may be used, as the situation could change with an increase in tick infestation resulting in severe outbreaks.

Smith[17] developed the first model which included a description of the tick vector population by superimposing *Babesia* transmission on a simple tick model. The model had a fixed number of identical (equilibrium) and discrete generations of ticks per year and a fixed generation time. Components contributing to the survival of the *B. bovis* in its tick vector, *Boophilus microplus*, and in the bovine host were identified and their relationships simulated with an interactive basic computer program. Estimates for major independent variables (host-finding success, feeding success, filial infection rate, and recovery rate) were derived from published laboratory and field data. Values for dependent variables (tick burden, inoculation rate, tick and bovine infection rates) were calculated during successive tick generations until equilibrium conditions were reached. The simulation was used to predict the effects of alternative disease control strategies.

B. Simulation Modeling of the Epidemiology of Babesiosis

It is hoped that by applying this new approach for accurately simulating the epidemiology of babesiosis and other tick-borne diseases, a much more accurate epidemiological model can be obtained.[6] A useful first step is the estimation of the potential distributions of different disease vectors using the climatic matching program, "Climex" of Sutherst and Maywald.[25]

They used the model to describe the potential distribution of *B. microplus*, *Rhipicephalus appendiculatus*, and *Amblyomma variegatum*. Low Climex-derived climatic indexes indicated zones in which transmission may be interrupted because the size of the vector population is often inadequate to maintain transmission.

The first step in producing a generalized tick-borne disease model has been to develop a realistic tick population model which is applicable to one-, two-, three-host ticks.[6] It increments events daily and is climate-driven to make it applicable in different locations. The model describes the number of ticks or individual animals rather than average hosts, and it has a smoothly distributed transaction between each life cycle state. In addition, the model has maximum flexibility to describe various management options including chemical control of parasitic ticks, animal movement, host resistance, and application of chemicals to pastures. The model also includes the selection of acaricide-resistant strains of ticks.

The prevalence of disease-causing organisms within ticks is represented as the proportion of ticks of each age class that are uninfected, lightly infected, or heavily infected. Host animals are treated as individuals, and infective bites are received at random. The level of infection of each host is determined by its history of infective bites. Different parasites may be modeled one at a time by setting appropriate values for various parameters, such as the stages of feeding at which transmission occurs and the probabilities of transmission. The development of infective stages in the tick and the effect of environmental temperature on infectivity of the tick can be included in the model as can the dose of parasites inoculated into the host.

For the purpose of describing transmission, the disease is represented as the ''infectivity'' of that host to ticks feeding on it. The infectivity determines both the probability that a tick feeding at a particular stage of its life cycle, usually during engorgement, will become infected and the level of that infection. The infectivity can be related to observable phenomena in the host, such as density of parasites in the blood or its health or loss of production, but knowledge of these relationships is not essential for modeling the transmission of the disease unless the host dies.

V. EPIDEMIOLOGY AND CONTROL OF BABESIOSIS

The reason for the study of epidemiology of babesiosis in the field is to understand the disease and to introduce improved methods of control into the field. Also, techniques to evaluate how control measures work in the field can be assessed in the situations reviewed. On the British scene, babesiosis causes a minor problem, but perhaps this has been underestimated, considering the work of Donnelly, Purnell and colleagues[58,60,66-68] Britain exports cattle to many areas of the world and has been responsible for developing many breeds. The U.S. scene is of interest since that country was the first to recognize the *Babesia* problem and managed to eradicate *Boophilus microplus* and *Babesia bigemina*, showing what can be done if the epidemiological problem is recognized and if control measures are introduced.

Australia, on the other hand, has decided to live with its tick and *Babesia* problem because the required level of management for eradication in infected or infested areas does not exist. However, they have developed the epidemiological techniques and are developing the control techniques to allow them to manage the tick and *Babesia* parasites. We hope we can be equally successful in Africa where a much more complex situation exists with many tick species, alternative hosts for ticks, and a variety of tick-borne diseases occur. I believe that a more integrated approach to control, as suggested by the Australian workers, will be applicable in Africa, and we are pursuing this goal.

REFERENCES

1. **Mehlhorn, H. and Schein, E.,** Piroplasms: life cycle and sexual stages, *Adv. Parasitol.*, 23, 38, 1984.
2. **Leitch, B. L.,** The Effect of Acquired Resistance to *Rhipicephalus appendiculatus* Neumman 1901 on the Transmission of *Theileria parva* in Cattle, M.Sc. thesis, University of Salford, England, 1986.
3. **Mahoney, D. F.,** *Babesia* of domestic animals, in *Parasitic Protozoa*, Vol. 4, Kreier, J., Ed., Academic Press, New York, 1977, 1.
4. **Joyner, L. P. and Donnelly, J.,** The epidemiology of babesial infections, *Adv. Parasitol.*, 17, 115, 1979.
5. **Friedhoff, K. T. and Smith, R. D.,** Transmission of *Babesia* by ticks, in *Babesiosis*, Ristic, M. and Kreier, J. P., Eds., Academic Press, New York, 1981, 267.
6. **Dallwitz, M. J., Young, A. S., Mahoney, D. F., and Sutherst, R. W.,** Comparative epidemiology of tick-borne diseases of cattle with emphasis on modeling, *Parasitology—Quo Vadit*, Proc. 6th Int. Congr. Parasitology, Howell, M. J., Ed., Australian Academy of Sciences, Brisbane, 1986, 629.
7. **Schwabe, C. W., Riemann, H. P., and Franti, C. E.,** *Epidemioogy in Veterinary Practice*, Lea & Febiger, Philadelphia, 1977.
8. **McDonald, G.,** *Epidemiology and Control of Malaria*, Oxford University Press, London, 1957.
9. **Moll, G., Lohding, A., and Young, A. S.,** Epidemiology of theileriosis in the Trans-Mara Division, Kenya. Husbandry and disease background and preliminary investigations on theileriosis in calves, *Prev. Vet. Med.*, 2, 802, 1984.
10. **Moll, G., Lohding, A., Young, A. S., and Leitch, B. L.,** Epidemiology of theileriosis in calves in an endemic area of Kenya, *Vet. Parasitol.*, 19, 255, 1986.
11. **Smith, T. and Kilborne, F. L.,** Annual Report, Bureau of Animal Health, U.S. Department of Agriculture, Washington, D.C., 1893.
12. **Mahoney, D. F., Wright, I. G., Goodger, B. V., Mirre, G. B., Sutherst, R. W., and Utech, K. B. W.,** The transmission of *Babesia bovis* in herds of European and Zebu × European cattle infected with the tick *Boophilus microplus*, Aust. Vet. J., 57, 461, 1981.
13. **Young, A. S., Brown, C. G. D., Burridge, M. J., Cunningham, M. P., Payne, R. C., and Purnell, R. E.,** Establishment of an experimental field population of *Theileria lawrencei* infected ticks maintained by African buffalo *(Syncerus caffer)*, *J. Parasitol.*, 65, 903, 1977.
14. **Young, A. S., Leitch, B. L., Dolan, T. T., Newson, R. M., Ngumi, P. N., and Omwoyo, P. L.,** Transmission of *Theileria parva* by a population of *Rhipicephalus appendiculatus* under simulated natural conditions, *Parasitology*, 86, 255, 1983.
15. **Newson, R. M., Cheira, J. W., Young, A. S., Dolan, T. T., Cunninghma, M. P., and Radley, D. E.,** Survival of *Rhipicephalus appendiculatus* (Acarina: Ixodidae) and persistence of *Theileria parva* (Apicomplexa: Theileriidae) in the field, Int. J. Parasitol., 14, 483, 1984.
16. **Young, A. S. and Morzaria, S. P.,** Biology of *Babesia* species, *Parasitol. Today*, 2, 211, 1986.
17. **Smith, R. D.,** *Babesia bovis:* computer simulation of the relationship between the tick vector, parasite and bovine host, *Exp. Parasitol.*, 56, 27, 1983.
18. **Callow, L. L.,** Vaccination against bovine babesiosis, in *Immunity to Blood Parasites of Animals and Man*, Miller, L. H., Pino, J. A., and McKelvey, J. J., Jr., Eds., Plenum Press, New York, 1977, 121.
19. **Mahoney, D. F., Wright, I. G., and Goodger, B. V.,** Immunization against babesiosis: current studies and future outlook, in *Impact of Disease on Livestock Produciton in the Tropics*, Riemann, H. P. and Burridge, M. J., Eds., Elsevier, Amsterdam, 1984, 401.
20. **Kuttler, K. L.,** Chemotherapy of babesiosis. A review, in *Babesiosis*, Ristic, M. and Kreier, J. P., Eds., Academic Press, New York, 1981, 65.
21. **Sutherst, R. W.,** Management of arthropod parasitism in livestock, in *Tropical Parasites and Parasitic Zoonoses*, Dunsmore, J. D., Ed., Murdoch University, Perth, 1983, 41.
22. **Seifert, G. W.,** Selection of beef cattle in Northern Australia for resistance to the cattle tick (*Boophilus microplus*): research and application, in *Impact of Diseases on Livestock Production in the Tropics*, Riemann, H. P. and Burridge, M. J., Eds., Elsevier, Amsterdam, 1984, 553.
23. **Neitz, W. O.,** Classification, transmission and biology of piroplasms of domestic animals, *Ann. N.Y. Acad. Sci.*, 64, 56, 1956.
24. **Morzaria, S. P., Young, A. S., and Hudson, E. B.,** *Babesia bigemina* in Kenya: experimental transmission by *Boophilus decoloratus* and production of tick-derived stabilates, *Parasitology*, 74, 291, 1977.
25. **Sutherst, R. W. and Maywald, G. F.,** A computerized system for matching climates in ecology, *Agric. Ecosyst. Environ.*, 13, 281, 1985.
26. **Sutherst, R. W., Wharton, R. H., and Utech, K. B. W.,** Guide to Studies on Tick Ecology, Tech. paper, No. 14, CSIRO Division of Entomology, Council of Science and Industrial Research Organizaiton, Australia, 1978, 59.
27. **Sutherst, R. W.,** Is the Austrialian pest management approach to tick control relevant to Africa?, in *Tick Biology and Control*, Whitehead, G. B. and Gibson, J. D., Eds., Rhodes University, Grahamstown, Australia, 1981, 79.

28. **Mahoney, D. F. and Mirre, G. B.,** Bovine babesiosis: estimation of infection rates in the tick vector *Boophilus microplus* (Canestrini), *Ann. Trop. Med. Parasitol.*, 65, 309, 1971.
29. **Mahoney, D. F.,** Bovine babesiosis: a study of factors concerned in transmission, *Ann. Trop. Med. Parasitol.*, 63, 1, 1969.
30. **Mahoney, D. F., Wright, I. G., and Mirre, G. B.,** Bovine babesiosis: the persistence of immunity to *Babesia argentina* and *B. bigemina* in calves (*Bos taurus*) after naturally acquired infection, *Ann. Trop. Med. Parasitol.*, 67, 197, 1973.
31. **Curnow, J. A.,** *In vitro* agglutination of bovine erythrocytes infected with *Babesia argentina, Nature (London)*, 217, 267, 1968.
32. **Curnow, J. A.,** Use of a slide agglutination test to demonstrate antigenic differences between *Babesia bigemina* parasites, *Aust. Vet. J.*, 49, 290, 1973.
33. **Mahoney, D. F.,** Bovine babesiosis: an assessment of the significance of the complement fixing antibodies based on experimental infections, *Aust. Vet. J.*, 40, 369, 1964.
34. **Curnow, J. A.,** Studies on the epizootiology of bovine babesiosis in northeastern New South Wales, *Aust. Vet. J.*, 49, 284, 1973.
35. **Ross, J. P. J. and Lohr, K. F.,** Serological diagnosis of *Babesia bigemina* infection in cattle by the indirect fluorescent antibody test, *Res. Vet. Sci.*, 9, 557, 1968.
36. **Johnston, L. A. Y., Pearson, R. D., and Leatch, G.,** Evaluation of an indirect fluorescent antibody test for detecting *Babesia argentina* infection in cattle, *Aust. Vet. J.*, 49, 373, 1973.
37. **Goodger, B. V. and Mahoney, D. F.,** Evaluation of the passive hemagglutination test for the diagnosis of *Babesia argentia* infection in cattle, *Aust. Vet. J.*, 50, 246, 1974.
38. **Cowman, A. F., Kemp, D. J., Timms, P., and Dalgleish, B.,** cDNA clones from Poly A$^+$ RNAs which differ between virulent and avirulent strains of the protozoan parasite *Babesia bovis, Proc. Int. Conf.* on *Manipulatoin and Expression of Genes in Eukaryotes,* Academic Press, Australia, 1983, 31.
39. **Timms, P., de Vos, A. J., and Dalgleish, R. J.,** Molecular vaccines against parasite bovine babesiosis, *Parasitol. Today*, 2, S11, 1986.
40. **MacKerras, I. M., Waterhouse, D. F., Maiden, A. C. B., and Edgar, G.,** The Cattle Tick Problem in NSW, Sci. Bull. No. 78, Department of Agriculture, New South Wales, Sydney, 1961.
41. **Sutherst, R. W. and Utech, K. B. W.,** Controlling livestock parasites with host resistance, in *CRC Handbook of Pest Management in Agriculture,* Vl. 2, Pimentel, D., Ed., CRC Press, Boca Raton, Fla., 1981, 385.
42. **Wharton, R. H.,** Tick-borne diseases and the vectors. V. Acaricide resistance and alternative methods of tick control, *World Anim. Rev.*, 20, 8, 1976.
43. **Wilkinson, P. R.,** The spelling of pastures in cattle tick control, *Aust. J. Agric. Res.*, 8, 414, 1957.
44. **Francis, J. and Little, D. A.,** Resistance of Drought master cattle to tick infestation and babesiosis, *Aust. Vet. J.*, 40, 247, 1964.
45. **Wharton, R. H., Utech, K. W. B., and Sutherst, R. W.,** Tick resistant cattle for the control of *Boophilus microplus,* Proc. Third Int. Cong. Acarology, 1973, 697.
46. **Hewetson, R. W.,** Results of selection for cattle tick resistance in cattle, in *Tick Biology and Control,* Whitehead, G. B. and Gibson, J. D., Eds., Rhodes University, Grahamstown, Australia, 1981, 91.
47. **Anon.,** Cattle Tick in Australia. Inquiry by the Cattle Tick Control Commission, Australian Government Publishing Service, Canberra, 1975.
48. **Hewetson, R. W.,** Resistance of cattle to cattle tick, *Boophilus microplus.* The inheritance of resistance to experimental infestations, *Aust. J. Agric. Res.*, 19, 497, 1968.
49. **Wharton, R. H., Utech, K. B. W., and Turner, H. G.,** Resistance of the cattle tick *Boophilus microplus* in a herd of Australian Illawarra short cattle: its assessment and heritability, *Aust. J. Agric. Res.*, 21, 103, 1970.
50. **Siefert, G. W.,** Variations between and within breeds of cattle in resistance to field infestation of the cattle tick *(Boophilus microplus), Aust. J. Agric. Res.*, 22, 159, 1971.
51. **Utech, K. B. W. and Wharton, R. H.,** Breeding for resistance to *Boophilus microplus* in Australia Illawarra shorthorn and Brahman × Australian Illawarra shorthorn cattle, *Aust. Vet. J.*, 58, 41, 1982.
52. **Young, A. S.,** The epidemiology of theileriosis in East Africa, in *Advances in the Control of Theileriosis,* Irvin, A. D., Cunningham, M. P., and Young, A. S., Eds., Martinus Nijhoff, Dordrecht, Netherlands, 1981, 38.
53. **Gray, M. A. and Brown, C. G. D.,** *In vitro* neutralization of *Theileria* sporozoites, in *Advances in the Control of Theileriosis*, Irvin, A. D., Cunningham, M. P., and Young, A. S., Eds., Martinus Nijhoff, Dordrecht, Netherlands, 1981, 125.
54. **Winger, C. M. and Canning, E. U.,** Monoclonal antibodies to merozoites of *Babesia divergens,* Proc. 7th Int. Congr. Protozoology, Nairobi, 1985, 96.
55. **Johnston, L. A. Y., Kemp, D. H., and Pearson, R. D.,** Immunization of cattle against *Boophilus microplus* using extracts derived from adult female tick. Effects of induced immunity on tick populations, Int. J. Parasitol., 16, 27, 1986.

56. **Milne, A.**, The ecology of the sheep tick *Ixodes ricinus* L. Host relationships of the tick. II. Observation on hill and moorland grazings in northern England, *Parasitology*, 39, 173, 1949.
57. **McLeod, J.**, The bionomics of *Ixodes ricinus* L., the "sheep tick" of Scotland, *Parasitology*, 24, 382, 1932.
58. **Donnelly, J., Crossman, P. J., and McKendrick, M. D.**, An outbreak of redwater on a farm in Sussex, *Vet. Rec.*, 87, 729, 1970.
59. **McLeod, J.**, The ticks of domestic animals in Britain, *Emp. J. Exp. Agric.*, 7, 97, 1939.
60. **Donnelly, J. and McKellar, J. C.**, The effect of weather and season on the incidence of redwater fever in cattle in Britain, *Agric. Meterol.*, 7, 5, 1970.
61. **Edwards, E. E. and Arthur, D. R.**, Seasonal activity of the tick *Ixodes ricinus* L. in Wales, *Parasitology*, 38, 72, 1947.
62. **Arthur, D. R.**, Some aspects of the ecology of the tick, *Ixodes ricinus* L. in Wales, *Bull. Entomol. Res.*, 39, 321, 1948.
63. **Campbell, J. A.**, Recent work on the ecology of the pasture tick, *Isodes ricinus* in Britain, Rep. 14th Int. Vet. Congr., 1950, 113.
64. **Chmela, J.**, On the development cycle of the common tick (*Ixodes ricinus* L.) in the north Moravian natural focus of tick-borne encephalitis, *Folia Parasitol. (Prague)*, 16, 313, 1969.
65. **Lees, A. D. and Milne, A.**, Seasonal and diurnal activities of individual sheep ticks (Ixodes ricinus L.), *Parasitology*, 41, 189, 1951.
66. **Donnelly, J., Joyner, L. P., and Crossman, P. J.**, Incidence of *Babesia divergens* infection in a herd of cattle as measured by the indirect immunofluorescent antibody test, *Res. Vet. Sci.*, 13, 511, 1972.
67. **Donnelly, J.**, Epidemiology of *Babesia* infection in cattle, *Proc. R. Soc. Med.*, 66, 774, 1973.
68. **Purnell, R. E.**, Vaccines against piroplasms, in *Vaccine Against Parasites*, Taylor, A. E. R. and Muller, R., Eds., Blackwell Scientific, Oxford, 1980, 25.
69. **Mason, C. A. and Norval, R. A. I.**, Ticks of Zimbabwe. I. The genus, *Boophilus*, *Zimbabwe Vet. J.*, 11, 36, 1980.
70. **Anon.**, Annual Report of the Department of Agricultural Technical Services (1970—1971), Republic of South Africa, 1972, 177.
71. **de Vos, A. J.**, Epidemiology and control of bovine babesiosis in South Africa, *J. S. Afr. Vet. Assoc.*, 50, 1979.
72. **Howell, C. J., de Vos, A. J., Beziridenbait, J. D., Potgieter, F. T., and Barrowman, P. R.**, The role of chemical tick-eradication in the control or prevention of tick-transmitted diseases of cattle, in *Tick Biology and Control*, Whitehead, G. B. and Gibson, J. P., Eds., Rhodes University, Grahamstown, Australia, 1981, 61.
73. **Howell, C. J., Walker, J. B., and Nevill, E. M.**, Ticks, Mites and Insect of Domestic Animals in South Africa. I. Description and Biology, Sci. Pamph. No. 393, Tech. serv., Departemnt of Agriculture, Republic of South Africa, 1978.
74. **de Vos, A. J. and Every, R.**, Epidemiology and control of bovine babesiosis in the natal midlands, in *Tick Biology and Control*, Whitehead, G. B. and Gibson, J. D., Eds., Rhodes University, Grahamstown, Australia, 1981, 57.
75. **Bigalke, R. D.**, Thoughts on the future control of ticks and tick-borne diseases of cattle in South Africa, *J. S. Afr. Biol. Soc.*, 17, 7, 1976.
76. **Nari, A., Cardozo, A., Berdie, J., Canabez, F., and Bawden, R.**, unpublished report.
77. **Estrada Correa, A., Chapa Ruiz, R., Gallo de la Torre, M., Bautista Garfias, C. R., and Morilla Gonzalez, M.**, Indice de infeccion de babesiosis bovina causada par *Babesia bovis* en el centro Experimental Pecuariode, Haeytamulco, Puebla, Mexico, Vet. Mex., 11, 9, 1980.
78. **Payne, R. C. and Scott, J. M.**, Anaplasmosis and Babesiosis in El Salvador, *Trop. Anim. Health Prod.*, 14, 75, 1982.
79. **Berry, S., Ibata, G., and Edwards, S.**, Antibody formation to *Babesia bovis* and *Anaplasma marginale* in calves in Bolivia, *Trop. Anim. Health Prod.*, 13, 240, 1981.
80. **Applewhaite, L. M., Craig, T. M., and Wagner, G. G.**, Serological prevalence of bovine babesiosis in Guyana, *Trop. Anim. Health Prod.*, 13, 13, 1981.
81. **Bundy, D. A., Hylton, G. A., and Wagner, G. G.**, Bovine haemoparasitic diseases in Jamaica, *Trop. Anim. Health Prod.*, 15, 47, 1983.
82. **James, M. A., Coronado, A., Lopez, W., Melendez, R., and Ristic, M.**, Seroepidemiology of bovine anaplasmosis and babesiosis in Venezuela, *Trop. Anim. Health Prod.*, 17, 9, 1984.
83. **Madruga, C. R., Aycardi, E., Kessler, R. H., Schenk, M. A. M., de Figueiredo, G. R., and Cuirvo, J. B. E.**, Niveis de anticorpos anti-*Babesia bigemina* e *Babesia bovis* em bezerros da raca Nejore, Ibage' e cruzamentos de Nelore, *Pesq. Agropecu. Bras.*, 19, 1163, 1983.
84. **Corrier, D. E., Gonzalex, E. F., and Betancourt, A.**, Current information on the epidemiology of bovine anaplasmosis and babesiosis in Colombia, in *Tick-Borne Diseases and Their Vectors*, Wilde, J. K. H., Ed., University of Edinburgh, Scotland, 1978, 114.,

85. **Healy, G. L. and Ristic, M.,** Human Babesiosis, in *Babesiosis of Domestic Animals and Man,* Vol. 2, Ristic, M., Ed., CRC Press, Boca Raton, Fla., in press.
86. **Young, A. S.,** Investigations on the Epidemiology of Blood Parasites of Small Mammals with Special Reference to Piroplasma, Ph.D. thesis, University of London, 1970.
87. **Shortt, H. E. and Blackie, E. J.,** An account of the genus *Babesia* as found in certain small animals in Britain, *J. Trop. Med. Hyg.*, 68, k37, 1965.
88. **Cole, A. C.,** Blood parasties, found in mammals, birds and fishes in England, *Parasitology,* 7, k17, 1914.
89. **Baker, J. R., Chitty, D., and Phipps, E.,** Blood parasites of wild voles in England, *Parasitology,* 53, 297, 1963.
90. **Healing, T. D.,** Infection with blood parasites in British small rodents, *Apodemus sylvaticus, Clethrionomys glareolus* and *Microtus agrestis, Parasitology,* 83, 179, 1981.
91. **Kirner, S. H., Barbehenn, K. R., and Travis, B. V.,** A summer survey of two *Microtus pennsylvanicus* p. (Ord.) population, *J. Parasitol.*, 44, 103, 1958.
92. **Blewett, D. A., Beasly, G. J., Campbell, J. A., and Turnbull, D. Mc. D.,** A simple model for the interpretation of serological data for tick-borne diseases, in *Tick-Borne Diseases and Their Vectors,* Wilde, J. K. H., Ed., Centre for Tropical Veterinary Medicine, University of Edinburgh, Scotland, 1978, 130.
93. **Kostitzin, V. A.,** *Symbiose, Parasitisme et Evolution,* Hermann, Paris, 1934.
94. **Crofton, H. D.,** A model of host-parasite relationships, *Parasitology,* 63, 343, 1971.
95. **Anderson, R. M.,** Mathematical models of host-helminth parasite interactions, in *Ecological Stability,* Usher, M. B. and Williamson, M. H., Eds., Chapman & Hall, London, 1974, 43.
96. **McDonald, G.,** *Dynamics of Tropical Disease,* Bruce-Chwatt, L. J. and Granville, V. J., Eds., Oxford University Press, London, 1973.

Chapter 6

PATHOGENESIS OF BABESIOSIS

I. G. Wright and B. V. Goodger

TABLE OF CONTENTS

I. INTRODUCTION

This review is intended to elucidate the pathogenic mechanisms of acute babesial infections. Aspects such as hematological and biochemical changes per se are not included as these have been extensively reviewed elsewhere.[1-3]

Acute babesiosis, particularly that caused by *Babesia bovis*, induces a hypotensive shock syndrome characterized by low circulating parasite numbers, but with a high percentage of parasitized red cells in the capillaries of brain, muscle, and kidney. This shock commences when as few as 1×10^9 organisms are present in the circulation and is progressive until death. A generalized nonspecific inflammation is initiated by this disease process and is manifested by activation of the coagulation-complement-kallikrein cascade and by the release of numerous vasoactive compounds. As a result of this vasodilatation and circulatory stasis, generalized organ damage due to anoxia and toxic products of both parasite and damaged host tissue origin occurs. Generalized nonspecific damage does not occur with less pathogenic strains or species of *Babesia* in which the most common signs are hemolysis directly proportional to the often high parasitemia and elevated body temperatures.

In the assessment of strain and/or species virulence, the physiologic status of the infected host is of major importance. For instance, splenectomy can greatly exacerbate symptoms with less pathogenic species such as *B. bigemina*. On the other hand, with highly virulent organisms such as *B. bovis*, splenectomized calves are not as severely affected as nonsplenectomized adult animals, even though the parasitemia is significantly lower, rarely exceeding 1% (5 to 20% in splenectomized calves). The reason for this difference in susceptibility is unknown. Thus, in any discussion on the relative virulence of various *Babesia* species and strains the type of host must be taken into consideration. The most reliable comparison can be made using the natural host, i.e., nonsplenectomized weaned animals, because unpredictable results are obtained with young animals due to the possible presence of protective colostral antibodies in their serum for up to 6 months after birth[4] and also to the presence of a dialysable inhibitory factor in serum.[5] Unfortunately, most studies on the pathogenesis of the various babesia species have been made using splenectomized hosts, a factor which modifies their true pathogenic effects.

II. VIRULENCE

Babesia parasites vary greatly in their degree of virulence. In general, small babesias, such as *B. bovis*, *B. ovis*, and *B. gibsoni*, are highly pathogenic, while larger babesias, such as *B. major*, *B. motasi*, and *B. ovata*, are much less virulent. However, some species, such as *B. bigemina* and *B. canis*, exhibit variable virulence, especially those strains found in South Africa which may be highly virulent, while others, such as those found in Australia, are of low to moderate virulence. The nature of this variability in virulence is not clearly understood, with the exception of *B. bovis* which has been the subject of intensive study. Virulence also depends on the nature of the host, and it is a consistent finding that splenectomized hosts are generally affected much more severely than those with intact spleens. Table 1 summarizes a classification of virulence of various *Babesia* species in nonsplenectomized hosts under natural or experimental conditions. Reactions to organisms are influenced not only by splenectomy, but by age and possibly by breed. For instance, young animals are reported to be more resistant to challenge to *B. bovis* than older ones, though this effect has now been shown to be related to the presence of colostral antibodies.[4,6] Likewise, *Bos indicus* animals have been shown to be less susceptible to the same parasite than *Bos taurus* breeds.[6] Unfortunately no comparable studies have been made with other *Babesia* spp.

Because the nature of virulence has only been thoroughly investigated with one *Babesia*

Table 1
CATEGORIES OF VIRULENCE STATUS OF *BABESIA* SPECIES IN NONSPLENECTOMIZED NATURAL HOSTS

Species	Host
Virulent	
B. bovis	Bovine
B. gibsoni	Canine
B. ovis	Ovine
Moderately high virulence	
B. bigemina	Bovine
B. canis	Canine
Moderate virulence	
B. caballi	Equine
B. divergens	Bovine
B. equi	Equine
B. felis	Feline
B. motasi	Ovine/caprine
B. trautmanni	Porcine
Low virulence	
B. crassa	Ovine/caprine
B. foliata	Ovine
B. major	Bovine
B. microti	Rodent/human
B. occultans	Bovine
B. ovata	Bovine
B. perroncitoi	Porcine
B. taylori	Caprine

species, *B. bovis*, the following discussion will relate entirely to the latter. Extrapolation of these findings to other species should not be attempted as the nature of most *Babesia* infections differs from that caused by *B. bovis*. A possible exception is *B. ovis* of sheep which has been shown to cause vascular stasis similar to that caused by *B. bovis*, and to some *B. canis* strains which also cause vascular stasis in cerebral vessels.[7]

Most studies on virulence have attempted to attenuate the particular organism to be used as an immunogen. It has long been recognized that rapid passage of *B. bovis* through a series of splenectomized calves reduces the virulence of this organism. In an effort to produce a relatively stable avirulent yet immunogenic strain, Callow and co-workers preformed a number of studies. These workers found that, depending on the strain, a variable number of passages was necessary to induce avirulence; 23 in the case of the K strain, 10 with the S strain of *B. bovis*.[8] However, they further found that this avirulence was not stable, for even after 34 passages, the K strain reverted to full virulence after passage in cattle. Furthermore, these strains retained their ability to be transmitted by their tick vector. Subsequent studies also confirmed that at least ten passages of a South African strain of *B. bovis* were necessary to reduce virulence. The South African strain and the Australian K strain cross protected against subsequent virulent heterologous challenge.[9]

Stable avirulent *B. bovis* organisms have been developed using ionizing radiation. In a series of studies, Wright and co-workers found that 350 Gy of gamma irradiation from a ^{60}Co source produced avirulent parasites which were not tick transmissible, but were strongly immunogenic.[10-12] Furthermore, after 1 year in a carrier host, these organisms maintained their avirulence and immunogenicity and were not tick transmissible. The parent virulent strain maintained its virulence, immunogenicity, and tick transmissibility under the same conditions. These workers concluded that irradiation was selecting a subpopulation from a heterogeneous population, rather than altering the whole population; both virulent and non-virulent organisms multiplied at the same rate (8 hr doubling time) and attained the same maximum parasitemia. This indicates that both virulent and avirulent organisms had similar metabolic characteristics in the vertebrate host, but that only those animals that received virulent organisms were severely affected. This relationship between virulence and tick transmissibility was first observed by Dalgliesh and Stewart[13] who found that after about 60 passages through splenectomized calves, the parasite was not longer tick transmissible.

Irradiation has also been used to attenuate some other species of *Babesia* including *B. divergens*, *B. major*, and *B. bigemina*.[14-16] While the effect was clear cut with both *B. divergens* and *B. major*, data from our laboratory with *B. bigemina* suggest that strain selection is important, as with one strain no attenuation was attained, whereas with a second strain marked attenuation was apparent. This suggested, as with the *B. bovis* attenuation studies, using either irradiation or passage through splenectomized calves, that attentuation was achieved by selecting a subpopulation from an otherwise heterogeneous population. Callow et al.[8] thought that less virulent strains were less likely to be sequestered in internal organs and hence by rapid passaging, the selection pressure would result in an enrichment of less of the virulent forms in circulating blood. This is in agreement with electron microscopic studies,[17-19] which demonstrated that red cells parasitized by virulent strains of *B. bovis* had altered external membranes which attached to capillary walls in the brain and kidney resulting in sequestration. Such changes were not demonstrated in infections with attenuated strains. All irradiation studies on *B. bovis*[10-12] indicated that selection of substrains was occurring without any evidence of permanent mutation. More recently, in vitro cloning techniques have shown that *B. bovis* is composed of a number of subpopulations,[20] and one of these populations which was not virulent was selected for use as an attenuated vaccine.

The best evidence to date that *Babesia* strains are in fact heterogeneous populations has resulted from studies on the DNA of various strains.[21,22] these studies found that for a particular gene (designated Bab R) up to three different sized transcripts were found in different strains, both virulent and avirulent, but they were unable to elucidate the mechanism that caused subpopulations to be either virulent or avirulent.

In earlier studies,[23-26] however, a proteolytic enzyme was isolated from virulent organisms which was not present in avirulent forms. In strains that had been syringe passaged in splenectomized calves 10 to 12 times, this enzyme was present in only small amounts, even though it was abundant in the virulent parent strain. In the first of these studies,[23] it was shown that this trypsin-like enzyme was most active at neutral pH and cleaved a broad spectrum of substrates including hemoglobin. In a later study,[24] it was demonstrated that this enzyme formed fibrin gels when reacted with normal plasma in vitro. This enzyme was also shown to be hypotensive, causing an immediate fall in systolic blood pressure after intravenous injection.[27] Mahoney and Wright[28] also showed that upon intravenous injection of 2×10^9 disrupted *B. bovis* parasites, the hematocrit fell and plasma kallikrein (hypotensive serum enzyme) was activated, presumably by this protease. Concurrent studies also demonstrated that the alternative pathway of complement was activated, either by this enzyme or by the cascade of plasma enzyme activation initiated by the same protease.[28]

This work culminated in the isolation of the protease and in a detailed study of its biological properties. The enzyme was then purified by affinity chromatography techniques using soya bean trypsin inhibitor as a ligand.[26] On sodium dodecyl sulfate (SDS) electrophoresis under denaturing and reducing conditions, it was found to have a molecular weight of 20 kdaltons. The antigen was found by indirect fluorescent antibody (IFA) studies to be largely secreted by the parasite into the red cell cytoplasm, with the highest antigen concentration being produced by recently divided parasites. This strongly suggests that the enzyme has no intracellular function, but may be involved in some escape mechanism from the parasitized cell. However, avirulent parasites, which do not contain any detectable enzyme, also manage to divide at the same rate and presumably to escape from red cells as effectively as enzyme-containing virulent strains.[25] This indicates that the enzyme serves no essential metabolic function within the mammalian host cell nor does it facilitate the escape mechanism. The likely function of this enzyme became apparent when it was found that irradiated avirulent *B. bovis* could not be transmitted by its tick vector *Boophilus microplus,* even when 100,000 larval progeny of adult females previously fed on infected blood were placed on a susceptible splenectomized calf. Under the same conditions, 100 larval progeny of ticks previously fed on the virulent parent strain were able to transmit the infection.[12] These experiments, while not conclusive, offer substantial support to the view that the proteolytic enzyme is involved only in infectivity of the organism for its tick vector. This is supported by findings using passage in splenectomized calves which indicate that once a strain is rendered completely avirulent, it is also nontransmissible by the tick vector.[13] The relevance of this enzyme to the acute disease syndrome is discussed more fully in Section III.

In limited studies in our laboratory, we have not been able to demonstrate this enzyme in *Babesia bigemina* parasites, but its existence in *B. equi* has been verified.[29] The enzyme is certainly a "marker protein" of virulent *B. bovis* strains, and it could prove to be a useful criterion when selecting various *Babesia* strains for use as attenuated vaccines.

III. THE COAGULATION-COMPLEMENT-KALLIKREIN CASCADE

A. Coagulation Disturbances

Severe coagulation disturbances have been reported in both acute *B. bovis* and acute *B. canis* infections.[30,31] At first it was thought that the coagulation disturbance in sera of cattle severely affected with *B. bovis* was due to disseminated intravascular coagulation (DIC), based on the detection of fibrinogen-like protein (FLP).[32] This view was supported by Dalgliesh and co-workers[33] who reported the presence of thrombi in splenectomized calves acutely infected with *B. bovis*. However, in previous studies, neither thrombi nor deposited fibrin could be demonstrated using electron microscopic, conventional histochemical, and immune staining techniques in acute *B. bovis* infections, and it was considered unlikely that DIC was occurring[17,34,35] (Figures 1, 2). Studies on endotoxemia in pigs demonstrated that "thrombi" which were detected by histochemical staining were in fact hyaline deposits of degenerating red blood cells in the absence of cross-linked fibrin.[36,37] Electron microscopy was needed to establish whether fibrin filaments were present. More recent studies on canine endotoxin shock[38] and on human DIC have confirmed these earlier findings.[39] The histochemical study on DIC in canine babesiosis must therefore be viewed in the same light.[31]

While classical DIC does not occur during acute *B. bovis* infections, at least, profound disturbances to fibrinogen metabolism have been demonstrated. In acute *B. bovis* infections, changes indicative of hypercoagulability have been observed.[33,40] These changes included an initial fall in both reptilase time (RT) and thrombin time (TT), suggesting activation of the coagulation system, with prolonged terminal RT and TT along with prolonged prothrombin time (PT), partial thromboplastin time (PTT), and whole blood coagulation time (WBCT). Corresponding to these prolonged times was the acute thrombocytopenia and dysfibrinogenemia. Prolonged RT and TT are indicative of thrombin activation and consumption and/or dysfibrinogenemia.[41] In addition, there is marked consumption of antithrombin activity consistent with neutralization of activated thrombin.

In a series of studies on fibrinogen metabolism, it was shown that a whole spectrum of altered fibrin (ogen) products were formed during acute *B. bovis* infections. In general, these represented soluble intermediates in fibrinogen to fibrin conversion. Of particular interest

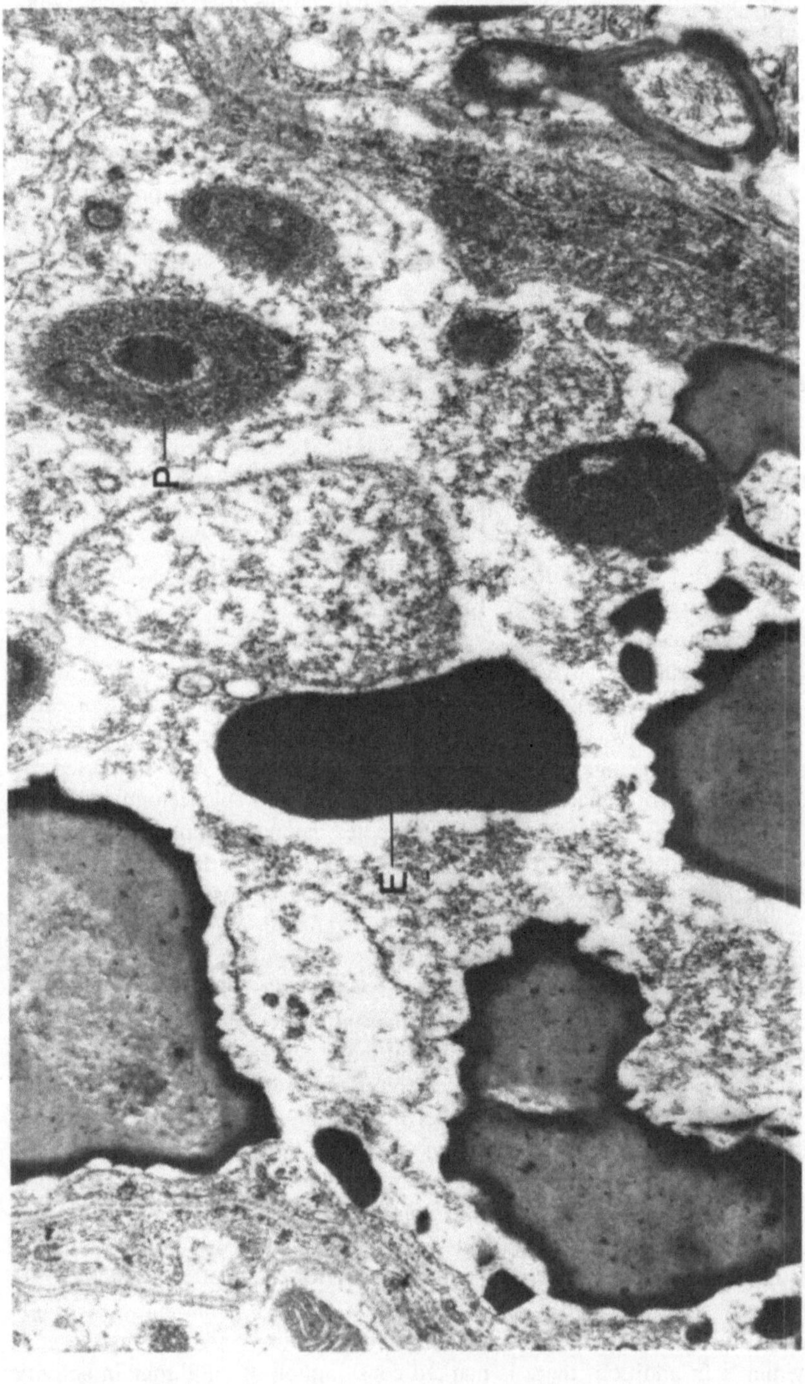

FIGURE 1. Stasis of brain capillary in acute *B. bovis* infection. The capillary contains numerous infected erythrocytes, both lysed and intact. All infected erythrocytes have a stellate appearance. An infected erythrocyte (e) has normal biconcave outline. Parasite (P) in lysed erythrocyte. (Magnification × 30,000.)

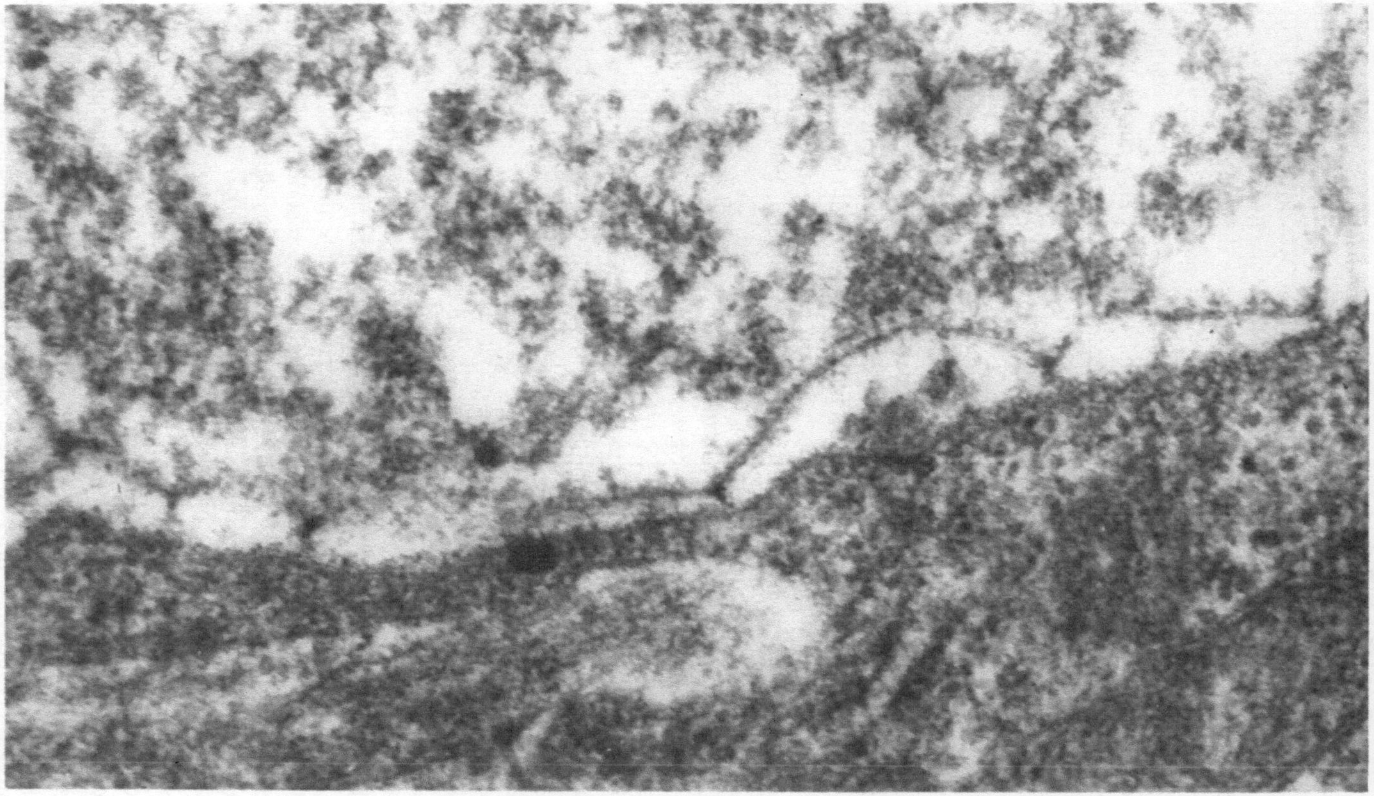

FIGURE 2. A lysed infected erythrocyte attached by strands to the capillary endothelium. The strands appear to be an integral part of the erythrocyte membrane. Histochemical and immunostaining techniques failed to demonstrate evidence of disseminated intravascular coagulation. (Magnification × 120,000.)

was the rapid formation of cryoprecipitates in plasma.[42] The gel only partially dissolved after heating at 37°C and upon analysis was found to be composed of fibrinogen and soluble fibrin, the latter in both monomer and complexed forms. Traces of γ-chain dimers were found in these high molecular weight complexes indicating a slight degree of interaction with factor XIII. In addition, plasma cryofibrinogen was present in the last 1 to 3 days and represented as much as one third of the total plasma fibrinogen. Further studies demonstrated the presence of a noncoagulable fibrinogen-like protein (FLP) in sera obtained from clotted blood in terminal infections.[40]

Electrophoretic analysis of the chain structure of the complexes showed close similarities with noncross-linked fibrin. The major polypeptide chains of fibronogen (Aα, Bα, γ) showed little or no change during infection, and it was concluded that the complexes were composed predominantly of fibrinogen and noncross-linked fibrin. There was also no evidence of fibrinolysis as the Aα chain was unaltered[43] and the plasma plasminogen levels remained relatively unaltered.[44] Moreover, activated plasmin could not be detected in plasma of acutely infected cattle. The total antiplasmin activity and the concentration of α_1 antiplasmin actually increased during infection[45] which is further evidence of lack of plasminogen activation. In addition, factor XIII levels increased during acute *B. bovis* infections, which again indicates that it was not being consumed to a major degree in fibrin cross linking.

Although all these data indicate the lack of classical DIC, as evidence by the virtual absence of fibrin-split products, particularly D dimer, use of the Thrombo-Wellco test which utilizes antisera to human FDP to detect FDP gives a positive reaction with serum. Likewise, in canine babesiosis, a similar positive reaction has been observed.[31] In our hands, this reagent failed to react with degradation products of bovine fibrinogen, but did react with soluble fibrin. These data should therefore be cautiously interpreted.

The bulk of FLP in infected sera has a molecular size greater than fibrinogen, but its chain structure is similar to fibrinogen or noncross-linked fibrin and is a product of coagulation and not of fibrinolysis. The serum FLP represent noncoagulable intermediates in fibrinogen-to-fibrin conversion whose role is to prevent fibrin deposition in the vascular system. This noncoagulable product may in part be the result of the direct action of babesial esterase which has been shown to form FLP in vitro[24] independently of the intrinsic coagulation system.

In less acute infections with *B. bigemina*, and with avirulent strains of *B. bovis*[46,47] only trace amounts of FLP and cryofibrinogen are observed. Likewise, a transient decline in plasma fibrinogen occurs during these infections, which suggests only a limited degree of thrombin activation and subsequent RES removal. On the other hand, small increases in fibrinogen occur during infections with the moderately virulent infections of *B. canis, B. caballi*, and *B. equi*.[48-50] Cryofibrinogen has not been reported in infections other than those caused by virulent *B. bovis*. Moreover, the FLP-forming esterase reported in virulent but not avirulent strains of *B. bovis*[24,25] have not been recorded in less virulent species, with the possible exception of *B. equi*.[29] This probably accounts for the lack of involvement of the coagulation system in those infections, especially since this esterase activates kallikrein and possibly complement.[24,28,43]

Recently, work in our laboratories has shown that erythrocytes infected with virulent strains of *B. bovis* have enhanced procoagulant activity.[51] The normal bovine plasma coagulation time in the presence of infected erythrocytes is reduced by 33 to 66% depending on the strain of parasite. This, together with earlier findings[44,52] that fibrinogen has an affinity for *B. bovis* infected erythrocytes, offers one explanation of the mechanism of capillary stasis. It is likely that infected red cells, because of their enhanced procoagulation activity, can activate the coagulation system *in situ* causing the strand-like formations[17] (see Figure 2). It is of interest that Wright[17,18] only observed this phenomenon with erythrocytes infected with virulent parasites.

B. Complement Activation

Involvement of the complement system in acute babesiosis has been demonstrated by a number of workers. Annable and Ward[53] first showed a depression of serum complement levels during *B. rodhaini* infections of mice, although a later study of rodent babesiosis found no evidence for the depletion of components of the alternate pathway.[54] However, in acute *B. bovis* infections, it was found that both the classical and the alternate pathways of complement were activated and that hemolytic complement levels fell to zero and remained at that level long after parasites had disapperaed from the blood stream.[55] Levels commenced falling when parasites were barely detectable, at which time both the coagulation and kallikrein systems were activated. This activation had been previously shown to be due to parasite proteases and occurred before antibody levels were detectable.[28] In a further study on acute *B. bovis* infections, a marked depletion of C3 and of the natural nonantibody opsonin, conglutinin, was demonstrated.[55] Bovine conglutinin reacts specifically with C3b[56] and aggregates particles to which complement is fixed. Its role in the stasis of capillaries with infected red cells has not been established, but is highly likely.

Associated with complement activation is the production of anaphlytoxins C3a and C5a. These are extremely potent vasodilators in their own right, as well as potentiators of bradykinin and kallikrein.[57] Consequently, complement activation would augment the shock syndrome occurring during acute *B. bovis* infection. It has been shown that in acute *B. bovis* infections, serum carboxypeptidase B (SCPB) levels fall markedly. As this enzyme inactivates both kinins and anaphylatoxins, depression of this enzyme would lead to vasodilation since control of SCPB appears necessary for the maintenance of a stable circulation, and decreases of 29 to 50% in SCPB activity would result in shock.[58] Falls of 50% and greater were observed with acute *B. bovis* infections caused by three virulent strains, although in avirulent *B. bovis* and *B. bigemina* infections, little change was observed.[59] Marked reductions in SCPB have also been recorded in infections with the moderately virulent parasite, *B. divergens*.[60]

A further role for complement in *Babesia* infections was proposed by Chapman and Ward[54] who believed that components of the alternate pathway were involved in the invasion process of the red cell by babesial parasites. This work was supported by studies demonstrating the role of C3b in the invasion mechanism of red cells by *B. rodhaini*.[61] However, this work has been invalidated by other studies using the same model.[62] More recently, it has been demonstrated that complement does not facilitate the invasion of erythrocytes by *B. bovis*.[63] Thus, the role of complement as a chemoreceptor for parasite invasion is now considered unlikely in babesiosis.

C. Kallikrein Activation and Hypotension

The hypotensive, vasodilatory serum enzyme, kallikrein, and the vasoactive peptide, bradykinin, which is generated after cleavage from its precursor protein kininogen by kallikrein, have both been incriminated in numerous shock syndromes, including those caused by endotoxin, bacteria, dengue fever, trypanosomiasis, malaria, and babesiosis.[64-67] The role of kinins and kallikrein in hemoprotozoan infections, and especially in babesiosis, has been extensively reviewed by Wright.[1,68] Consequently, the general pharmacology of these substances will not be reviewed here. In summary, during acute *B. bovis* (and presumably *B. ovis* and *B. canis* infections), there is a large increase in circulating activated kallikrein and a large decrease in plasma kininogen, possibly involving free kinin itself.[67,69] Such changes are not demonstrable with either avirulent strains of *B. bovis* or with the less virulent *B. bigemina*. Kallikrein activation by *B. bovis* esterase was demonstrated both in vitro and in vivo.[24,28] In experiments performed in vivo, it was shown that as few as 1×10^9 disrupted organisms injected intraveneously caused activation of kallikrein and a fall in PCV. An estimated 2×10^9 replicating organisms caused similar changes.[28] During a fulminating infection, it was demonstrated that a significant fall in systolic blood pressure was also initiated by a similar parasite load and that these changes were correlated with hematocrit reduction.[70] Observations in our laboratory have consistently shown that the hematocrit falls disproportionately to the estimated red cell destruction caused by replicating parasites during acute *B. bovis* infections; this disproportionate fall is not apparent with infections by avirulent strains. This fall in the hematocrit is due in part to aggregation of infected and uninfected red cells in capillary beds due to sludging. This aspect is dealt with in Section IV. One of the reasons for capillary sludging is vasodilatation, coupled with diminished blood pressure as a result of the synergistic effect of kallikrein, kinin, and other vasoactive agents. Another action of kallikrein is chemoattraction. This may be especially relevant in the shock-lung syndrome described by Wright et al.[35] in which massive accumulation of neutrophils, as well as edema, was demonstrated in that organ. The reason for this is unknown, but in view of the chemotactic effect of kallikrein on these cells, it is highly possible that kallikrein is at least partly responsible. Indeed, Clark et al.[71] have implicated C5a, proteases, and leukocytes in the adult respiratory distress syndrome (ARDS) of acute malaria and considered that a similar mechanism was operative in *B. bovis* infections. Both parasite derived proteases and plasma kallikrein as well as C5a could initiate this pulmonary edema as a result of this action on leukocytes causing them to release free oxygen radicles and peroxide,[71] Toxic products of lipid peroxidation, a process for which indirect evidence exists in acute *B. bovis* infection (see Sections IV and V), would then cause endothelial cell damage in the lung. A similar mechanism has been shown to exist with endotoxin, and although endotoxin is not involved in acute babesiosis,[1] the similarities between the two conditions are obvious.

Another role for plasma kallikrein and cleavage products of kininogen is their positive feedback on the activation of the coagulation pathway via factor XII.[72,73] As there is a large increase in activated kallikrein and a subsequent produciton of bradykinin from high-molecular- weight kininogen during acute *B. bovis* infections, the observed hypercoagulability of plasma is due, in part at least, to these compounds directly activating Hageman factor (factor XII). Hageman factor activation by the trypsin-like enzymes of *B. bovis* is also likely, especially as in vitro studies demonstrated such a role.[24]

D. Other Pharmacological Agents

Changes in plasma levels of the biogenic amines histamine, and 5-hydroxytryptamine (5-HT) have been reported in acute *B. bovis* infections.[74] Both histamine and 5-HT levels rose markedly during infection, the rise corresponding to a severe thrombocytopenia. Levels of norepinephrine and its precursor, dopamine, remained unaffected. Although both histamine and 5-HT are not the major mediators of bovine anaphylaxis,[75] both potentiate the effect of kinins and kallikrein.[76] Thus, the likely sequence of events in babesiosis is the initial release of kinins, which triggers the release of biogenic amines from thrombocytes. These agents act both independently and synergistically to produce profound vasodilatory hypotensive shock.

Although endotoxin has been incriminated in acute malaria, Wright[1] was unable to demonstrate its presence either in sera from *B. bovis* infected animals, or in the parasites themselves. It is thus unlikely to have any role in acute babesiosis. The role of biogenic amines and endotoxin in other babesia infections has not been studied.

IV. STASIS

A. Introduction

Probably the most deleterious pathophysiological lesion in certain acute babesial infections is the stasis induced in the microvascular system by aggregated infected erythrocytes. The infected erythrocytes adhere not only to each other, but also to the endothelium, particularly of the brain (Figures 1, 2), kidney, and skeletal muscle.[17,18,35] The lesion occurs in *B. bovis* infections and has been postulated in *B. ovis* and *B. canis* infections and is partly responsible for muscle damage and increases in creatine kinase and lactate-dehydrogenases.[78] While morphological changes to *B. bovis*-infected erythrocyte membrane have been detected at the electron microscope level, and while these changes are obviously physically involved in cell-cell and cell-endothelium attachments,[18,19] the exact mechanisms have not been elucidated by immunological, biological, or pharmacological analyses. Presumably the mechanisms are multifactorial and probably cascadial, but three main facets must be involved, these being (1) the endothelium, (2) changes to the erythrocytic membrane, and (3) host contribution via plasma constituents.

B. The Endothelium

The contribution of the endothelium fibrin lining to stasis has been reviewed by Copley[78] who refers to the term "cement fibrin" which can form in close proximity to the endothelial cell. This consists of fibrin monomers and fibrin intermediates which can "anchor" fibrinogen and other substances by hydrophobic reactions. In *B. bovis* infection, not only are there substantial increases in soluble fibrin,[42] but there are also increases in erythrocytic-bound fibrinogen, especially in infected erythrocytes.[79] Thus, an important contributor to stasis could be the interaction between endothelial fibrin and erythrocytic fibrinogen. Interestingly, binding of erythrocytes to endothelium of the microvascular system occurs in diabetes and is due in part to increased concentration and interaction of erythrocytic fibrinogen with endothelium.[80,81] This phenomenon might be occurring in acute babesiosis.

The endothelium also contains the major antiprotease α_2 macroglobulin.[82] The reaction between this globulin and babesial proteases, although not initiating stasis, could exacerbate the process. The two identified babesial proteases[23,24] are partly membrane bound and, if exposed, could interact with the antiprotease.

Heparin and its analogues are also contained in the endothelium and are highly mobilized during tissue damage and inflammation.[84] *Babesia bovis*-infected erythrocytes bind to heparin in vitro.[86] Assuming this reaction occurs in vivo, then infected erythrocytes could bind to endothelium via heparin. The heparin-binding proteins of infected erythrocytes have been partially purified and have been shown to be highly effective as a vaccine.[87] It is reasonable to speculate that this protection was partly due to antibody neutralization of adhesive antigens by antibody.

Hemostasis associated with increased levels of fibrinogen also occurs in dogs with *B. canis* infection.[48] Plasma fibrinogen levels also increase in both *B. caballi*[50] and *B. equi*[51] infections, but there is no definitive data incriminating stasis.

C. Erythrocytic Changes

One possible explanation for aggregation of erythrocytes in acute *B. bovis* infections would be the presence of babesial antigen on the surface of the infected erythrocyte. However, if present, such an antigen must be in low concentration as it cannot be detected by fluorescent antibody analysis of intact infected erythrocytes,[87] and it may even be strain specific.[88] Howard et al.[89] used surface labeling and identified three proteins specific for the surface of infected erythrocytes, but were unable to designate them as being of babesial origin. Studies by Goodger et al.,[90] using rabbit antiserum to partly purified babesial antigen, indicated the infected erythrocyte membrane contained modified host antigen or babesial antigen complexed with host serum protein, suggesting that host membrane components are involved in aggregation.

Malonyldialdehyde (MDA) is a sensitive indicator of lipid peroxidation, and increased amounts of MDA and decreased Vitamin E have been detected in *B. bovis*-infected erythrocytes.[91] Lipid peroxidation predisposes toward stasis by causing increased rigidity and decreased deformability of erythrocytes.[92,93] Vitamin E is consumed via its antioxidant role, and its decrease leads to an increased binding of hemoglobin to the erythrocytic membrane which also produces increased membrane viscosity.[94] Lipid peroxidation is influenced by glucose metabolism,[95] and as there is a large metabolic consumption of glucose by the parasite,[96,97] this may modify peroxidation in some manner. Thus, it seems that lipid peroxidation contributes to the stasis of *B. bovis*-infected erythrocytes by producing an increased viscosity and a decreased deformability of the erythrocytes, thereby enhancing contact not only between altered erythrocytes, but also between the endothelium and erythrocytes. Beta blockers have been shown to reduce the adhesion of erythrocytes to the endothelium by diminishing their viscosity and increasing their deformability,[99] and it is interesting to speculate that such compounds may have a similar effect in acute babesiosis and hence have a therapeutic value.

D. Plasma Constituents

Increased plasma concentrations of fibrinogen, soluble fibrin, and fibrinogen-fibrin complexes occur in *B. bovis* infections. These proteins have been shown in other studies to contribute to increased viscosity.[99] Increased plasma viscosity, plus the interaction of infected erythrocytes with the endothelium, appear to play a crucial role in the development of intravascular stasis. In diabetes, fibrinogen binds erythrocytes to endothelium, and this binding is enhanced by fibronectin.[80,81] With *B. bovis* infection, fibronectin has also been detected in lesions by fluorescent antibody studies[100] and may play a secondary role in the development of stasis. Additionally, since stasis with concommitant endothelial injury caused an increased release of fibronectin,[101] then in *B. bovis* infections, fibronectin could act as an additional molecular bridge between the endothelium and fibrinogen/fibrin of the infected erythrocyte, thus aggravating stasis. Another contributing factor to stasis is free cholesterol, which has been shown to increase significantly in *B. bovis* infection.[102] Elevated levels of free cholesterol are also related to decreased erythrocyte fluidity in diabetes[103] and may play a similar role in acute babesiosis.

Soluble fibrin levels are associated with increased blood viscosity and decreased erythrocyte deformability,[99] and thus the plasma-soluble fibrin seen during *B. bovis* infections should also predispose toward microvascular sludging. During stasis, fibrinogen adheres to vessel walls, and this adhesion is exacerbated by fibrinolytic inhibition.[104] Fibrinolytic inhibition occurs during *B. bovis* infections.[105] An increased concentration of plasma fibrinogen also causes increased erythrocytic "stickiness",[106] and subsequently, cell-cell and cell-endothelium reactions occur in the microvascular system. Such changes in fibrinogen levels also have been observed in acute babesiosis and may result in increased erythrocyte "stickiness".[40,43]

V. INFLAMMATION

A. Introduction

Inflammation, as evidenced by the changes in acute phase proteins, occurs during acute babesial infections. This nonspecific inflammation is probably initiated by bradykinin.[69] Probably the most sensitive indicator of inflammation in cattle is the level of haptoglobin, which is a protein not normally detected in bovine serum, but is readily detected during acute babesial infection.[107] Indeed, the elevated haptoglobin levels associated with low concentrations of fibronectin observed during infection[107] are useful indicators of reticuloendothelial system (RES) dysfunction.[108] It is uncertain as to whether inflammation is beneficial or detrimental to the host. With malarial parasites, which are closely related to babesia, Maegraith[109] stated that human beings typically react to infection by an acute inflammatory response, while Rest[110] claimed that the inflammatory response aids the host

Table 2
CHANGES IN SERUM PROTEINS AND RELATED COMPOUNDS IN ANIMALS ACUTELY INFECTED WITH *BABESIA BOVIS*

Not normally present	Expected increase	Expected decrease
Soluble fibrin	Fibrinogen	Fibronectin
Cryofibrinogen	α_1-Antitrypsin	Conglutinin
Serum fibrinogen-like proteins	Total antiplasmin	α_2-Macroglobulin
Haptoglobin	α_1-Antiplasmin	Phospholipid
Immunoconglutinin	Free cholesterol	Esterified cholesterol
Vasoactive peptides	Creatine kinase	Lipoprotein
Circulating cryoprecipitable immune complexes	Lactate dehydrogenase	Total complement
	Creatinine	Complement C3
	MDA	Total antithrombin
		Antithrombin III
		Serum carboxypeptidase D
		Sialic acid
		Vitamin E

and prolongs survival. Conversely, uncontrolled inflammation in response to an infectious agent can cause death due to "endogenous suicide".[111] Moreover, vaccination of mice with killed *Corynebacterium* can protect against subsequent *B. rodhaini* challenge, and while this protection is claimed to be due to nonspecific immunity,[112] it is presumably associated with a strong inflammatory response. Corticosteroid therapy exacerbates infection due to *B. bovis*.[33] This effect could be attributed to the anti-inflammatory and/or immunosuppressive effects of corticosteroids.

Detailed studies of many acute phase proteins during babesial infection show that some, such as the major antiprotease, α_2 macroglobulin, actually decrease during infection.[105] The consumption of this protein by neutralization of babesial proteases and by activated host enzymes is apparently in gross excess of its production. By contrast, another major antiprotease, α_1 antitrypsin, increases during infection and may remain high, long after the parasitemia has waned.[113] This indicates that inflammation continues for some period after elimination of peripheral parasites, and it also suggests that some important babesial proteases are not trypsin-like. Table 2 lists likely changes in those acute phase proteins which have been studied during *B. bovis* infections.

An interesting but disturbing inflammatory response has been observed after challenging cattle previously vaccinated with partly purified babesial antigens; 2 to 3 days after challenge, there is a rapid rise in temperature, most likely associated with an inflammatory reaction induced by immunocomplexes.[113] The ideal vaccine must thus provoke a specific immunoresponse without predisposing the host to hypotensive shock. A resolution to this is required, as an exacerbated inflammatory response to challenge could negate an otherwise protective immunity. Indeed, many of the cellular and humoral components of the acute phase reaction can modulate the immune response by eliciting an immunodepressive effect.[114,115] Additionally, the peak parasitemia also induces an immune suppression,[116] and one could argue that the onset of inflammation followed by the peak parasitemia potentiates immune depression at that period when high concentrations of parasite antigen are being presented to the immune system. This might explain in part why protective antibody to babesosis is found only in sera of chronic carriers and is lacking in sera of cattle recently recovered from a primary infection.[117]

Inflammation also causes a reduction in plasminogen activator activity as well as an increase in antiplasmin levels,[118] which might explain, in part, why *B. bovis* infections are characterized by high levels of soluble fibrin, but only small amounts of classical fibrin/fibrinogen degradation products.[40]

B. Reticuloendothelial System (RES) Depression and Fibronectin

The RES should play an active role in babesial infection by removing erythrocytic debris, soluble fibrin, and infected erythrocytes. However, the evidence suggests that the RES functions poorly during acute infection and may in fact be temporarily saturated. The main indicator for this reduced RES activity is plasma fibronectin, the concentration of which is thought to be related directly to RES efficacy.[119] During *B. bovis* infections, plasma fibronectin levels decrease significantly,[55] indicating dysfunction or impairment of the RES. In addition, at this time there are significant amounts of haptoglobin-hemoglobin complexes in the plasma, and this, combined with a decrease in plasma fibronectin, is indicative of RES blockade.[108] The situation is further compounded by the erythrocytic debris resulting from parasite invasion, since such debris causes RES depression and a decrease in total hemolytic complement.[55,119] Hemoglobin released from infected erythrocytes induces RES suppression, and thrombin activation that occurs during *B. bovis* infections[28] leads to hepatic RES depression, thus presumably prolonging the half-life of particulate material.[121] The cumulative effects of the RES depressants correlate with increased fibrinogen concentrations[122] and the massive amounts of soluble immune complexes present in plasma soon after parasitemia wanes.[45]

Under normal circumstances, fibronectin should increase in concentration during inflammation, but during *B. bovis* infections, there is a massive consumption.[55] It has been postulated that fibronectin may be intimately involved in removing infected erythrocytes as (1) its concentration rapidly returns to normal levels as parasitemia wanes, and (2) following challenge, the fibronectin level in vaccinated cattle falls 2 to 3 days earlier than in control animals.[55] In vaccinated animals it is conceivable that infected erythrocytes, subtly changed by antibody, are recognized as "foreign" by fibronectin and are thus subsequently coded for RES removal.

C. Immune Complexes

Immune complexes (IC) were demonstrated in kidney sections of mice and rats infected with *B. rodhaini*.[53,54] Babesial antigen and C3 were detected within these complexes, suggesting that infection produced an immune complex disease with consumption of complement. In *B. bovis* infections, activation and consumption of both complement pathways has been demonstrated,[28,43] with massive decreases of C3 occurring.[55] The C3 levels remained low long after the parasitemia had waned, and immune complexes were subsequently detected at this time.[45] Surprisingly, in contrast to *B. rodhaini* infections, the complexes formed during *B. bovis* infections were cryoprecipitable and formed in antibody excess, with the major antibody being IgM. Babesial antigen was detected in these complexes, but it was not protective when used in vaccination experiments.[123] Normal erythrocyte antigens were also detected in the complex, suggesting that an autoimmune hemolytic anemia may occur during infection, as has recently been demonstrated in human infections.[124] The detection of IC in *B. bovis* infections helps explain some of the pathology of infection. The complex contains C3, and such complexes may produce severe inflammation in vascular beds, as well as adhering to blood vessel walls,[125] thereby prolonging stasis. Current studies of microvascular lesions in *B. bovis* infections at our laboratory, using fluorescent antibody techniques, have detected components of IC intimately associated with aggregated infected erythrocytes.[126] A neutropenia, due to their sequestration in the lung,[127] is produced by IC, and as both neutropenia and lung sequestration occur in acute *B. bovis* infections, IC must contribute to pulmonary edema.[35] IC can suppress or modulate the immune response, which does, in part, explain the slow production of protective antibodies following *B. bovis* infection. Immune modulation by complexes has also been suggested in infections caused by the related malarial parasite.[128]

A puzzling aspect of the presence of IC in *B. bovis* infections is their duration in plasma.

Such large complexes should be rapidly cleared form the circulation, and their prolonged presence requires some comment. The complexes contain C3 and should be ideally and rapidly removed via appropriate receptors on the erythrocyte. However, a reduction of complement decreases binding of immune complexes to erythrocytes,[129] which provides one explanation.[129,130] An alternate explanation may be that the erythrocytic C3 receptors could be saturated following RES depression. In addition, immune complexes can produce loss of receptor function especially on Kupffer, cells which would result in a diminished clearance.[131]

VI. LIPIDS

Lipid metabolism has been studied in some detail in both *B. bovis* and *B. bigemina* infections,[133] and the presence of fatty infiltration in liver parenchymal cells in *B. bovis* infection has been reported by Wright.[134] In *B. bovis* infections, there are significant decreases in total plasma lipid, total cholesterol, and phospholipids. In contrast, free cholesterol levels increase while esterified-cholesterol levels decrease, possibly due to reductions in lecithin-cholesterol acyltransferase or degradation of lipoproteins which result in nonmetabolizable cholesterol.[103] In addition, corticosteroids also decrease during *B. bovis* infections,[132] leading to activation of both prostaglandins and complement with the subsequent activation of the coagulation system, thus enhancing the hypercoagulable state seen during acute infections. The decrease in steroids must be due in part to neutralization of inflammation, while the associated decrease in endothelial membrane integrity appears to predispose toward microvascular stasis. Decreases in serum lipoproteins and the changes in electrophoretic heterogeneity also occur, and these may be partly the result of protease inhibition.[103] Since there is an equilibrium between plasma and erythrocytic lipids, any changes in plasma lipids invariably affect the integrity of the erythrocytes. This could, in part, explain the differences in osmotic fragility observed in infected as compared to noninfected erythrocytes.[135]

VII. CONCLUSIONS

The pathogenesis of *Babesia* infections is highly complex, as evidenced by multisystemic involvement described in this review and elsewhere.[136] The bulk of evidence indicates that the pathogenesis is immunologically mediated and the role of complement in invasion remains putative for most *Babesia* species. The clinical outcome following exposure to *Babesia* is clearly dependent on the genetic, physiologic, and immunologic status of the host. The major system involved in modulating the pathogenesis of babesiosis is the reticuloendothelial system, particularly the spleen. Another important parameter in the pathogenesis of babesiosis is the virulence of the etiologic agent. Clinicopathologic manifestations characterized by the appearance of abnormal quantities of cellular and humoral elements in the blood of affected animals should be considered when selecting appropriate immunogens against specific *Babesia* species. *Babesia bovis* offers the best model for the study of pathogenesis of diseases caused by *Babesia* species.

REFERENCES

1. **Wright, I. G.,** Biochemical characteristics of *Babesia* and physiochemical reactions in the host, in *Babesiosis*, Ristic, M. and Kreier, J. P., Eds., Academic Press, New York, 1981, 171.
2. **Zwart, D. and Brocklesby, D. W.,** Babesiosis: non-specific resistance, immunological factors and pathogenesis, *Adv. Parasitol.*, 17, 49, 1979.
3. **Callow, L. L. and Dalgliesh, R. J.,** Immunity and immunopathology in babesiosis, in *Immunology of Parasitic Infections*, Cohen, S. and Warren, K. S., Eds., Blackwell, Oxford, 1982, 475.

4. **Tracey-Patte, P. D. and Wright, I. G.,** unpublished data.
5. **Levy, M. G., Clabaugh, G., and Ristic, M.,** Age resistance in bovine babesiosis: role of blood factors in resistance to *Babesia bovis*, *Infect. Immun.*, 37, 1127, 1982.
6. **Johnston, L. A. Y. and Sinclair, D. F.,** Differences in response to experimental primary infection with *Babesia bovis* in Hereford, Droughtmaster and Brahman cattle, in *Ticks and Tick-Borne Diseases*, Johnston, L. A. Y. and Cooper, M. G., Eds., Australian Veterinary Association, Sydney, 1980, 18.
7. **Purchase, H. S.,** Cerebral babesiosis in dogs, *Vet. Rec.*, 59, 269, 1947.
8. **Callow, L. L., Mellors, L. T., and McGregor, W.,** Reduction in virulence of *Babesia bovis* due to rapid passage in splenectomized cattle, *Int. J. Parasitol.*, 9, 333, 1979.
9. **de Vos, A. J., Bessenger, R., and Fourie, C. G.,** Virulence and heterologous strain immunity of South African and Australian *Babesia bovis* strains with reduced pathogenicity, *Onderstepoort J. Vet. Res.*, 49, 133, 1982.
10. **Wright, I. G., Goodger, B. V., and Mahoney, D. F.,** The irradiation of *Babesia bovis*. I. The difference in pathogenicity between irradiated and non-irradiated populations, *Z. Parasitenkd.*, 63, 47, 1980.
11. **Wright, I. G., Mahoney, D. F., Mirre, G. B., Goodger, B. V., and Kerr, J. D.,** The irradiation of *Babesia bovis*. II. The immunogencity of irradiated blood parasites for intact cattle and splenectomised calves, *Vet. Immunol. Immunopathol.*, 3, 591, 1982.
12. **Wright, I. G., Mirre, G. B., Mahoney, D. F., and Goodger, B. V.,** Failure of *Boophilus microplus* to transmit irradiated *Babesia bovis*, *Res. Vet. Sci.*, 34, 124, 1983.
13. **Dalgliesh, R. J. and Stewart, N. P.,** Failure of vaccine strains of *Babesia bovis* to regain infectivity for ticks during long-standing infections in cattle, *Aust. Vet. J.*, 53, 429, 1977.
14. **Purnell, R. E., Brocklesby, D. W., and Stark, A. J.,** *Babesia divergens*: reactions of splenectomized calves to the inoculation of titrated or irradiated piroplasms, *Parasitology*, 75, xxxi, 1977.
15. **Purnell, R. E., Brocklesby, D. W., Stark, A. J., and Young, E. R.,** Protection of cattle against *Babesia major* by the inoculation of irradiated piroplasms, *Res. Vet. Sci.*, 25, 388, 1978.
16. **Bishop, J. P. and Adams, L. G.,** *Babesia bigemina:* immune response of cattle inoculated with irradiated parasites, *Exp. Parasitol.*, 35, 35, 1974.
17. **Wright, I. G.,** An electron microscopic study of intravascular agglutination in the cerebral cortex due to *Babesia argentina* infection, *Int. J. Parasitol.*, 2, 209, 1972.
18. **Wright, I. G.,** Ultrastructural changes in *Babesia argentina*-infected erythrocytes in kidney capillaries, *J. Parasitol.*, 59, 735, 1973.
19. **Aikawa, M., Rabbege, J., Uni, S., Ristic, M., and Miller, L. H.,** Structural alterations of the membrane of erythrocytes infected with *Babesia bovis*, *Am. J. Trop. Med. Hyg.*, 34, 45, 1985.
20. **Rodriguez, S. D., Buening, G. M., Green, T. E. J., and Carson, C. A.,** Cloning of *Babesia bovis* by *in vitro* cultivation, *Infect. Immun.*, 42, 15, 1983.
21. **Cowman, A. F., Timms, P., and Kemp, D. J.,** DNA polymorphisms and subpopulations in *Babesia bovis*, *Molec. Biochem. Parasitol.*, 11, 91, 1984.
22. **Cowman, A. F., Bernard, O., Stewart, N., and Kemp, D. J.,** Genes of the protozoan parasite *Babesia bovis* that rearrange to produce RNA species with different sequences, *Cell*, 37, 653, 1984.
23. **Wright, I. G. and Goodger, B. V.,** Proteolytic enzyme activity in the intra-erythrocytic parasites *Babesia argentina* and *Babesia bigemina*, *Z. Parasitenkd.*, 42, 213, 1973.
24. **Wright, I. G.,** The probable role of *Babesia argentina* esterase in the in vitro activation of plasma prekallikrein, *Vet. Parasitol.*, 1, 91, 1975.
25. **Wright, I. G., Goodger, B. V., and Mahoney, D. F.,** Virulent and avirulent strains of *Babesia bovis:* the relationship between parasite protease content and pathophysiological effect of the strain, *J. Protozool.*, 28, 118, 1981.
26. **Wright, I. G., Goodger, B. V., Rode-Bramanis, K., Mattick, J. S., Mahoney, D. F., and Waltisbuhl, D. J.,** The characterisation of an esterase derived from *Babesia bovis* and its use as a vaccine, *Z. Parasitenkd.*, 69, 703, 1983.
27. **Wright, I. G.,** unpublished data.
28. **Mahoney, D. F. and Wright, I. G.,** *Babesia argentina:* immunization of cattle with a killed antigen against infection with a heterologous strain, *Vet. Parasitol.*, 2, 273, 1976.
29. **Commins, M. A.,** unpublished data.
30. **Goodger, B. V. and Wright, I. G.,** *Babesia bigemina* (*argentina*): observations of coagulation parameters, fibrinogen catabolism and fibrinolysis in intact and splenectomised cattle, *Z. Parasitenkd.*, 54, 9, 1977.
31. **Moore, D. J. and Williams, M. C.,** Disseminated intravascular coagulation: a complication of *Babesia canis* infection in the dog, *J. S. Afr. Vet. Assoc.*, 50, 265, 1979.
32. **Mahoney, D. F. and Goodger, B. V.,** *Babesia argentina:* serum changes in infected calves , *Exp. Parasitol.*, 24, 375, 1969.
33. **Dalgliesh, R. J., Dimmock, C. K., Hill, H. W. M., and Mellors, L. T.,** *Babesia argentina:* disseminated intravascular coagulation in acute infections in splenectomized calves, *Exp. Parasitol.*, 40, 124, 1976.

34. **Rogers, R. J.,** Observations on the pathology of *Babesia argentina* infections in cattle, *Aust. Vet. J.*, 47, 242, 1971.
35. **Wright, I. G., Goodger, B. V., McKenna, R. V., and Mahoney, D. F.,** Acute *Babesia bovis* infection: a study of the vascular lesions in kidney and lung, *Z. Parasitenkd.*, 60, 19, 1979.
36. **Nordstoga, K.,** Porcine salmonellosis: a counterpart to the generalized Schwartzman reaction, *Acta Pathol. Microbiol. Scand.*, 82, 690, 1974.
37. **Nordstoga, K.,** Disseminated intravascular coagulation: a pathologist's view, *Thromb. Haemostasis*, 37, 180, 1977.
38. **Dale, J., Ohlsson, K., Nordstoga, K., and Aasen, A. O.,** Intravascular hemolysis and ultrastructural changes of erythrocytes in lethal canine endotoxic shock, *Eur. Surg. Res.*, 12, 39, 1980.
39. **Zerbino, D. D. and Lukasevich, L. L.,** Disseminated intravascular blood coagulation syndrome: morphological criteria, *Arkh. Patol.*, 44, 29, 1982.
40. **Goodger, B. V. and Wright, I. G.,** *Babesia bovis (argentina)*: analysis of fibrinogen-like proteins during infection, *Z. Parasitenkd.*, 60, 211, 1979.
41. **Latallo, Z. S. and Teisseyre, E.,** Evaluation of reptilase-R and thrombin clotting time in the presence of fibrinogen degradation products and heparin, *Scand. J. Haematol. Suppl.*, 13, 219, 1971.
42. **Goodger, B. V.,** A cold preciptitable fibrinogen complex in the plasma of cattle dying from infection with *Babesia argentina*, *Z. Parasitenkd.*, 48, 1, 1975.
43. **Goodger, B. V., Wright, I. G., and Mahoney, D. F.,** *Babesia bovis (argentina)*: analysis of paracoagulation proteins in acutely infected cattle, *Aust. J. Exp. Biol. Med. Sci.*, 58, 179, 1980.
44. **Goodger, B. V.,** Erythrophilic proteins of bovine plasma, *Experientia*, 32, 518, 1976.
45. **Goodger, B. V., Wright, I. G., and Mahoney, D. F.,** Initial characterization of cryoprecipitates in cattle recovering from acute *Babesia bovis (argentina)* infection, *Aust. J. Exp. Biol. Med. Sci.*, 59, 521, 1981.
46. **Goodger, B. V. and Wright, I. G.,** Acute *Babesia bigemina* infection: changes in fibrinogen catabolism, *Z. Parasitenkd.*, 53, 53, 1977.
47. **Wright, I. G. and Goodger, B. V.,** Acute *Babesia bigemina* infection: changes in coagulation and kallikrein parameters, *Z. Parasitenkd.*, 53, 63, 1977.
48. **Guelfi, J. F., Dubois, P., and Boneu, B.,** Exploration of hemostasis on dogs affected by babesiasis, *Rev. Med. Vet.* (Toulouse), 135, 699, 1984.
49. **Allen, P., C., Frerichs, W. M., and Hobrook, A. A.,** Experimental acute *Babesia caballi* infections. II. Response of platelets and fibrinogen, *Exp. Parasitol.*, 37, 373, 1975.
50. **Rudolph, W., Correa, J., Zurita, L., and Manley, W.,** Equine piroplasmosis: leukocytic response to *Babesia equi* (Laveran, 1901) infection in Chile, *Br. Vet. J.*, 131, 601, 1975.
51. **Goodger, B. V., Rodebramanis, K., and Wright, I. G.,** Procoagulant activity of *Babesia bovis*-infected erythrocytes, *J. Parasitol.*, 73, 1052, 1987.
52. **Goodger, B. V.,** *Babesia argentina*: observations on the immunogenicity of the cryofibrinogen complex, *Z. Parasitenkd.*, 53, 47, 1977.
53. **Annable, C. R. and Ward, P. A.,** Immunopathology of the renal complications of babesiosis, *J. Immunol.*, 112, 1, 1974.
54. **Chapman, W. E. and Ward, P. A.,** The complement profile in babesiosis, *J. Immunol.*, 117, 935, 1976.
55. **Goodger, B. V., Wright, I. G., and Mahoney, D. F.,** Changes in conglutinin, immunocongulutinin, complement C3 and fibronetin concentrations in cattle acutely infected with *Babesia bovis*, *Aust. J. Exp. Biol. Med. Sci.*, 59, 531, 1981.
56. **Lachmann, P. J.,** Conglutinin and immunoconglutinins, *Adv. Immunol.*, 6, 479, 1967.
57. **Osler, R. G.,** *Complement Mechanisms and Functions*, Prentice-Hall, Englewood Cliffs, N.J., 1976.
58. **Corbin, N. C., Hugli, T. E., and Muller-Eberhard, H. J.,** Serum carboxypeptidase B: a spectrophotometric assay using protamine as substrate, *Anal. Biochem.*, 73, 41, 1976.
59. **Wright, I. G., Mahoney, D. F., and Goodger, B. V.,** Serum carboxypeptidase B levels during acute and mild *Babesia bovis* and acute *Babesia bigemina* infections of cattle, *Z. Parasitenkd.*, 63, 191, 1980.
60. **Taylor, S. M. and Elliott, C. T.,** Serum carboxypeptidase B levels during acute and mild *Babesia divergens* infections of adult cattle, *Res. Vet. Sci.*, 35, 370, 1983.
61. **Jack, R. M. and Ward, P. A.,** The role *in vivo* of C3 and the C3b receptor in babesial infection in the rat, *J. Immunol.*, 124, 1574, 1980.
62. **Seinen, W., Stegmann, T., and Kuil, H.,** Complement does not play a role in promoting *Babesia rodhaini* infections in Balb/C mice, *Z. Parasitenkd.*, 68, 249, 1982.
63. **Levy, M. G., Kakoma, I., Clabaugh, G. C., and Ristic, M.,** Complement does not facilitate *in vitro* invasion of bovine erythrocytes by *Babesia bovis*, *Ann. Trop. Med. Parasitol.*, 3, 377, 1986.
64. **Onabanjo, A. O. and Maegraith, B. G.,** Kallikrein as a pathogenic agent in *Plasmodium knowlesi* infection in *Macaca mulatta*, *Br. J. Exp. Pathol.*, 51, 523, 1970.
65. **Boreham, P. F. L.,** Immune reactions and kinin formation in chronic trypanosomiasis, *Br. J. Pharmacol. Chemother.*, 32, 493, 1968.

66. **Goodwin, L. G., and Richards, W. H. G.,** Pharmacologically active peptides in blood and urine of animals infected with *Babesia rodhaini* and other pathogenic organisms, *Br. J. Pharmacol. Chemother.*, 15, 152, 1960.
67. **Wright, I. G.,** Plasma kallikrein levels in acute *Babesia argentina* infections in splenectomised and intact calves, *Z. Parasitenkd.*, 41, 269, 1973.
68. **Wright, I. G.,** The kallikrein-kinin system and its role in the hypotensive shock syndrome of animals infected with the haemoprotozoan parasites *Babesia, Plasmodium* and *Trypanosoma, Gen. Pharmacol.*, 10, 319, 1979.
69. **Wright, I. G.,** Kinin, kininogen and kininase levels during acute *Babesia bovis* (= B. argentina) infection of cattle, *Br. J. Pharmacol.*, 61, 567, 1977.
70. **Wright, I. G. and Kerr, J. D.,** Hypotension in acute *Babesia bovis* (= *B. argentina*) infections of splenectomised calves, *J. Comp. Pathol.*, 87, 531, 1977.
71. **Clark, I. A., Hunt, N. H., and Cowden, W. B.,** Immunopathology of malaria, in *Immunology, Immunopathology and Immunoprophylaxis of Parasitic Infections*, Soulsby, E. J. L., Ed., CRC Press, Boca Raton, Fla., 1986.
72. **Griffin, J. H. and Cochrane, C. G.,** Mechanisms for the involvement of high molecular weight kininogen in surface- dependent reactions of Hageman Factor, *Proc. Natl. Acad. Sci. U.S.A.*, 73, 2554, 1976.
73. **Liu, C. Y., Scott, C. F., Bagdasarian, A., Pierce, J. V., Kaplan, A. P., and Colman, R. W.,** Potentiation of the function of Hageman Factor fragments by high molecular weight kininogen, *J. Clin. Invest.*, 60, 7, 1977.
74. **Wright, I. G.,** Biogenic amine levels in acute *Babesia bovis* infected cattle, *Vet. Parasitol.*, 4, 393, 1978.
75. **Eyre, P., Lewis, A. J., and Wells, P. W.,** Acute systemic anaphylaxis in the calf, *Br. J. Pharmacol.*, 47, 504, 1973.
76. **Mizuta, K.,** Synergistic effect of histamine and bradykinin on capillary permeability, *Mie Med. J.*, 20, 157, 1970.
77. **Wright, I. G., McKenna, R. V., and Goodger, B. V.,** Acute *Babesia bovis* infections: plasma creatine kinase, lactate dehydrogenase and creatinine levels and associated muscle damage, *Z. Parasitenkd.*, 64, 297, 1981.
78. **Copley, A. L.,** The endothelial fibrin lining, fibrinogen gel clotting and the endothelium-blood interface, *Ann. N.Y. Acad. Sci.*, 416, 377, 1983.
79. **Goodger, B. V.,** *Babesia bovis (= argentina)*: changes in erythrophylic and associated proteins during acute infection of splenectomized and intact calves, *Z. Parasitenkd.*, 55, 1, 1978.
80. **Wautier, J. L., Pintigny, D., Wautier, M. P., Paton, R. C., Galacteros, F., Passa, P., and Caen, J. P.,** Fibrinogen, a modulator of erythrocyte adhesion to vascular endothelium, *J. Lab. Clin. Med.*, 101, 911, 1983.
81. **Jones, R. L., Jovanovic, L., Forman, S., and Peterson, C. M.,** Time course of reversibility of accelerated fibrinogen disappearance in Diabetes Mellitus: association with intravascular volume shifts, *Blood*, 63, 22, 1984.
82. **Becker, C. G. and Harpel, P. C.,** α_2-Macroglobulin on human vascular endothelium, *J. Exp. Med.*, 144, 1, 1976.
83. **Commins, M. A., Goodger, B. V., and Wright, I. G.,** Proteinases in the lysate of bovine erythrocytes infected with *Babesia bovis*: initial vaccination studies, *Int. J. Parasitol.*, 15, 491, 1985.
84. **Anon.,** The endothelial cell surface in health and disease, *Lancet*, i, 1255, 1985.
85. **Goodger, B. V., Mahoney, D. F., and Wright, I. G.,** *Babesia bovis*: attachment of infected erythrocytes to heparin-sepharose columns, *J. Parasitol.*, 69, 248, 1983.
86. **Goodger, B. V., Commins, M. A., Wright, I. G., Waltisbuhl, D. J., and Mirre, G. B.,** *Babesia bovis*: successful homologous vaccination using a heparin-binding fraction of infected erythrocytes, *Int. J. Parasitol.*, 17, 935, 1987.
87. **Ludford, C. G.,** Fluorescent antibody staining of four *Babesia* species, *Exp. Parasitol.*, 24, 327, 1969.
88. **Curnow, J. A.,** *In vitro* agglutination of bovine erythrocytes infected with *Babesia argentina, Nature (London)*, 217, 267, 1968.
89. **Howard, R. J., Rodwell, B. J., Smith, P. M., Callow, L. L., and Mitchell, G. F.,** Comparison of the surface proteins and glycoproteins on erythrocytes of calves before and during infection with *Babesia bovis, J. Protozool.*, 27, 241, 1980.
90. **Goodger, B. V., Wright, I. G., Mahoney, D. F., and McKenna, R. V.,** *Babesia bovis (argentina)*: studies on the composition and location of antigen associated with infected erythrocytes, *Int. J. Parasitol.*, 10, 33, 1980.
91. **Commins, M. A., Goodger, B. V., Waltisbuhl, D. J., and Wright, I. G.,** *Babesia bovis:* studies of parameters influencing microvascular stasis of infected erythrocytes, *Res. Vet. Sci.*, 1988, in press.
92. **Pfafferott, C., Meiselman, H. J., and Hochstein, P.,** The effect of malonyldialdehyde on erythrocyte deformability, *Blood*, 59, 12, 1982.

93. **Stocks, J. and Dormandy, T. L.,** The autoxidation of human red cell lipids induced by hydrogen peroxide, *Br. J. Haematol.*, 20, 95, 1971.
94. **Flynn, T. P., Johnson, G. J., and Allen, D. W.,** Sucrose density gradient analysis of erythrocyte membranes in hemolytic anemias, *Blood*, 57, 59, 1981.
95. **Trotta, R. J., Sullivan, S. G., and Stern, A.,** Lipid peroxidation and hemoglobin degradation in red blood cells exposed to *t*-pbutyl hydroperoxide. Dependence on glucose metabolism and hemoglobin status, *Biochim. Biophys. Acta*, 678, 230, 1981.
96. **Richard, M. D.,** Carbohydrate metabolism in *Babesia rodhainin*: investigation with ^{14}C-labeled substrates and enzyme assays, *Exp. Parasitol.*, 28, 512, 1970.
97. **Farlow, G. E.,** Differences in the infectivity of *Babesia* incubated in plasma and serum and the role of glucose in prolonging viability, *Int. J. Parasitol.*, 6, 513, 1976.
98. **Boogaerts, M. A., Roelant, C., Temmerman, J., Goossens, W., and Verwilghen, R. L.,** Effect of beta-blocking drugs on red cell adhesive and rheological properties, *J. Lab. Clin. Med.*, 102, 899, 1983.
99. **Thorburn, J., Drummond, M. M., Whigham, K. A., Lowe, G. D. O., Forbes, C. D., Prentice, C. R. M., and Whitfield, C. R.,** Blood viscosity and haemostatic factors in late pregnancy, pre-eclampsia and fetal growth retardation, *Br. J. Obstet. Gynaecol.*, 89, 117, 1982.
100. **Waltisbuhl, D. J.,** unpublished data.
101. **Schwarz, H. P., Luger, A., Graf, H., Stummvoll, H. K., and Aiginger, P.,** Fibronectin and venostasis, *Thromb. Haemostasis*, 48, 345, 1982.
102. **Goodger, B. V., Wright, I. G., and Mahoney, D. F.,** *Babesia bovis (argentina)*: studies of plasma lipids and lipoproteins during acute infections in cattle, *Z. Parasitenkd.*, 65, 271, 1981.
103. **Otsuji, S., Baba, Y., and Kamada, T.,** Erythrocyte membrane microviscosity in diabetes, *Horm. Metab. Res. Suppl.*, 11, 97, 1981.
104. **Witte, S.,** The affinity of fibrinogen for vessel walls, *Immunol. Probl. Blutgerinnung von Willebrand-Juergens-Syndr. Verhandlungsber, 1976*, Heene, D. L., Ed., Schattauer, Stuttgart, 1978, 263.
105. **Goodger, B. V., Wright, I. G., and Mahoney, D. F.,** The use of pathophysiological reactions to assess the efficacy of the immune response to *Babesia bovis* in cattle, *Z. Parasitenkd.*, 66, 41, 1981.
106. **Marsh, N. A.,** The accelerating effect of fibrinogen and early fibrinogen degradation products on erythrocytes sedimentation, *Thromb. Haemostasis*, 42, 757, 1979.
107. **Goodger, B. V.,** Preliminary characterization of the bovine polymeric Hb binding protein and comparison of some properties with human haptoglobins, *Aust. J. Exp. Biol. Med. Sci.*, 50, 11, 1972.
108. **Brandslund, I., Svehag, S. E., Teisner, B., and Hyltoft Peterson, P.,** Activation of the complement system and accumulation of hemoglobin-haptoglobin complexes in plasma during an adverse reaction to penicillin treatment, *Int. Arch. Allerg. Appl. Immunol.*, 71, 173, 1983.
109. **Maegraith, B.,** Interdependence, *Am. J. Trop. Med. Hyg.*, 26, 344, 1977.
110. **Rest, J. R.,** Cerebral malaria in inbred mice. I. A new model and its pathology, *Trans. R. Soc. Trop. Med. Hyg.*, 76, 410, 1982.
111. **Kalter, E. S.,** Inflammatory mediators and acute infections, *Resuscitation*, 11, 133, 1984.
112. **Cox, H. W. and Saleh, S. M.,** Anemia and thrombocytopenia from *Corynebacterium parvum* stimulated resistance against malaria, trypanosomiasis, and babesiosis, *J. Parasitol.*, 69, 654, 1983.
113. **Goodger, B. V., Commins, M. A., Wright, I. G., and Mirre, G.B.,** *Babesia bovis:* vaccination of cattle against heterologous challenge with fractions of lysate from infected erythrocytes, *Z. Parasitenkd.*, 70, 321, 1984.
114. **Israel, L., Samak, R., Edelstein, R., Bogucki, D., and Breau, J. L.,** Mise en evidence du role immunodepresseur des proteines de l'inflammation. Leur role physiopathologique chez les cancereaux, *Ann. Med. Intern.*, 132, 26, 1981.
115. **Gauldie, J., Lamontagne, L., and Stadnyk, A.,** Acute phase response in infectious disease, *Surv. Synth. Pathol. Res.*, 4, 126, 1985.
116. **Gray, G. D. and Phillips, R. S.,** Suppression of primary and secondary antibody responses and inhibition of antigen priming during *Babesia microti* infections in mice, *Parasitol. Immunol.*, 5, 123, 1983.
117. **Mahoney, D. F.,** Bovine babesiosis: the passive immunization of calves against *Babesia argentina* with special reference to the role of complement fixing antibodies, *Exp. Parasitol.*, 20, 119, 1967.
118. **Hirsch, J.,** Review. Blood tests for the diagnosis of venous and arterial thrombosis, *Blood*, 57, 1, 1981.
119. **Saba, T. M. and Jaffe, E.,** Plasma fibronectin (opsonic glycoprotein): its synthesis by vascular endothelial cells and role in cardiopulmonary integrity after trauma as related to reticuloendothelial function, *Am. J. Med.*, 68, 577, 1980.
120. **Loegering, D. J., Kaplan, J. E., and Grover, G. J.,** Effect of erythrocyte stroma in reticuloendothelial system phagocytic function, *J. Lab. Clin. Med.*, 102, 231, 1983.
121. **Kaplan, J. E. and Saba, T. M.,** Enhancement of reticuloendothelial activity by low dose heparin during intravascular coagulation, *J. Reticulosendothel. Soc.*, 29, 381, 1981.

122. **Zubairov, D. M., Andrushko, I. A., Akhmetshina, M. K., Timerbaev. V. N., and Voronina, I. E.,** Role of the reticuloendothelial system in the catabolism of fibrinogen, *Fiziol. Zh. SSSR im. I. M. Sechenova*, 64, 524, 1978.
123. **Goodger, B. V., Waltisbuhl, D. J., Wright, I. G., Mahoney, D. F., and Commings, M. A.,** *Babesia bovis*: analysis and vaccination trial with the cryoprecipitable immune complex, *Vet. Immunol. Immunopathol.*, 14, 57, 1987.
124. **Wolf, C. F., Resnick, G., Marsh, W. L., Benech, J., and Habicht, G.,** Autoimmunity to red blood cells in babesiosis, *Transfusion*, 22, 533, 1982.
125. **Benacerraf, B. and Unanue, E. R.,** *Textbook of Immunology*, Williams & Wilkins, Baltimore, 1979, 239.
126. **Waltisbuhl, D. J.,** unpublished data.
127. **Camussi, G., Bussolino, F., Tetta, C., Caligaris Cappio, F., Coda, R., Macchiorlatti, E., Alberton, M., Roffinello, C., and Segoloni, G.,** Effect of prostaglandin (PGI_2) on immune-complex-induced neutropenia, *Immunology*, 48, 625, 1983.
128. **Shear, H. L.,** Murine malaria: immune complexes inhibit Fc receptor-mediated phogocytosis, *Infect. Immun.*, 44, 130, 1984.
129. **Waxman, F. J., Hebert, L. A., Cornacoff, J. B., VanAman, M. E., Smead, W. L., Kraut, E. H., Birmingham, D. J., and Taguiam, J. M.,** Complement depletion accelerates the clearance of immune complexes from the circulation of primates, *J. Clin. Invest.*, 74, 1329, 1984.
130. **Waxman, F. J., Hebert, L. A., Cosio, F. G., Smead, W. L., Vanaman, M. E., Taguiam, J. M., and Birmingham, D. J.,** Differential binding of immunoglobulin-A and immunoglobulin-G1 immune complexes to primate erythrocytes in vivo. Immunoglobulin-A immune complexes bind less well to erythrocytes and are preferentially deposited in glomeruli, *J. Clin. Invest.*, 77, 82, 1986.
131. **Nishi, T., Bhan, A. K., Collins, A. B., and McCluskey, R. T.,** Effect of circulating immune complexes on Fc and C3 receptors of Kupffer cells, *in vivo, Lab. Invest.*, 44, 442, 1981.
132. **Elissalde, G. S., Wagner, G. G., Craig, T. M., Elissalde, M. H., and Rowe, L.,** Hypocholesterolemia and hypocortisolemia in acute and terminal *Babesia bovis* infections, *Vet. Parasitol.*, 12, 1, 1983.
133. **Muley, A. K., Singh, B., Ghafoor, M. A., and Anantwar, L. G.,** Note on biochemical changes during experimental *Babesia bigemina* infection in splenectomized cross-bred calves, *Indian J. Anim. Sci.*, 50, 455, 1980.
134. **Wright, I. G.,** Studies on the pathogenesis of *Babesia argentina* and *Babesia bigemina* infections in splenectomised calves, *Z. Parasitenkd.*, 39, 85, 1972.
135. **Wright, I. G.,** Osmotic fragility of erythrocytes in acute *Babesia argentina* and *Babesia bigemina* infections in splenectomised *Bos taurus* calves, *Res. Vet. Sci.*, 15, 299, 1973.
136. **Kakoma, I. and Ristic, M.,** Pathogenesis of babesiosis, in *Malaria and Babesiosis*, Ristic, M., Ambroise-Thomas, P., and Kreier, J. P., Eds., Martinus Nijhoff Dodrecht, Netherlands, 1984.

Chapter 7

IMMUNOLOGY OF BABESIOSIS

Mark A. James

TABLE OF CONTENTS

I. INTRODUCTION

Immunity to *Babesia* infection depends on the innate resistance of the host in addition to specific responsiveness to babesial antigens. Activated macrophages and their soluble products (monokines) are effective in stimulating nonspecific immunity to babesiosis. Specific immune mechanisms include both humoral and cellular components. Protective antibodies induced primarily as a result of helper T lymphocyte activity neutralize merozoite invasion of erythrocytes and enhance phagocytosis of free parasites and infected cells. There is now good evidence that recovery from *Babesia* infections is T lymphocyte-dependent, with natural killer cells also playing a role in protective immune responses. Exoantigen-containing supernatant fluids from *Babesia* in vitro cultures are now being used as soluble immunogens for vaccination against bovine babesiosis and are efficacious in the induction of heterologous strain immunity in susceptible cattle.

II. MECHANISMS OF NONSPECIFIC RESISTANCE

The capacity of the host to overcome a babesial infection is determined not only by specific immunologic responsiveness, but also by certain innate characteristics of the host. The reticuloendothelial system and its components are important regulators of nonspecific resistance to *Babesia* spp. and may be activated by killed bacterial agents.[1,2] Clark[3] has demonstrated that a commercially available extract of *Coxiella burnetii* protects mice against *B. microti* and *B. rodhaini*, possibly by activating interferon production or natural killer cells. These agents may also activate macrophages with subsequent intraerythrocytic parasiticidal effects, e.g., crisis form induction, a consequence of soluble monokine activity. Similar pretreatment of calves with low doses (0.20 mg/kg) of killed *Corynebacterium parvum* has been reported to alter the course of acute *B. bigemina* infections.[4] Although *C. parvum* failed to stimulate significant protection against virulent challenge exposure, lower parasitemias and less severe decreases in hematocrit levels were noted.

Nonspecific immunity to *Babesia* spp. may be enhanced by concurrent infections with other parasites. Mzembe et al.[5] showed that macrophage-mediated resistance to *B. microti* in mice was enhanced during infection with the larval stages of *Nematospiroides dubius*, a gastrointestinal nematode. Peritoneal macrophages harvested from infected mice and supernatants collected from cultures of these mononuclear cells were both effective in suppressing *B. microti* development in vitro.

Other studies have also indicated that mononuclear phagocytes are active against *Babesia* parasites in vitro.[6,7] Decreased multiplication of *B. microti* in hamster erythrocytes was seen when the organisms were cocultivated with activated peritoneal macrophages. Montealegre et al.[8] studied the in vitro growth of *B. bovis* in bovine erythrocytes[9] following treatment with supernatants from bovine blood monocyte cultures. Briefly, phagocytic cell monolayers were treated with either culture-derived soluble *B. bovis* exoantigens, immune complexes, or bovine anti-*B. bovis* immune serum. Subsequently, the monolayers were washed and allowed to develop further in the presence of standard culture medium. Transfer of supernatant fluids from these cultures to those of *B. bovis* revealed the presence of growth-inhibiting factors. Supernatants from antigen- and immune complex-treated monolayers demonstrated greater inhibitory effects than did supernatants from antibody-treated or untreated monolayers.

Biochemically or physiologically based innate resistance may also play an important role in the protective immunity to babesiosis. In vitro cultivation of *B. bovis* in erythrocyte cultures showed that blood from young animals contains a factor(s) responsible for their resistance to severe disease.[10] The presence of this factor results in inhibition of parasite growth and the eventual intraerythrocytic death of the parasite. Cattle breeds also vary in their susceptibility to babesiosis. *Bos indicus* breeds of cattle are known to be more resistant than *Bos taurus* breeds,[11] with the *Bos indicus* breeds having a reportedly greater resistance to *Babesia bovis* than *B. bigemina.*[12,13]

III. ANTIBODY-MEDIATED IMMUNE RESPONSES

A. Nature of Humoral Immunity to *Babesia* Parasites

The role of antibody-mediated immune responses to *Babesia* parasites is well established.[14] Passive transfer of immune serum into susceptible animals has been shown to confer partial protection against *Babesia* infections,[15] with immunity predominantly species specific.[16] Protection against *B. microti* in mice may be transferred with either serum or cells. Meeusen et al.[17] reported that animals which received immune serum developed parasitemias at a much slower rate and reached a lower peak parasitemia (25%) as compared to that of untreated mice (48%). The effect of passively transferred immune serum on *B. microti* infection was immediate but short lived. Adoptive transfer experiments showed that only B lymphocyte-enriched spleen cell populations were effective in protecting mice against primary infection, whereas T cell-enriched subpopulations were best at transferring immunity to a second reinfection with *B. microti*. This suggested that primed helper T cells regulated B cell expression of the protective antibody response.

B. Antibody Levels During *Babesia* Infections

In general, antibody levels rise to an early peak and thereafter decline during the chronic stage of infection.[18] Meeusen et al.[19] observed that infection of adoptively immunized mice with *B. microti* induced high levels of IgG antibody as compared with that of nonimmunized control mice. Apparently, development of the anamnestic humoral response was dependent on the proliferation of memory B cells.

Similar types of antibody responses are seen in cattle raised in endemic areas for babesiosis. Calves are protected by maternal colostral antibodies until approximately 3 months of age.[20] At this time, animals usually become infected with peak antibody titers occurring between 4 to 5 months of age. Once infected, the resulting immunity lasts at least 4 years.[21]

O'Donoghue et al.[22] have studied the kinetics of the humoral response to *B. bigemina* infection using IgM- and IgG-specific enzyme immunoassays. IgG-antibodies first appeared at 7 days postinfection. Titers peaked by day 12, and then plateaued at that level for 7 weeks. IgM-antibodies appeared at 7 days postinfection, reached and persisted at maximal levels from days 12 to 22, and then declined to low levels by 28 days postinfection.

James et al.[23] studied the kinetics of specific indirect fluorescent antibody (IFA) titers (IgG) and total serum IgG and IgM levels in yearling Angus heifers vaccinated twice with culture-derived[9] soluble *B. bovis* antigens[24] and challenged with live *Babesia* organisms;[25] 19 days after challenge exposure, vaccinated cattle had a strong anamnestic response as mean IFA titers reached 1:48,710. Control animals demonstrated a quick response to challenge exposure; however, the mean antibody titers were considerably lower than those of the vaccinated heifers during the postchallenge exposure period. Vaccinated animals demonstrated significant increases in mean serum IgG values soon after each injection with soluble *B. bovis* vaccine. These values returned to near baseline values by day 54, then increased to a mean concentration of 32.8 mg/mℓ (49% increase from baseline value) on day 67 (19 days postchallenge). Prechallenge values in controls were constant, but quickly decreased to 17.3 mg/mℓ 6 days postchallenge, a time coincident with patent parasitemia in these animals. Slight increases in serum IgM values were noticed after each vaccination. A marked increase to a mean value of 9.7 mg/mℓ (149% increase from baseline value) was seen 19 days postchallenge. This increase in serum IgM was preceded by significant decreases in vaccinated and control heifers during the patent *B. bovis* infection. Again, low values were generally associated with peak parasitemias.

A nonspecific suppression of total IgG (controls) and IgM (vaccinated animals and controls) was observed, the depression coinciding with the period of peak parasitemia. Others have reported immunosuppresion to heterologous antigens in mice infected with *B. microti*. Purvis[26] found decreased antibody production to sheep erythrocytes in infected mice. Depression of IgM and IgG plaque-forming cells began 3 days after peak parasitemia and was maximal 4 days later. Specific antibabesial antibody levels continued to rise throughout the infection period. Moreover, normal immune responsiveness was correlated with the gradual disappearance of parasites from the blood. Similarly, Gray[27] observed the effect of *B. microti* on both primary and secondary responses to sheep erythrocytes with maximal depression associated with peak parasitemia.

IV. ROLE OF CELL-MEDIATED IMMUNITY

A. T Lymphocytes

1. Involvement in Effector Mechanisms

Ruebush and Hanson[28,29] have reported that lymph node and spleen cells from BALB/c mice infected with *B. microti* developed the ability to transfer adoptive immunity to naive mice. Results suggested that immunologic memory was modulated by the T lymphocyte. They found that resistance to primary *B. microti* infections is T cell-dependent with a concurrent depression in B cell function.

Zivkovic et al.[30] have shown that spleen cells from *B. rodhaini*-immune mice could transfer protection to naive animals pretreated with cyclophosphamide. Treatment of naive mice with cyclophosphamide (300 mg/kg) 5 days prior to a virulent *B. rodhaini* challenge exposure resulted in over 50% survival. The protective effect of cyclophosphamide was explained by its inhibition of suppressor T cell activity. Moreover, protection was unaffected by sublethal irradiation at 400 rad. Since B lymphocyte function is suppressed by this dosage, the role of antibodies in the protective immune response to *B. rodhaini* may be relatively minor. Considerable protection was still preserved after irradiation of immune animals with 875 rad. This suggests that highly radioresistant macrophages, and to a lesser extent, T lymphocytes, are required for complete protection.

Protective immunity against a primary *B. microti* infection in mice, induced by adoptive transfer of immune spleen cells, can be abolished if the cells are pretreated with mitomycin C prior to transfer.[31] Since mitomycin C treatment prevents lymphocyte replication without affecting other cell functions, this suggests that undifferentiated lymphoid cells require proliferation in recipient animals before they can exert a protective effect.

The role of T lymphocytes and macrophages in the specific immunity to lethal *B. rodhaini* infections in BALB/c mice has been confirmed in studies with selective immunosuppressive agents, e.g., antithymocyte serum (ATS) and corticosteroids.[32] Hydrocortisone acetate given intraperitoneally in six doses (100 mg/kg) during a 3-week postchallenge period caused increased parasitemias and mortalities as compared to that in untreated mice. Likewise, dexamethasone placed in drinking water at 20 mg/ℓ or 10 mg/ℓ doses for 22 days had a similar suppressive effect on protection against *B. rodhaini*. Mortality was found proportional to the dosage administered in both immunized and control animals. Mice treated with ATS also showed increased susceptibility to challenge infections. Peak parasitemias were significantly higher than in untreated immune mice, yet remained far below the levels observed in nonimmune, challenged mice. Mortality in the ATS-treated animals reached 78%. No deaths were recorded in untreated, immune mice following challenge infection.

2. Measurement of Cell-Mediated Immune (CMI) Responses in Bovine Babesiosis

Timms et al.[33] have studied the CMI response to *B. bovis* antigen in cattle vaccinated with live parasites or with culture-derived soluble exoantigens. Lymphocyte transformation

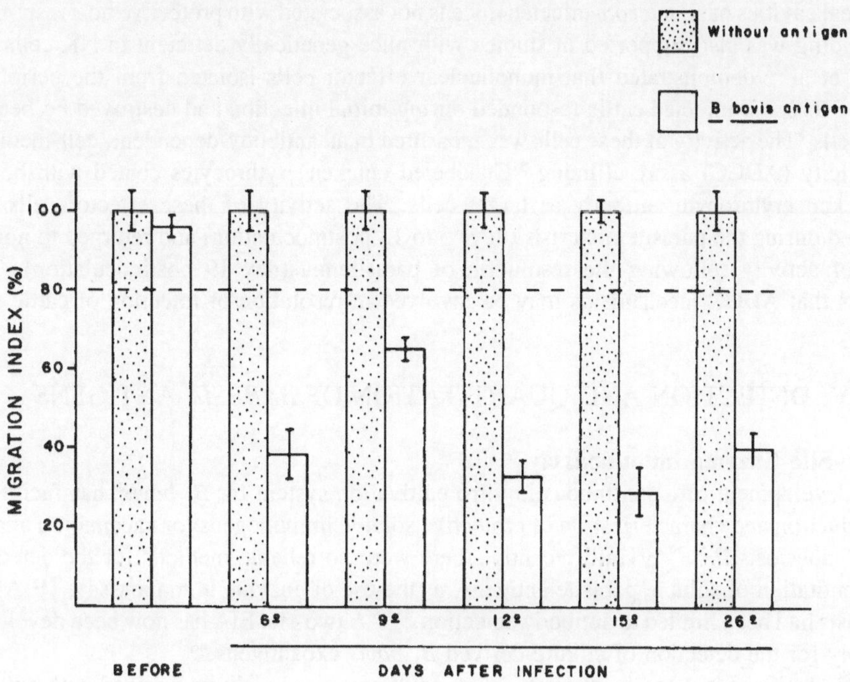

FIGURE 1. Peripheral leukocyte migration indexes in intact calves (n = 3) before and after experimental infection with *B. bovis*. Migration assays were performed in the presence or absence of *B. bovis* antigen. An average migration index (± S.D.) below the dashed line (80% level) was considered a significant inhibition of migration. (From Machado, R. Z., Pinto, A. A., and Valacao, C. A. A., Paper 152, Proc. 11th Conf. World Assoc. Adv. Vet. Parasitol., Rio de Janeiro, 1985.)

assays indicated that the response to live vaccine was short-lived, a relatively low stimulation index being recorded 10 to 18 days after vaccination. Conversely, cattle injected with soluble immunogens responded with positive stimulation indexes throughout a 6-month postvaccination period, with maximum values observed after approximately 18 weeks.

Machado et al.[34] utilized the leukocyte migration inhibition test (LMIT) to evaluate CMI responses to *B. bovis* infection. Samples of peripheral blood were collected periodically from susceptible, intact *Bos taurus* calves (n = 3) before and after experimental inoculation. Leukocytes were isolated and the test performed according to standardized procedures. Cells were incubated for 16 hr at 37°C in the presence or absence of solubilized *B. bovis* antigen, and then the migration area was measured. An average migration index of <80% was considered as a significant level of migration inhibition. The CMI response of calves to *B. bovis* infection is illustrated in Figure 1. Positive CMI responses were noted as early as 6 days postinoculation and continued at significant levels throughout the 26-day monitoring period.

B. Cell-Mediated Cytotoxicity: Killer Cells

Studies have focused on the role of the natural killer (NK) cell in cellular immune responses to babesial infection. Eugui and Allison[35] found that differences in susceptibility to *B. microti* infections among various mouse strains could be correlated with NK cell activity. NK cell activity is known to be high in strains of mice that are resistant to *B. microti*, such as C57B1 and CBA, and low in the susceptible A strain. Irvin et al.[36] suggested that NK cells may be important mediators of functional splenic activity during *Babesia* infections of mice. Conversely, Wood and Clark[37] have shown that increased NK activity in the spleens and peritoneal cavities of *B. microti*-infected mice is not associated with protective host responses. This finding was also supported in studies with mice genetically deficient in NK cells.

Goff et al.[38] demonstrated that mononuclear effector cells isolated from the peripheral blood of *B. bovis*-infected cattle responded during initial infection and destroyed Fc-bearing target cells. The activity of these cells was measured in an antibody-dependent, cell-mediated cytotoxicity (ADCC) assay utilizing ^{51}Cr-labeled chicken erythrocytes coated with bovine antichicken erythrocyte antibody as target cells. The activity of these effector cells was enhanced during the parasitemic crisis (days 5 to 13 postinoculation) and returned to normal levels of activity following the resolution of parasitemia (day 19 postinoculation). This suggests that ADCC mechanisms may be involved in resolution of infection of cattle with *B. bovis*.

V. DETECTION AND QUANTITATION OF *BABESIA* ANTIGENS

A. Two-Site Enzyme Immunoassay

The development of a continuous in vitro cultivation system for *B. bovis*[9] has facilitated the production and characterization of protective soluble immunogens for vaccination against bovine babesiosis.[24,25,39-42] Until recently, there were no reliable methods for the detection and quantitation of soluble parasite antigens, as the use of enzyme immunoassays (EIA) for babesiosis had been limited to antibody detection.[43-45] A two-site EIA has now been developed expressly for the detection of culture-derived *B. bovis* exoantigens.[46]

Bovine IgG was isolated from serum of an adult cow previously immunized with culture-derived *B. bovis* exoantigens and needle challenged with virulent parasites. The specific IgG was used as a capture antibody and as an enzyme-conjugated recognizing antibody. The test was found to be highly specific, reproducible, and sensitive for detection of antigenic cross reactivity between *B. bovis* and *B. bigemina* (Figure 2). The potential usefulness of this assay system is limitless in that the test can be employed to (1) monitor parasite antigens during purification procedures, (2) detect antigens in plasma of infected animals and in the tick vector, and (3) evaluate the standardization of vaccines, e.g., dose titration and antigenic stability.

VI. HETEROLOGOUS STRAIN IMMUNITY IN BOVINE BABESIOSIS

Interstrain antigenic differences exist within the *B. bovis* species, and have been well documented.[47-50] Strains maintain their antigenic identity, reverting to a common antigenic type, even after tick transmission.[50] Callow et al.[51] have shown that the virulence of *B. bovis* organisms may be reduced with their rapid passage in splenectomized calves. Attenuated and avirulent parasites are apparently able to induce protective strain immunity to heterologous challenge.[41,51,52]

An important consideration in the development of effective babesiosis vaccines is the ability to induce heterologous immunity to geographically isolated *B. bovis* strains. Montenegro-James et al.[42] studied the cross protection between *B. bovis* isolates from various regions (Venezuela, Mexico, Argentina, Ecuador, and Colombia) of Latin America. Susceptible yearling cattle were immunized with culture-derived soluble immunogens from Venezuelan or Mexican strains. Cattle were then challenged with virulent *B. bovis* organisms of the Argentinian, Colombian, Ecuadorean, Mexican, and Venezuelan strains. The Venezuelan immunogen was effective in inducing heterologous strain immunity to challenge with the Ecuadorean and Argentinian strains. However, vaccination with the Venezuelan immunogen failed to protect against challenge with the Mexican and Colombian strains, as indicated by most clinical parameters. Heterologous antibody levels were lower for all strains, with the Mexican strain demonstrating the greatest difference during the vaccination period.

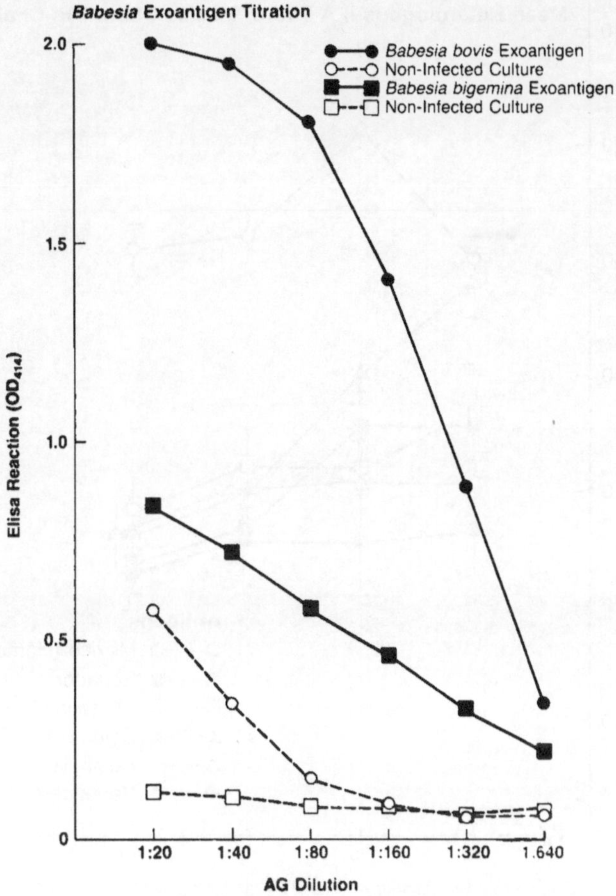

FIGURE 2. Titration of culture-derived *Babesia* exoantigens by the two-site EIA using bovine anti-*B. bovis* (Mexican strain) IgG.

Conversely, animals administered the Mexican *B. bovis* immunogen responded poorly to heterologous challenge exposure, as vaccinated cattle showed clinical reactions similar to those found in control animals. Homologous and heterologous IFA titers produced in response to vaccination are depicted in Figure 3. Maximal homologous titers occurred 2 weeks after the second vaccinal dose and were followed by a transient decline in antibody levels until challenge. Heterologous responses were once again consistently lower, especially those specific to the Venezuelan strain.

Thus, the widespread use of soluble *Babesia* immunogens as vaccines may ultimately depend on the cross-protective capacity between different strains. In the above study, a good degree of cross protection was shown when cattle vaccinated with the Venezuelan *B. bovis* strain were challenged with virulent strains from Ecuador and Argentina, respectively. This indicated that common protective antigens exist among these strains. The degree of antigenic activity contained in culture supernatants of the respective strains has also been analyzed by the two-site EIA (Figure 4). Antigenic cross reactivity was monitored with anti-*B. bovis* antibodies to the homologous Venezuelan strain. Data from the EIA experiments confirmed serological and clinicopathophysiological observations.

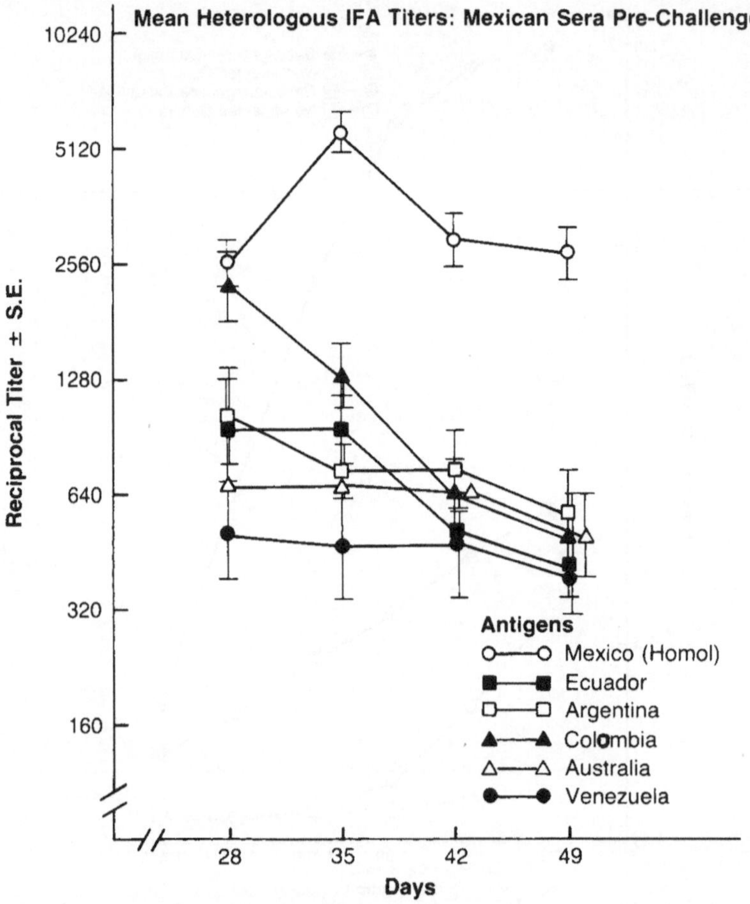

FIGURE 3. Mean homologous and heterologous anti-*B. bovis* IFA titers following vaccination with a culture-derived soluble immunogen of the Mexican *B. bovis strain*. (From Montenegro-James, S., Toro Benitez, M., Leon, E., Lopez, R., and Ristic, M., *Vet. Parasitol.*, 18, 321, 1985. With permission.)

VII. ANTIGENIC RELATIONSHIPS BETWEEN *BABESIA* AND *PLASMODIUM* SPECIES

Much of the humoral immune response to both *Babesia* and *Plasmodium* spp. is directed against asexual blood stage antigens, particularly against those of the merozoite. Numerous antigens, some of which may be involved in the induction of immunity, have been variously identified as being associated with the organism, the erythrocyte membrane, or as exoantigens in blood plasma or culture supernatant fluids.[53,54]

Previous studies have demonstrated antigenic similarities between *Babesia* and *Plasmodium* parasites. Cox et al.[55] detected serological cross reactions by gel precipitation and bentonite flocculation tests. Cox and Turner[56] and Ludford et al.[57] used the IFA test to determine antigenic relationships between *Babesia* and *Plasmodium* species. Cross-protection studies in mice[58] showed that *B. rodhaini*-immunized animals were protected against *P. vinckei* infection, although mice were susceptible to *P. berghei*.

James et al.[59] have studied the antigenic relationship between *B. bovis* and the human malaria parasite, *P. falciparum*, using rabbit monospecific antibodies produced against individual culture-derived *P. falciparum* polypeptides and bovine polyspecific antibodies to

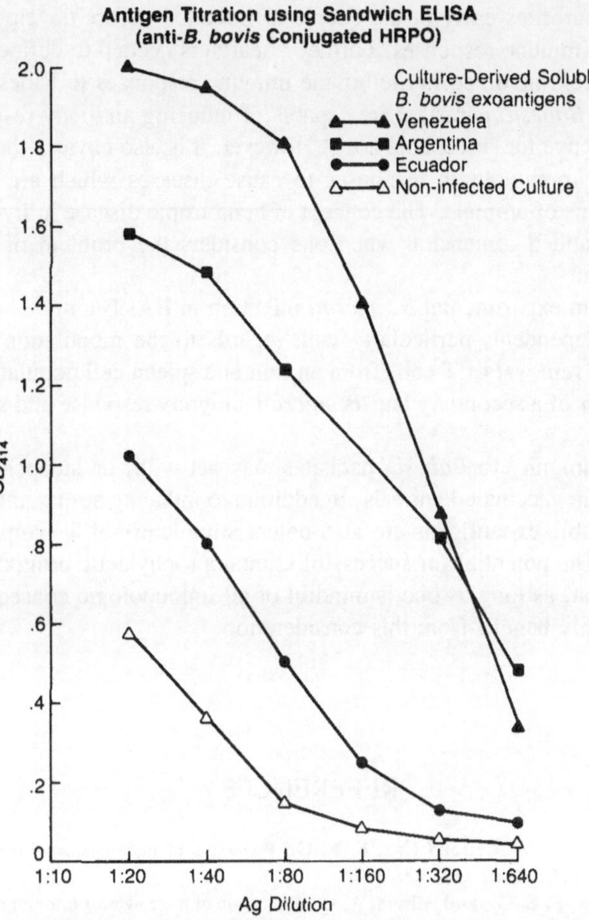

FIGURE 4. Two-site EIA titration of culture-derived *B. bovis* exoantigens from strains isolated in Venezuela, Argentina, and Ecuador. Capture antibody and conjugate were prepared from an IgG fraction of hyperimmune anti-*B. bovis* (Venezuelan strain) serum.

B. bovis exoantigens. These respective antibodies were found to be distinctly cross reactive in the IFA test using infected erythrocytes as antigen substrates. Immunofluorescence was shown to be highly specific for parasite surfaces. Additionally, the degree of reactivity with soluble exoantigens contained in *Plasmodium* and *Babesia* culture supernatants was monitored by a two-site EIA employing the cross reactive antibodies. Further evidence for antigenic cross reactivity between *B. bovis* and *P. falciparum* was shown with an in vitro inhibition assay. Antibodies to *P. falciparum* and *B. bovis* were found to be highly inhibitory for the in vitro growth of *P. falciparum* in human erythrocytes. These data suggest that *B. bovis* and *P. falciparum* share common epitopes of surface antigens.

VIII. SUMMARY AND CONCLUSIONS

The study of immunity to babesial infections has elucidated specific and nonspecific, humoral, and cell-mediated mechanisms of resistance. Soluble products secreted by bovine blood mononuclear phagocytes activated by soluble antigens or immune complexes inhibit the in vitro growth of *B. bovis*. Secretory activity of these phagocytes might be important in organs such as the spleen where large numbers of these cells reside. The products may aid in controlling parasites early in the course of infection before development of specific antibody-mediated immune responses. Further research is needed to define and characterize the nonspecific factors involved in the bovine immune responses to babesiosis.

It is evident that *Babesia* parasites are capable of inducing antibody responses which are in many cases protective for susceptible hosts. However, it is also obvious that these organisms often evade potent immunologic responses to cause diseases which are among the most devastating infections of animals. The concept of hemotropic disease in livestock-producing areas takes on an added dimension when one considers the problem of parasite-induced immunosuppression.

The recovery from experimental *B. microti* infection in BALB/c mice has been shown to be T lymphocyte-dependent, particularly with regards to the modulation of immunologic memory. Complete removal of T cells from an immune spleen cell population will abrogate the adoptive transfer of a secondary anti-*B. microti* antibody response and protection against infection.

Cell-mediated immunity to *Babesia* parasites may act with, or independently from, humoral mechanisms in vaccinated animals. In addition to inducing strong antibody responses, culture-derived soluble exoantigens are also potent stimulators of T lymphocyte and macrophage function. The potential for successful immunoprophylactic programs in the control of babesiosis is great, as long as one is mindful of all immunologic consequences. Vaccine development can only benefit from this consideration.

REFERENCES

1. **Clark, I. A., Allison, A. C., and Cox, F. E. G.,** Protection of mice against *Babesia* and *Plasmodium* with BCG, *Nature (London),* 259, 309, 1976.
2. **Clark, I. A., Cox, F. E. G., and Allison, A. C.,** Protection of mice against *Babesia* spp. and *Plasmodium* spp. with killed *Corynebacterium parvum, Parasitology,* 74, 9, 1977.
3. **Clark, I. A.,** Resistance to *Babesia* spp. and *Plasmodium* spp. in mice pretreated with an extract of *Coxiella burnetii, Infect. Immun.,* 24, 319, 1979.
4. **Corrier, D. E. and Wagner, G. G.,** The protective effect of pretreatment with killed *Corynebacterium parvum* against acute babesiosis in calves, *Vet. Parasitol.,* 15, 165, 1984.
5. **Mzembe, S. A. T., Lloyd, S., and Soulsby, E. J. L.,** Macrophage mediated resistance to *Babesia microti* in *Nematospiroides dubius*-infected mice, *Z. Parasitenkd.,* 70, 753, 1984.
6. **Bautista, C. R. and Kreier, J. P.,** Effect of immune serum on the growth of *Babesia* in hamster erythrocytes in short-term culture, *Infect. Immun.,* 23, 470, 1979.
7. **Bautista, C. R. and Kreier, J. P.,** The action of macrophages and immune serum on growth of *Babesia microti* in short-term cultures, *Tropenmed. Parasitol.,* 31, 313, 1980.
8. **Montealegre, F., Levy, M. G., Ristic, M., and James, M. A.,** Growth inhibition of *Babesia bovis* in culture by secretions from bovine mononuclear phagocytes, *Infect. Immun.,* 50, 523, 1985.
9. **Levy, M. G. and Ristic, M.,** *Babesia bovis:* continuous cultivation in a microaerophilous stationary phase culture, *Science,* 207, 1218, 1980.
10. **Levy, M. G., Clabaugh, G., and Ristic, M.,** Age resistance in bovine babesiosis: role of blood factors in resistance to *Babesia bovis, Infect. Immun.,* 37, 1127, 1982.
11. **Lohr, K. -F., Otieno, P. S., and Gacanga, W.,** Susceptibility of Boran cattle to experimental infections with *Anaplasma marginale* and *Babesia bigemina, Zentralbl. Veterinaermed. Reihe* B., 22, 842, 1975.
12. **Rogers, R. J.,** An evaluation of tick fever outbreaks in Northern Queensland in recent years, *Aust. Vet. J.,* 47, 415, 1971.
13. **Johnston, L. A. Y., Leatch, G., and Jones, P. N.,** The duration of latent infection and functional immunity in droughtmaster and hereford cattle following natural infection with *Babesia argentina* and *Babesia bigemina, Aust. Vet. J.,* 54, 14, 1978.
14. **Carson, C. A. and Phillips, R. S.,** Immunologic response of the vetebrate to *Babesia,* in *Babesiosis,* Ristic, M. and Kreier, J. P., Eds., Academic Press, New York, 1981, 411.
15. **Zwart, D. and Brockelsby, D. W.,** Babesiosis: non-specific resistance, immunological factors and pathogenesis, *Adv. Parasitol.,* 17, 50, 1979.

16. **Mahoney, D. F., Kerry, J. D., Goodger, B. V., and Wright, I. G.,** The immune response of cattle to *Babesia bovis (B. argentina)*. Studies on the nature and specificity of protection, *Int. J. Parasitol.*, 9, 297, 1979.
17. **Meeusen, E., Lloyd, S., and Soulsby, E. J. L.,** *Babesia microti* in mice. Adoptive transfer of immunity with serum and cells, *Aust. J. Exp. Biol. Med. Sci.*, 62, 551, 1984.
18. **Smith, R. D., Osorno, B. M., Brener, J., De la Rosa, R., and Ristic, M.,** Bovine babesiosis: severity and reproducibility of *Babesia bovis* infections induced by *Boophilus microplus* under laboratory conditions, *Res. Vet. Sci.*, 24, 287, 1978.
19. **Meeusen, E., Lloyd, S., and Soulsby, E. J. L.,** Antibody levels in adoptively immunized mice after infection with *Babesia microti* or injection with antigen fractions, *Aust. J. Exp. Biol. Med. Sci.*, 63, 261, 1985.
20. **James, M. A., Coronado, A., Lopez, W., Melendez, R., and Ristic, M.,** Seroepidemiology of bovine anaplasmosis and babesiosis in Venezuela, *Trop. Anim. Health Prod.*, 17, 9, 1985.
21. **Mahoney, D. F., Wright, I. G., and Mirre, G. B.,** Bovine babesiosis: the persistence of immunity to *Babesia argentina* and *B. bigemina* in calves *(Bos taurus)* after naturally acquired infection, *Ann. Trop. Med. Parasitol.*, 67, 197, 1973.
22. **O'Donoghue, P. J., Friedhoff, K. T., Vizcaino, O. G., and Weyreter, H.,** The detection of IgM and IgG antibodies against *Babesia bigemina* in bovine sera using semi-defined antigens in enzyme immunoassays, *Vet. Parasitol.*, 18, 1, 1985.
23. **James, M. A., Kuttler, K. L., Levy, M. G., and Ristic, M.,** Antibody kinetics in response to vaccination against *Babesia bovis*, *Am. J. Vet. Res.*, 42, 1999, 1981.
24. **James, M. A., Levy, M. G., and Ristic, M.,** Isolation and partial characterization of culture-derived soluble *Babesia bovis* antigens, *Infect. Immun.*, 31, 358, 1981.
25. **Kuttler, K. L., Levy, M. G., James, M. A., and Ristic, M.,** Efficacy of a nonviable culture-derived *Babesia bovis* vaccine, *Am. J. Vet. Res.*, 43, 281, 1982.
26. **Purvis, A. C.,** Immunodepression in *Babesia microti* infections, *Parasitology*, 75, 197, 1977.
27. **Gray, G. B.,** The effect of *Babesia microti* on the immune response of mice to sheep erythrocytes, *Parasitology*, 77, xliv, 1978.
28. **Ruebush, M. J. and Hanson, W. L.,** Transfer of immunity to *Babesia microti* of human origin using T lymphocytes in mice, *Cell. Immunol.*, 52, 255, 1980.
29. **Ruebush, M. J. and Hanson, W. L.,** Thymus dependence of resistance to infection with *Babesia microti* of human origin in mice, *Am. J. Trop. Med. Hyg.*, 29, 507, 1980.
30. **Zivkovic, D., Seinen, W., Kuil, H., Albers-Van Bemmel, C. M. G. and Speksnijder, J. E.,** Immunity to *Babesia* in mice. I. Adoptive transfer of immunity to *Babesia rodhaini* with immune spleen cells and the effect of irradiation on the protection of immune mice, *Vet. Immunol. Immunopathol.*, 5, 343, 1984.
31. **Meeusen, E., Lloyd, S., and Soulsby, E. J. L.,** *Babesia microti* in mice. Subpopulations of cells involved in the adoptive transfer of immunity with immune spleen cells, *Aust. J. Exp. Biol. Med. Sci.*, 62, 567, 1984.
32. **Zivkovic, D., Speksnijder, J. E., Kuil, H., and Seinen, W.,** Immunity to *Babesia* in mice. III. The effects of corticosteroids and anti-thymocyte serum on mice immune to *Babesia rodhaini*, *Vet. Immunol. Immunopathol.*, 9, 131, 1985.
33. **Timms, P., Stewart, N. P., Rodwell, B. J., and Barry, D. N.,** Immune responses of cattle following vaccination with living and non-living *Babesia bovis* antigens, *Vet. Parasitol.*, 16, 243, 1984.
34. **Machado, R. Z., Pinto, A. A., and Valacao, C. A. A.,** Cell mediated immune response in cattle babesiosis, Paper 152, Proc. 11th Conf. World Assoc. Adv. Vet. Parasitol., Rio de Janeiro, 1985.
35. **Eugui, E. M. and Allison, A. C.,** Differences in susceptibility of various mouse strains to haemoprotozoan infections: possible correlation with natural killer activity, *Parasite Immunol.*, 2, 277, 1980.
36. **Irvin, A. D., Young, E. R., Osborn, G. D., and Francis, L. M. A.,** A comparison of *Babesia* infections in intact, surgically splenectomized, and congenitally asplenic (Dh/+) mice, *Int. J. Parasitol.*, 11, 251, 1981.
37. **Wood, P. R. and Clark, I. A.,** Apparent irrelevance of NK cells to resolution of infections with *Babesia microti* and *Plasmodium vinckei petteri* in mice, *Parasite Immunol.*, 4, 319, 1982.
38. **Goff, W. L., Wagner, G. G., and Craig, T. M.,** Increased activity of bovine ADCC effector cells during acute *Babesia bovis* infection, *Vet. Parasitol.*, 16, 5, 1984.
39. **Smith, R. D., Carpenter, J., Cabrera, A., Gravely, S., Erp, E., Osorno, M. B., and Ristic, M.,** Bovine babesiosis: vaccination against tick-borne challenge exposure with culture-derived *Babesia bovis* immunogens, *Am. J. Vet. Res.*, 40, 1678, 1979.
40. **Smith, R. D., James, M. A., Ristic, M., Aikawa, M., and Vega, C.,** Bovine babesiosis: protection of cattle with culture-derived soluble *Babesia bovis* antigen, *Science*, 212, 335, 1981.
41. **Timms, P., Dalgliesh, R., Barry, D., Dimmock, C., and Rodwell, B.,** *Babesia bovis:* comparison of culture-derived, non-living antigen and conventional vaccine in the protection of cattle against heterologous challenge, *Aust. Vet. J.*, 60, 75, 1983.

42. **Montenegro-James, S., Toro Benitez, M., Leon, E., Lopez, R., and Ristic, M.,** Heterologous strain immunity in bovine babesiosis using a culture-derived soluble *Babesia bovis* immunogen, *Vet. Parasitol.*, 18, 321, 1985.
43. **Purnell, R. E., and Henry, D. J.,** Microplate enzyme linked immunosorbent assay for antibody to *Babesia divergens* in cattle, *Vet. Rec.*, 99, 102, 1976.
44. **Young, E. R. and Purnell, R. E.,** Evaluation of dried samples as a source of antibody in the micro ELISA test for *Babesia divergens, Vet. Rec.*, 106, 60, 1980.
45. **Barry, D. N., Rodwell, B. J., Timms, P., and McGregor, W.,** A microplate enzyme immunoassay for detecting and measuring antibodies to *Babesia bovis* in cattle serum, *Aust. Vet. J.*, 59, 136, 1982.
46. **Montealegre, F., Montenegro-James, S., Kakoma, I., and Ristic, M.,** Detection of culture-derived *Babesia bovis* exoantigen using a two-site enzyme immunoassay *J. Clin. Microbiol.*, 25, 1648, 1987.
47. **Callow, L. L.,** Strain immunity in babesiosis, *Nature (London)*, 204, 1213, 1964.
48. **Callow, L. L.,** A note on homologous strain immunity in *Babesia argentina* infections, *Aust. Vet. J.*, 44, 268, 1968.
49. **Curnow, J. A.,** *In vitro* agglutination of bovine erythrocytes infected with *Babesia argentina, Nature (London)*, 217, 267, 1968.
50. **Curnow, J. A.,** Studies on antigenic changes and strain differences in *Babesia argentina* infections, *Aust. Vet. J.*, 49, 279, 1968.
51. **Callow, L. L., Mellors, L. T., and McGregor, W.,** Reduction in virulence of *Babesia bovis* due to rapid passage in splenectomized cattle, *Int. J. Parasitol.*, 9, 333, 1979.
52. **De Vos, A. J., Bessinger, R., and Fourie, C. G.,** Virulence and heterologous strain immunity of South African and Australian *Babesia bovis* strains with reduced pathogenicity, *Onderstepoort J. Vet. Res.*, 49, 133, 1982.
53. **Perrin, L. H. and Dayal, R.,** Immunity to asexual erythrocytic stages of *Plasmodium falciparum:* role of defined antigens in the humoral response, *Immunol. Rev.*, 61, 245, 1982.
54. **James, M. A.,** An update of the isolation and characterization of culture-derived soluble antigens of *Babesia bovis, Vet. Parasitol.*, 14, 231, 1984.
55. **Cox, H. W., Milar, R., and Patterson, S.,** Serologic cross-reactions of serum antigens associated with acute *Plasmodium* and *Babesia* infections, *Am. J. Trop. Med. Hyg.*, 17, 13, 1968.
56. **Cox, F. E. G. and Turner, S. A.,** Antigenic relationships between the malaria parasites and piroplasms of mice as determined by the fluorescent-antibody technique, *Bull. W.H.O.*, 43, 337, 1970.
57. **Ludford, C. G., Hall, W. T. K., Sulzer, A. J., and Wilson, M.,** *Babesia argentina, Plasmodium vivax*, and *P. falciparum:* antigenic cross-reactions, *Exp. Parasitol.*, 32, 317, 1972.
58. **Zivkovic, D., Speksnijder, J. E., Kuil, H., and Seinen, W.,** Immunity to *Babesia* in mice. II. Cross protection between various *Babesia* and *Plasmodium* species and its relevance to the nature of *Babesia* immunity, *Vet. Immunol. Immunopathol.*, 5, 359, 1984.
59. **James, M. A., Montenegro-James S., Fajfar-Whetstone, C., Montealegre, F., Erickson, J., and Ristic, M.,** Antigenic relationship between *Plasmodium falciparum* and *Babesia bovis:* reactivity with antibodies to culture-derived soluble exoantigens, *J. Protozool.*, 34, 328, 1987.

Chapter 8

EXOANTIGENS OF *BABESIA*

Miodrag Ristic and Ibulaimu Kakoma

TABLE OF CONTENTS

I. INTRODUCTION

Use of culture and blood-derived merozoites as immunogens for the development of vaccines for malaria and babesiosis has shown promise.[1] However, even if large quantities of parasites can be produced by in vitro cultivation, it would be difficult, if not impossible, to produce quantities of erythrocyte-free merozoites suitable for industrial use. This difficulty was overcome when it was established that cultures of various *Babesia* species generate both intraerythrocytic organisms and masses of extraerythrocytic soluble antigens termed exoantigens.[2,3] These antigens are defined as soluble proteinaceous moieties naturally released in the blood plasma of animals infected with plasmodia or babesias and in the supernatant medium of in vitro cultures of these organisms. Cattle vaccinated with exoantigens of *Babesia bovis*[4,5] and *B. bigemina*[6] and dogs vaccinated with *B. canis*[7] exoantigens produced strong immunologic responses and were clinically protected against tick and needle challenge. Consequently, an immunoprophylactic system which uses an organism-free immunogen for immunization of animals against erythrocytic protozoa has been developed.

Prior to the development of cell culture systems, the only source of exoantigens was blood serum of animals acutely infected with various *Babesia* species.[8] It was in the course of studies of serum-derived exoantigens that their immunogenic potential was first recognized. Protective serum exoantigens were demonstrated in dogs infected with *B. canis*, horses infected with a mixture of *B. equi* and *B. caballi*, and rats infected with *B. rodhaini*.[9] Each of these antigens reacted with sera of recovered animals. Antigens were detected in the blood plasma of infected animals soon after the patent period, persisted throughout the acute phase, and disappeared with the development of latency.

II. GENESIS AND CHARACTERISTICS OF EXOANTIGENS

A. Blood Plasma Exoantigens

Several experiments have been conducted to ascertain the origin of the serum antigens. Fluorescent antibody studies indicated that these antigens are associated with the infected erythrocyte.[9,10] Since the maximal concentration of these antigens was observed at the peak of parasitemia, at which time the erythrocytes become increasingly fragile and hemolysis usually occurs, it is reasonable to assume that the origin of the serum antigens is the infected erythrocyte. It is suggested that exoantigens are metabolites intimately associated with the developmental cycle of the *Babesia* parasites.

Isolation studies using sucrose density gradient centrifugation, paper electrophoresis, and analytical ultracentrifugation revealed that each antigen was composed of two fractions.[8] Fraction A was located in a zone of 25% sucrose; it migrated electrophoretically in the gamma region and had a sedimentation coefficient of approximately 8 S. Fraction B, found in the 40% sucrose layer, migrated electrophoretically in the beta globulin region and had a sedimentation coefficient of 20 to 23 S. All the fractions reacted in a gel precipitation test with the respective specific sera of horses, dogs, and rats that had recovered from acute babesiosis.

Biophysical properties such as absorption at 280 nm, thermolability, precipitation by ammonium or sodium sulfate by low pH buffers, and osmotic liability indicated that the antigens were of proteinaceous nature. Amino acid analysis demonstrated that the composition of the antigens was distinct from that of normal serum proteins and most other common proteins since proline was absent. Methionine, the amino acid known to be essential for the metabolism of malaria parasites, was absent in the A fractions, and histidine was absent in the B fractions.[8]

Immunogenic studies revealed that the level of antibody directed against the serum antigens was inversely related to parasitemia in infected animals, indicating that the occurrence of antibody in the serum facilitated clearance of the parasite from the blood, and vice versa. These findings have been further substantiated by various experiments described herein.

The immunogenic activity of serum exoantigens was tested extensively in dogs and rats.[9] When dogs were immunized with *B. canis* exoantigens, they were clinically protected against challenge for at least 9 months. The exact mechanism by which the antibodies to exoantigens were able to induce protective immunity was not clear. It appears, however, that the inoculation of serum antigens induced production of antibodies that interfered with the completion of the developmental cycle of the parasite. Inoculation of exoantigens into immunologically competent, *Babesia*-free animals was shown to induce transitory anemia in the animals.[11] Further studies revealed that the inoculated antigens rapidly attach to the surface of erythrocytes. These antigen-sensitized erythrocytes were rapidly removed by the reticuloendothelial system from peripheral blood circulation. From these studies, it was concluded that the anemia associated with acute babesiosis was at least in part mediated by beta-globulin-associated antigen (antigen B) which is released into serum, particularly at peak parasitemia.

B. Cell Culture-Derived Exoantigens

The highlight of a series of important findings that followed development of the microaerophilous stationary phase (MASP) method for continuous in vitro propagation of *B. bovis*[3] is the abundant quantity of soluble exoantigens in the culture medium. These antigens, combined with appropriate adjuvants, have been extensively used as immunogens for the preventioin of clinical babesiosis caused by *B. bovis*.[1,4,5] More recently, similar antigens have been generated in short-term cultures of *B. bigemina*[6] and *B. canis*[7] and shown to be equally immunogenic in bovine and canine hosts, respectively. The principle of *B. canis* exoantigens was effectively utilized in the development of a commercial vaccine trademarked "Pirodog®" for canine babesiosis, as described elsewhere in this book.

Based upon results of various immunochemical studies, it is evident that many of the exoantigens originate from the merozoite surface coat. These antigens are shed into the culture medium at the time of parasite entry into the host erythrocytes, although a portion of the antigen may be liberated at the time of merozoite exit from these cells. Similar to plasma exoantigens, cell culture-derived antigens have strong affinity for the surfaces of cultured erythrocytes.

Exoantigens have been isolated and generally characterized as being proteinaceous moieties with a molecular wight of 37 to 40 kdaltons, are degraded by papain and trypsin, are stable at 60°C for 30 min, have fast electrophoretic mobility, and an isoelectric point (pi) 5.0 to 5.5. At least three antigens have been clearly identified and their location with reference to the infected erythrocyte determined.[10] Antigen 1 has an electrophoretic mobility in the albumin range and is bound to the erythrocyte membrane. Antigen 2 was found in the cytoplasm of infected erythrocytes, and antigen 3 was shown to be directly associated with the parasite. Electrophoretically, the latter two antigens migrated to the alpha-1 region. The methods used for detection, isolation, and characterization of culture-derived exoantigens are presented below.

1. Detection and Quantitation Methods

Methods for qualitative demonstration of plasma soluble exoantigens in various hemotropic diseases of man and animals have existed for more than 20 years. Ristic and associates reported the presence of exoantigens in the plasma of horses suffering from babesiosis,[12] in birds infected with *Plasmodium gallinaceum*,[13] and in dogs infected with *B. canis*.[8] Apparently related antigens were also reported by Wilson et al.[14] in human malaria. Similarly, Molinar et al.[7] demonstrated exoantigens in plasma of dogs infected with *B. canis* and in short-term cultures of this organism. Until recently, investigators relied on somewhat insensitive techniques such as double diffusion in gel for the detection of exoantigens. More sensitive techniques are now available, particularly following the establishment of the MASP technique[3] for continuous cultivation of *B. bovis*. To that effect, Montenegro[15] developed the latex agglutination test (LAT) and the protein A bacterial adsorbent coagglutination test for detection of both exoantigens and their antibodies. These tests were highly specific, inexpensive, and simple to use, particularly under field conditions. However, the level of sensitivity was low, requiring concentration of test material. The coagglutination test, for example, was only applicable after two times the concentration of *B. bovis* crude culture supernatants. The sensitivity of the LAT was not significantly higher than that of the coagglutination test. Another analytical, qualitative tool was provided by the crossed immunoelectrophoresis (CIE), which was successfully used by James et al.[16] to demonstrate three major exoantigens of *B. bovis*. This technique had the advantage of assaying for host-derived contaminant proteins and of recognizing unique parasite-specific antigen peaks. The latter peaks could be isolated and used in immunization studies for the production of monospecific antibodies.

With the advent of enzyme immunosorbent assays (EIA), highly sensitive, innovative methods have been developed by Montealegre et al.[17] and Wanduragala,[6] utilizing the two-site enzyme immunoassay for detection of exoantigens. These investigators used purified polyclonal bovine IgG antibodies derived from cattle vaccinated with exoantigens and subsequently challenged with virulent organisms. The purified IgG was used as a capture and recognizing antibody in EIA carried out on crude supernatants of in vitro cultures of *B. bovis*[17] and *B. bigemina*.[6] The technique was shown to be highly specific, 20 times more sensitive than conventional EIAs,[18] and could detect subtle differences between various *B. bovis* strains and cross reacting antigenic determinants of *B. bovis* and *B. bigemina*. The specificity of the EIA was validated by unlabeled antibody-mediated inhibition of the reaction between captured exoantigen and enzyme-labeled recognizing antibody.[18]

Similar studies were carried out by Wanduragala[6] using *B. bigemina* exoantigens. The latter studies were also extended to include measurement of soluble exoantigens in plasma of acutely infected animals, thus showing that EIA has a diagnostic potential based on detection of parasite-specific circulating exoantigens. In the course of these studies, it was clearly demonstrated that the concentration of exoantigens in the plasma of infected splenectomized cattle was consistently higher than that observed in control animals with the spleen *in situ*.[6] These data emphasized the importance of the spleen in hemotropic infections.

2. *Isolation and Purification*

A major constraint in the isolation and purification of *Babesia* exoantigens is the high concentration (40%) of bovine serum contained in the MASP culture medium.[3] Consequently, experiments were conducted to adapt the MASP system to utilization of rabbit serum substituting for bovine serum. The rabbit serum system, though giving fewer host-serum contaminants, yielded very low percent parasitized erythrocytes (PPE) and relatively low levels of exoantigens as compared to the standard MASP system. However, the rabbit serum system enabled James et al.[16] to raise rabbit polyvalent anti-*B. bovis* antisera, which were useful in analysis of antigens derived from the conventional MASP system. These workers reported that only 8 to 9 bovine serum proteins were detected in the rabbit culture system as compared to 40 proteins in normal bovine serum used in the standard MASP culture.

Prior to the development of the MASP system, Gravely et al.,[19] using differential centrifugation and ammonium sulfate precipitation, succeeded in partially purifying *B. bovis* exoantigens. Later, G-200 molecular sieve chromatography was applied by Molinar et al.[7] in the purification of *B. canis* exoantigens. Atyabi[20] was able to precipitate the major antigens of *B. canis* with 40, 50, 55, 60, and 65% saturated ammonium sulfate. All the antigens were completely precipitated in the 65% ammonium sulfate. Using the latter concentration, four antigen-antibody systems were demonstrated. A further refinement involved the use of conventional anion-exchange chromatography for the isolation of *B. bovis* exoantigens by Montenegro.[15] Parasite-specific exoantigens were first isolated in the 60 to 70% ammonium sulfate fractions. The remaining precipitate was then processed in a stepwise gradient (0.1 to 0.4 *M* Tris-HCl buffer pH 6.5) over diethylaminoethyl (DEAE) cellulose. In spite of the rigorous procedures applied in the above methods, host protein contaminants continued to cause problems. Hence, Montenegro[15] resorted to affinity purification methods on cyanogen-bromide-activated Sepharose® for analytical purification of soluble *B. bovis* antigens from crude supernatant fluids. Optimal yields were obtained with protein A-Sepharose® adsorbents which purified as much as 1.3 mg/mℓ or 2.85% of the original crude supernatant fluid.

The development of the two-site EIA for *B. bovis*[17,18] and *B. bigemina*[6] in our laboratory has facilitated more accurate monitoring procedures during the purification of exoantigens. Thus, using a combination of pH gradient and the high-performance liquid chromatography (HPLC) size exclusion technique, Montealegre[17] was able to isolate a relatively pure *B. bovis* fraction. By western immunoblotting using bovine anti-*B. bovis* IgG, the latter fraction was shown to contain three unique bands of relative molecular weight between 40 and 50 kdaltons.

Exoantigens from *B. bigemina* have also been recently purified by cationic exchange chromatography.[6] The antigenic activity was exclusively found in the first peak. By Sephadex® G-200 molecular sieve chromatography, it was demonstrated that the bulk of the *B. bigemina* antigenic activity was contained in the third peak with a relative molecular weight of 63 kdaltons. In similar fractionation procedures, Molinar et al.[7] found that culture-derived exoantigens of *B. canis* eluted in the void volume of Sephadex® G-200, whereas corresponding antigens from acute serum of a dog infected with *B. canis* coeluted with albumin. The significance of the affinity of exoantigens for albumin in modulating the course of hemotropic diseases will be discussed in a later section.

3. *Physical and Immunochemical Properties*

In order for a given moeity to be considered an exoantigen, it must be a product freely released from infected erythrocytes into plasma in vivo or into the supernatant medium under in vitro conditions. These characteristics distinguish exoantigens from lysate antigens and artificially extracted (solubilized) from parasitized erythrocytes. As a group, the exoantigens are highly heterogeneous with respect to composition and stability.

These antigens are relatively stable at 37 to 65°C, but are heat labile at 100°C for 15 min, thus making them different from analogous heat stable exoantigens observed in plasma of human beings suffering from malaria.[14]

In general, exoantigens of *Babesia* are proteinaceous in nature as determined by sensitivity to proteolytic enzymes trypsin and papain.[16] The polysaccharide or carbohydrate composition has not been definitely determined. However, Atyabi[20] reported that *B. canis* exoantigens were degradable by alpha-amylase. Tests involving sensitivity to 2-mercaptoethanol, which is targeted to disulfide bonds, demonstrated that the major antigenic activity depended on the integrity of these bonds and that 2-ME treatment completely eliminated the antigenicity.[7]

A unique property of exoantigens is their affinity for various serum proteins, particularly fibrinogen and immunoglobulins.[17] These features have potential effects on the protective and immunopathological role of exoantigens. Of special interest is the apparent heterogeneity of exoantigens which have been designated 1, 2, and 3 by James et al.[16] Antigen 1 was associated with the erythrocyte membrane, antigen 2 was intracytoplasmic, and antigen 3 was parasite specific. The parasite-specific Ag appeared to be immunodominant as revealed by its ability to interact with specific anti-*B. bovis* antibodies. The individual relevance of the three exoantigens in protective immunity has not been established.

III. ROLE OF EXOANTIGENS IN HOST-PARASITE INTERACTION

A. Immunogenic Properties

Since the development of the MASP culture system for *B. bovis*,[3] the exoantigens present in the supernatant medium have been subjected to various in vitro and in vivo studies for the purpose of investigating their immunogenic properties. These studies revealed that exoantigens interact with humoral and cell-mediated immune responses generated by *Babesia*-infected animals, and that inoculation of *Babesia* exoantigens in combination with suitable adjuvants into *Babesia*-free (susceptible animals) induces both humoral and cell-mediated immune responses. This combined immune effector mechanism is the basis of protective immunity following vaccination with exoantigens.

1. Humoral Immunity

The MASP culture system for *B. bovis* that ingeniously utilizes an optimal concentration of O_2 has also revealed that CO_2 is an essential factor required for invasion and entry of merozoites into erythrocytes. Consequently, withholding the CO_2 supply from *B. bovis* cultures allows merozoites to exit from erythrocytes, but interferes with their entry into new erythrocytes. Under these conditions, free merozoites accumulate in cell culture medium. When maintained at 4°C these merozoites retain their ability to invade and grow in erythrocytes for 16 to 20 hr. The availability of such invasive cell-free *B. bovis* merozoites has provided a basis for development of the merozoite serum neutralization (MN) test.[21] When cell-free merozoites were preincubated with sera of recovered or immunized animals and then introduced into the MASP culture system, they failed to invade erythrocytes. In contrast, similar merozoites incubated in sera of susceptible, nonimmunized animals remained fully functional as judged by their ability to invade and multiply in erythrocytes of MASP cultures. The MN test has provided the first in vitro system for detection and measurement of the biological activity of antibody and other humoral factors against epicomplexa. Using the MN test, we have been able to generate valuable information regarding the protective mechanism of antibodies produced in response to vaccination with the soluble cell culture-derived *B. bovis* antigen.

Antibodies to soluble antigen have been shown to block, slow down, or completely inhibit penetration of host erythrocytes by merozoites. When fresh, uninactivated antiserum was utilized, electron microscopic examination demonstrated that such antibodies actually exert a complement-dependent lytic effect on the merozoites.[22] There is also preliminary evidence suggesting an opsonizing effect of antibodies generated in response to inoculation of soluble *B. bovis* antigens into susceptible animals. The interaction of these antibodies with the merozoites is manifested by enhanced phagocytic uptake of these organisms.[23]

It has also been demonstrated that the *B. bovis* surface coat antigens react in vitro with antibodies of cattle convalescing from an active infection of cattle immunized with soluble MASP culture-derived antigens. For this purpose, a batch of culture-derived, cell-free *B. bovis* merozoites was divided into three equal portions. Fresh blood serum from a *Babesia*-free cow was added to one portion of merozoites, and a fresh serum from an animal recovered from *B. bovis* infection and an immunized calf was added to the second and third portion of merozoites, respectively. All the three systems were incubated at 37°C for 30 min, then fixed and processed for electron microscopic examination. Merozoites treated with normal bovine serum were morphologically normal with the surface coat intact. In contrast, merozoites treated with sera of recovered or immunized animals showed strong swelling and disruption of the surface coat.[1]

When cattle immunized with the soluble antigen are challenged with virulent *B. bovis* at variable time sequences (ranging from 2 to 6 months) after inoculation, they were consistently capable of mounting a strong anamnestic immune response.[1,4,5] This response was characterized by the development of high antibody titers within 10 to 14 days following challenge. This condition indicates that initial inoculation of the antigen-induced immunologic stimulus can be recalled rapidly for at least 6 months after primary vaccination.

2. *Proposed Mechanism of Protection by Antibodies to Exoantigens*

Following an experimental infection of a susceptible animal with *Babesia*-infected blood, two interdependent events are being initiated, namely, the growth and development of the organism in the erythrocytes and concomitant elaboration and release of soluble antigens into the plasma of the infected animals. The quantity of the exoantigens seems to be directly proportional to the level of parasitemia. In response to stimulation induced by these antigens, the animal begins to mount an immune response. At the humoral level of immunity, constantly rising soluble antigens efficiently neutralize the effectiveness of protective antibodies by complex formation. These antigen-antibody complexes appear to benefit the parasite by serving as a protective shield associated with immunopathologic changes described later in this chapter.

In an animal immunized with soluble antigens, the immunopathologic events are similar to those observed in the nonimmunized animal. However, because of the excess of the antiexoantigen antibodies, the shielding effect of the parasite by complexes is greatly minimized, thus permitting residual antibodies to exert protective antiparasite effect. Much of the plasma soluble antigens originate from the surface coat layer of merozoites whose existence has been clearly documented in babesias. Antibodies directed to this surface antigen prevent the attachment to and penetration of erythrocytes by invading merozoites by means of direct coating of the surface of these organisms and by exerting opsonizing and hemolytic antiparasite effects in association with phagocytic and complement systems, respectively. These antiparasite effects of antibodies are evidenced by the appearance of deformed, apparently nonfunctional parasites in peripheral circulation of immunized animals. The humoral protective effects appear to allow the animal to mount a vigorous hematopoietic response and maintain a relatively stable hematocrit level. Thus, the presence of antibodies in gross excess relative to soluble antigen protects the animal against pathohematologic effects of the critical phase of the disease and allows gradual recovery uncomplicated by major clinical manifestations.

3. *Cell-Mediated Immunity (CMI)*

Very little information is available on cell-mediated immunity (CMI) in bovine babesiosis. Most recently, however, in vitro studies in our laboratory have demonstrated that bovine blood macrophages would respond in a secretory and phagocytic manner when stimulated by culture-derived soluble *B. bovis* antigens.[17] Briefly, bovine blood monocytes from apparently normal cattle were first cultured in vitro. Cell monolayers were then treated with *B. bovis* culture-derived soluble antigen. The monolayers were washed free of the antigen and allowed to develop further in the presence of the standard MASP culture medium. Addition of this culture medium to *B. bovis* cultures revealed presence of a growth-inhibiting factor. It remains to be determined whether this *Babesia* growth-inhibiting factor directly interacts with invading merozoites or simply blocks erythrocyte receptor sites, thus preventing the attachment and penetration by the merozoites. Further in vitro studies revealed that immune cattle serum contains an antibody capable of enhancing phagocytosis of *B. bovis* by cultured blood mononuclear cells. It is proposed that antibodies to culture-derived exoantigen may act as cytophilic antibodies, thereby enhancing phagocytosis and that the antigen itself stimulates macrophages to secrete an as yet uncharacterized *B. bovis* growth inhibitory factor.

The evidence of induction of CMI responses in cattle immunized with *B. bovis* exoantigens was also produced. Timms et al.[24] demonstrated strong lymphocyte blast-transformation response, which persisted for more than 6 months after vaccination. The role of CMI in the protection against and pathogenesis of babesiosis warrants further investigation.

B. Immunopathological Manifestations

The pathophysiology of babesial infections has been reviewed extensively, and the evidence that exoantigens play a role in these processes is unequivocal.[25-28] Further, circulating soluble antigen and antigen-antibody complexes have been described in various babesial infections and shown to contribute to the pathogenesis of the disease. Fortuitous adherence of circulating exoantigens onto normal erythrocytes renders the latter vulnerable to indiscriminate reticuloendothelial removal,[11,27] and in rodent babesiosis circulating soluble antigen, immune complexes and complement are sequestered in the kidneys and are associated with glomerulonephritis.[26] Considering that exoantigens are highly anticomplementary in vitro,[29] it is not surprising that complement levels are depleted in infected animals,[30] thus exacerbating the shock syndrome.[31,32] In addition to uncomplexed circulating antigen and immune complexes, phospholipids from erythrocyte lysates and damaged tissues lead to reticuloendothelial blockade, causing enormous stress to infected animals.

C. Evasion of Immunologic Recognition by the Host

The recovery from the acute phase of babesiosis is followed by a carrier state. Affected animals, however, develop humoral and cell-mediated immune responses as a principal defense mechanism. While induction of an immune response is essential for the survival of the host, it also ironically provides for continuous survival and transmission of the parasite. This state of biological host-parasite balance has been described as a ''tolerant symbiosis''[33] which allows the parasite to persist and, in exchange, the host develops lasting protection against homologous virulent organisms. There are different escape mechanisms that the parasite uses to avoid the destructive effects of the immunologically hostile host. We will briefly discuss possible means by which *Babesia* parasite evade immunologic destruction.

Babesial exoantigens have been shown to attach to parasitized and nonparasitized erythrocytes. The resulting complex on the erythrocyte surface may serve as an additional mode of evasion of parasite destruction by humoral and other mechanisms.

Research efforts directed toward isolation and purification of exoantigen from babesial and plasmodial cultures which are supplemented by normal bovine or human sera, respectively, have shown that some of these exoantigens are difficult to separate from normal serum albumin. The interaction between *B. bovis*, *P. falciparum*, and possibly *B. bigemina* exoantigens with blood plasma albumin is probably beneficial to the parasite and detrimental to the host. The masking effect of certain important parasite antigens by albumin is of immediate benefit to the parasite through evasion of recognition of the host immune response (Figure 1). The consequence of this phenomenon is likely to be a delayed recognition of protective parasite antigens by the immune system of the host. It is not surprising, therefore, that repeated episodes of infection are required for the development of protective immunity in babesiosis and malaria.

Babesia bovis seems to be able to avoid splenic filtration by sequestration of infected erythrocytes in the small capillaries of the brain and other organs.[34] Thus, in infections with *B. bovis*, unlike those caused by *B. bigemina*, there is a poor correlation between parasitemia, anemia, and other clinical manifestations. In the acute form of babesiosis caused by *B. bovis*, parasitemia may be 1 to 2%, while the parasite rate in blood of brain capillaries may exceed 90%. An earlier belief maintained that the coating of erythrocytes by soluble parasite antigens neutralized the normal surface charge of these cells, thus favoring autoagglutination of erythrocytes in the capillaries. Recent studies produced convincing evidence on the mechanisms by which *B. bovis* escapes splenic filtration by sequestering in the blood capillaries.[34] These studies revealed that, similar to *P. falciparum*,[35] *B. bovis* possesses the

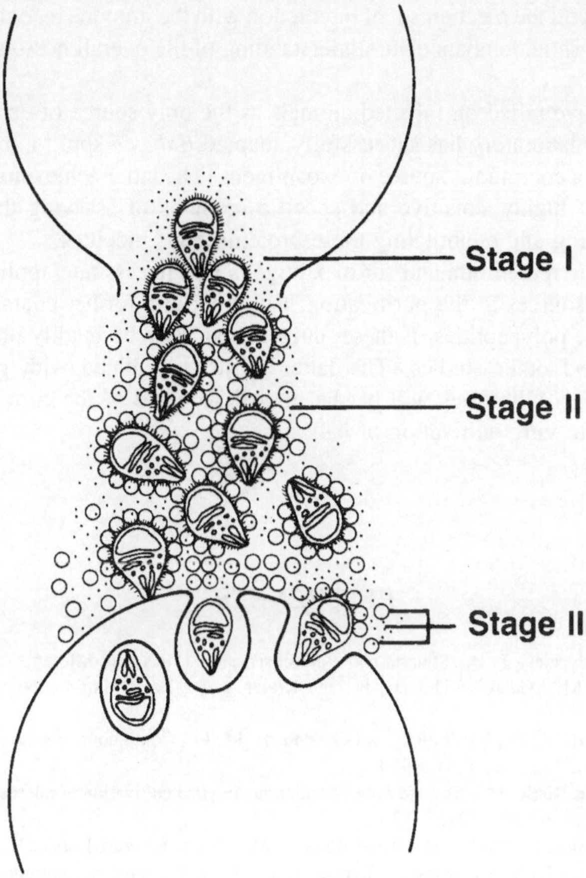

FIGURE 1. Schematic representation of the hypothesis for the mechanisms whereby hemoparasites (e.g., *Babesia, Plasmodia*) utilize their affinity for the albumin and other proteins of the host to evade or delay specific antiparasite immune recognition and elimination.

ability to induce changes in the membrane of the host erythrocytes.[36] These changes, appearing in the form of electron-dense spikes, seem to be specific antigens, which find corresponding receptor sites on the endothelial cells of blood capillaries facilitating adhesion of the altered erythrocytes to the endothelium. The resulting sequestration of erythrocytes in the narrow capillary passages impairs blood circulation in the vital organs such as the brain, bringing about anoxia and various clinicopathologic manifestations.

Preliminary studies using antibodies to exoantigens in the IFA test revealed specific staining of interacting tissue component between erythrocytic projections and endothelial surface. Further studies of this early finding may elucidate the role of exoantigens in the sequestration phenomenon of infected erythrocytes and their association with morphologically and antigenically altered cell membrane.

IV. CONCLUSIONS AND FUTURE PERSPECTIVES

Over the last 20 years, massive data have accumulated on the role of exoantigens in various infections caused by *Babesia* species. Their usefulness as immunogens and serodiagnostic reagents makes them valuable products. Exoantigens are also very intimately associated with the pathogenesis of babesial infections, thus a better understanding of their biologic properties and the mechanism of interaction with the immune recognition and defense systems of the host should enhance our understanding of the overall host-parasite interaction in babesiosis.

In the past we have relied on infected animals as the only source of exoantigens. During the last 6 years our laboratory has successfully adapted *Babesia* spp. to grow under in vitro systems, providing a continuous source of exoantigens. The latter achievement has facilitated the development of highly sensitive and specific methods of assaying these antigens and subsequently isolating and quantitating these proteinaceous moeities.

Ultimately, the increased demand for exoantigens will necessitate application of modern biotechnological resources in the purification, isolation, and further characterization of exoantigens. Purified, polypeptides of these antigens can now be readily obtained for amino acid sequencing and other studies. The latter studies combined with gene cloning and establishment of cDNA libraries will be the logical sequelae of the currently standardized techniques for the in vitro cultivation of babesias.

REFERENCES

1. **Ristic, M. and Kreier, J. P.,** Malaria and Babesiosis: similarities and differences, in *Malaria and Babesiosis,* Ristic, M., Ambroise-Thomas, P., and Kreier, J. P., Eds., Martinus Nijhoff Dordrecht, Netherlands, 1984, 3.
2. **Erp, E. E., Smith, R. D., Ristic, M., and Osorno, B. M. O.,** Continuous *in vitro* cultivation of *Babesia bovis, Am. J. Vet. Res.,* 41, 1141, 1980.
3. **Levy, M. G. and Ristic, M.,** *Babesia bovis*: continuous *in vitro* cultivation in microaerophilous stationary phase culture, *Science,* 207, 1218, 1980.
4. **Montenegro-James, S., Ristic, M., Toro Benitez, M., Leon, E., and Lopez, R.,** Heterologous strain immunity in bovine babesiosis using a culture-derived soluble *Babesia bovis* immunogen, *Vet. Parasitol.,* 18, 321, 1985.
5. **Smith, R. D., James, M. A., Ristic, M., Aikawa, M., and Vega, C.,** Bovine babesiosis: protection of cattle with culture-deprived soluble *Babesia bovis* antigen, *Science,* 212, 335, 1981.
6. **Wanduragala, L.,** Identification, Quantitation and Immunogenicity of Culture-Derived Soluble *Babesia bigemina* Antigens, Ph.D. thesis, University of Illinois, Urbana, 1986, in preparation.
7. **Molinar, E., James, M. A., Kakoma, I., Holland, C. J., and Ristic, M.,** Antigenic and immunogenic studies on cell culture-derived *Babesia canis, Vet. Parasitol.,* 10, 29, 1982.
8. **Sibinovic, K. H., MacLeod, R., Ristic, M., Sibinovic, S., and Cox, H. W.,** A study of some of the physical, chemical and serologic properties of antigens from sera of horses, dogs, and rats with acute babesiosis, *J. Parasitol.,* 53, 919, 1967.
9. **Sibinovic, K. H., Sibinovic, S., Ristic, M., and Cox, H. W.,** Immunogenic properties of babesial serum antigens, *J. Parasitol.,* 53, 1121, 1967.
10. **Montenegro-James, S., James, M. A., and Ristic, M.,** Localization of culture-derived soluble *Babesia bovis* antigens in the infected erythrocyte, *Vet. Parasitol.,* 13, 311, 1983.
11. **Sibinovic, K. H., Milar, R., Ristic, M., and Cox, H. W.,** *In vivo* and *in vitro* effects of serum antigens of babesial infection and their antibodies on parasitized and normal erythrocytes, *Ann. Trop. Med. Parasitol.,* 63, 327, 1969.
12. **Sibinovic, K. H., Ristic, M., Sibinovic, S., and Phillips, T. N.,** Equine babesiosis: isolation and serologic characterization of a blood serum antigen from acutely infected horses, *Am. J. Vet. Res.,* 26, 147, 1965.
13. **Smith, A. R., Lykins, D., Voss, E. W., and Ristic, M.,** Identification of an antigen and specific antibody in the sera of chickens infected with *Plasmodium gallinaceum, J. Immunol.,* 103, 6, 1969.
14. **Wilson, R. J. M., McGregor, I. A., and Williams, K.,** Occurrence of S-antigens in serum in *Plasmodium falciparum* infections in man, *Trans. R. Soc. Trop. Med. Hyg.,* 69, 453, 1975.
15. **Montenegro, S.,** Studies of Culture-Derived Soluble *Babesia bovis* Antigens: Purification, Characterization and Application in Serodiagnosis, Ph.D. thesis, University of Illinois, Urbana, 1982.
16. **James, M. A., Levy, M. G., and Ristic, M.,** Isolation and partial characterization of culture-derived soluble *Babesia bovis* antigens, *Infect. Immun.,* 31, 358, 1981.

17. **Montealegre, F. H., Montenegro-James, S., Kakoma, I., and Ristic, M.**, Detection of culture-derived *Babesia bovis* exoantigens using a two-site enzyme immunoassay, 1986, in preparation.
18. **Montealegre, F. H.**, Babesiosis: Detection and Characterization of Culture-Derived Exoantigens, Ph.D. thesis, University of Illinois, Urbana, 1986.
19. **Gravely, S. M., Smith, R. D., Erp, E. E., Canto, G. J., Aikawa, M., Osorno, B. M., and Ristic, M.**, Bovine babesiosis: partial purification and characterization of blood culture-derived *Babesia bovis, Int. J. Parasitol.*, 9, 591, 1979.
20. **Atyabi, N.**, Characterization of Culture-Derived Soluble Antigens of *Babesia canis*, M.S. thesis, University of Illinois, Urbana, 1979.
21. **Ristic, M. and Levy, M. G.**, unpublished observations, 1984.
22. **Ristic, M.**, Babesiosis, in *Diseases of Cattle in the Tropics*, Ristic, M. and McIntyre, I., Eds., Martinus Nijhoff, Dordrecht, Netherlands, 1981, 443.
23. **Montealegre, F., Levy, M. G., Ristic, M., and James, M. A.**, Growth inhibition of *Babesia bovis* by secretions from bovine mononuclear phagocytes, *Infect. Immun.*, 50, 523, 1985.
24. **Timms, P., Dalgliesh, R. J., Barry, D. N., Dimmock, C. K., and Rodwell, B. J.**, *Babesia bovis*: comparison of culture-derived parasites, nonliving antigen and conventional vaccine in the protection of cattle against heterologous challenge, *Aust. Vet. J.*, 60, 75, 1983.
25. **Mahoney, D. F.**, Circulating antigens in cattle infected with *Babesia bigemina* or *B. argentina*, *Nature (London)*, 211, 422, 1966.
26. **Annable, C. R. and Ward, P. A.**, Immunopathology of the renal complications of babesiosis, *J. Immunol.*, 112, 1, 1974.
27. **Goodger, B. V., Wright, I. G., Mahoney, D. F., and McKenna, R. V.**, *Babesia bovis (argentina)*: studies on the composition and location of antigen associated with infected erythrocytes, *Int. J. Parasitol.*, 10, 33, 1980.
28. **Kakoma, I. and Ristic, M.**, Pathogenesis of Babesiosis, in *Malaria and Babesiosis*, Ristic, M., Ambroise-Thomas, P., and Kreier, J. P., Eds., Martinus Nijhoff, Dordrecht, Netherlands, 1984, 85.
29. **Levy, M. G., Kakoma, I., Clabaugh, G. W., and Ristic, M.**, Complement does not facilitate *in vitro* invasion of bovine erythrocytes by *Babesia bovis*, *Ann. Trop. Med. Parasitol.*, 80, 377, 1986.
30. **Goodger, B. V., Wright, I. G., and Mahoney, D. F.**, Changes in conglutinin, immunoconglutinin, complement C3 and fibrinectin concentrations in cattle acutely infected with *Babesia bovis*, *Aust. J. Exp. Biol. Med. Sci.*, 59, 531, 1981.
31. **Wright, I. G.**, The kallekrein-kinin system and its role in the hypotensive shock syndrome of animals infected with the haemoprotozoan parasites *Babesia*, *Plasmodium* and *Trypanosoma*, *Gen. Pharmacol.*, 10, 319, 1979.
32. **Wright, I. G. and Kerr, J. D.**, Hypotension in acute *Babesia bigemina (argentina)* infections of splenectomized calves, *J. Comp. Pathol.*, 87, 531, 1977.
33. **Moulder, J. W.**, Comparative biology of intracellular parasitism, *Microbiol. Rev.*, 49, 298, 1985.
34. **Hildebrandt, P. K.**, The organ and vascular pathology of babesiosis, in *Babesiosis*, Ristic, M. and Kreier, J. P., Eds., Academic Press, New York, 1981, 459.
35. **Aikawa, M., Rabbege, J., Udeinya, I., and Miller, L. H.**, Electron microscopy of knobs in *P. falciparum*-infected erythrocytes, *J. Parasitol.*, 62, 435, 1983.
36. **Aikawa, M., Rabegge, J., Uni, S., Ristic, M., and Miller, L. H.**, Structural alterations of the membrane of erythrocytes infected with *Babesia bovis*, *Am. J. Trop. Med. Hyg.*, 34, 45, 1985.

Chapter 9

METHODS FOR THE MEASUREMENT OF THE SEROLOGICAL RESPONSE TO *BABESIA*

Georg Weiland and Ingrid Reiter

TABLE OF CONTENTS

I. INTRODUCTION

Most conventional immunodiagnostic methods are used for the measurement of the humoral response following natural or experimental babesiosis.

For epidemiological surveys of babesiosis, immunoserology is irreplaceable due to the difficulty of detecting the parasite in blood smears prepared from blood of chronically infected animals. A further aim of serological tests in babesiosis is to monitor therapeutic efficacy of antibabesial drugs for the treatment of acute *Babesia* infections. Furthermore, the detection of the humoral response in animals after specific vaccination is a useful tool for assessment of the immune status and efficacy of immunogens.

Each serological assay has some advantages or disadvantages depending on its level of sensitivity, specificity, simplicity, and cost effectiveness. The area of application — endemic areas or individual diagnosis — is essential for a final evaluation of the test system. At present, most workers propose the combination of at least two different assays to increase the diagnostic reliability of the serodiagnosis of *Babesia* infections.

In this review, data on established and modern techniques for the serodiagnosis of babesiosis will be discussed with special emphasis on current and recent developments.

II. LITERATURE REVIEW

A. Antigen

The extensive use of various serological assays revealed that the sensitivity and/or specificity of the tests depends on the antigen preparation used. The primary source of antigen for most of the serological assays used today is whole blood from infected animals collected at peak parasitemia. The parasite strain used for experimental infection should be as close as possible to the natural isolate. An increasing number of passages may reduce the specificity of the antigens.[1,2] On the other hand, high parasitemias are necessary for antigen production and are obtained for some *Babesia* species (e.g., *B. equi*, *B. canis*, *B. bigemina*) only in splenectomized, chemically immunosuppressed individuals or by serial passages in splenectomized animals.[3] Antigen harvested from blood with a parasitemia of less than 2% shows poor reactivity in the complement fixation test (CFT).[4] The formation of immune complexes attached to the host erythrocytic membrane can cause false positive or doubtful reactions in the immunoassay.[5] Animals should be bled during the ascending parasitemia for antigen production. Plasma factors must be removed from whole blood by several washes, and the remaining cell suspension should be depleted of leukocytes.[6] The isolation and purification of parasitized erythrocytes or red blood cells (RBC), the preparation of crude particulate, and soluble extract antigens are detailed elsewhere and summarized in Figure 1.

Briefly, infected erythrocytes are enriched by hypotonic lysis due to the fact that parasitized cells are more resistant to hypotonic lysis than normal cells.[5] A final parasitemia of 99% can routinely be achieved by this concentration method.[6] Infected blood cells can also be isolated by other concentration methods like Concanavalin A agglutination[7] or density gradient centrifugation.

Crude corpuscular antigen can be freed from concentrated, parasitized RBCs by controlled hypotonic lysis, whereas the crude soluble or insoluble extract antigen is generally prepared by means of a single or a combination of various physicochemical methods (ultrasonic disruption, deep freezing and thawing, hypotonic salt solutions, detergents, and mechanical stirring). The resulting soluble extract is harvested from the remaining membrane components by centrifugation. Recent developments in the purification and characterization of antigens have facilitated the isolation of specific babesial antigens. The purification and partial characterization of crude soluble extract antigen has been achieved by gel fractionation on

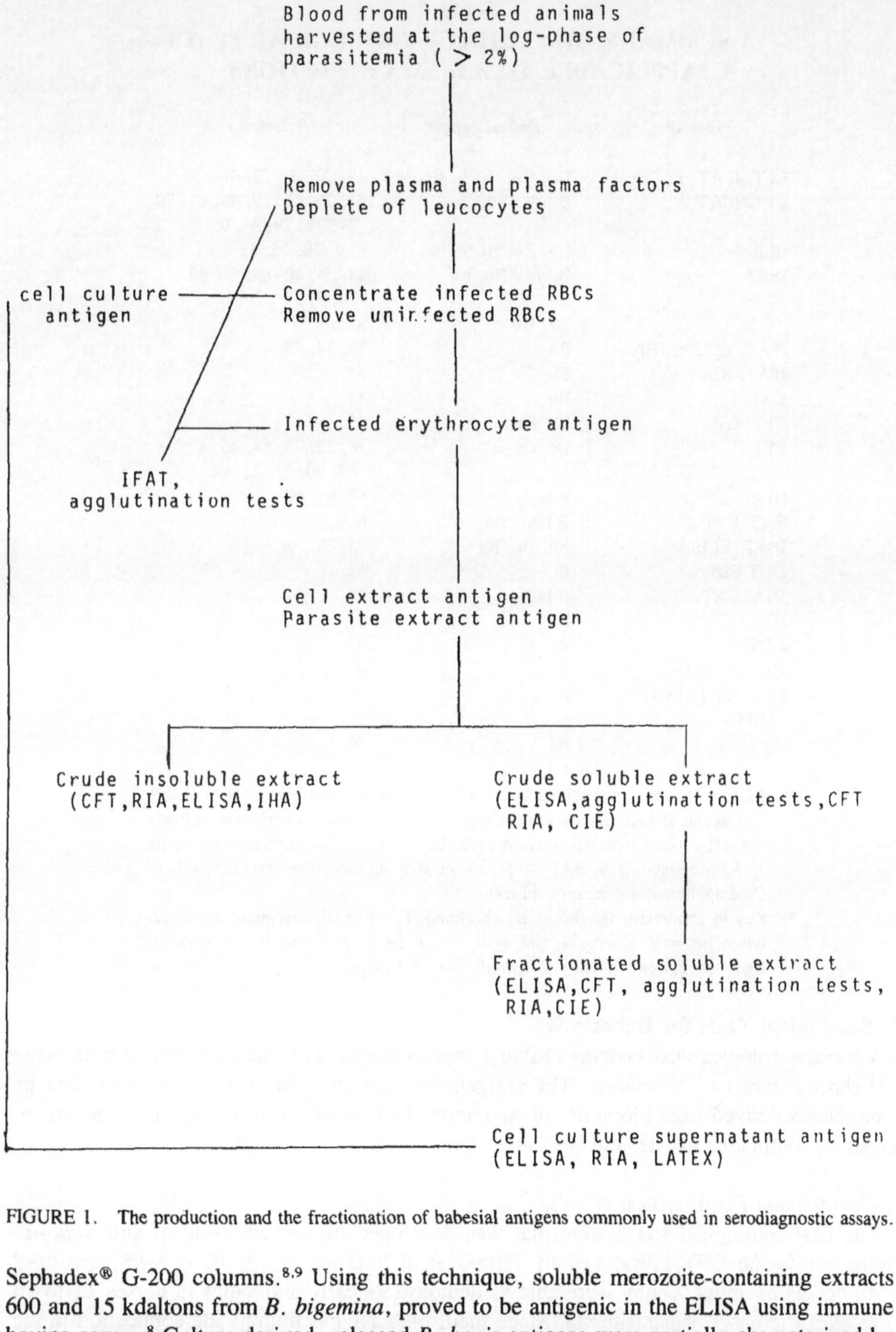

FIGURE 1. The production and the fractionation of babesial antigens commonly used in serodiagnostic assays.

Sephadex® G-200 columns.[8,9] Using this technique, soluble merozoite-containing extracts 600 and 15 kdaltons from *B. bigemina*, proved to be antigenic in the ELISA using immune bovine serum.[8] Culture-derived, released *B. bovis* antigens were partially characterized by Montenegro-James et al.[10] who described the presence of at least three soluble antigens of low molecular weight (37 to 50 kdaltons) isolated from culture supernatant. Antibodies to soluble antigens have also been described as blocking antibodies, and culture-derived soluble antigens have been successfully utilized in the serodiagnosis of *Babesia* infections in animals.[11]

Table 1
A SUMMARY OF VARIOUS SEROLOGICAL TESTS APPLICABLE TO *BABESIA* INFECTIONS

Test(s)[a]	*Babesia* spp.[b]	Selected refs.
CFT, IFAT, ELISA	Bd, Bm, BCb, Be	19, 20, 24, 32, 48
CFT, IFAT	B.big., Bb, Bo	21, 26, 28, 29, 45, 67, 70, 74, 78, 79, 94, 95
ELISA	Bd, Bb, BCb	8, 9, 50, 72, 82
IFAT	B.big., Bb, Bd	33, 34, 40, 44, 51, 68, 71, 76, 81
	Bg, Bc	93
IFAT, ELISA, IHA	Bd	36, 74, 75
IFAT, RIA	Bb	6
LAT	Bb	11
CFT, CT	Bc, Be	65
CFT	Bc, Be	18, 27, 77, 84, 86, 87, 88, 89, 90, 91, 92, 94
IHA	B.big.	57
SAT, CFT	B.big., Ba	66
IFAT, ELISA	Bd, Bc, Bg	35, 73, 83
DOT-EIA	Bc	52
IHA, BAT	B.big.	58
ID	Bc	80
GDT	Bc, Be	97
CA, GP, PHA	Be	62
CFT, IFAT, IHA, ELISA	Be	18
CAT	Be	96

[a] BAT = Bentonite agglutination test. CFT = Complement fixation test. ELISA = Enzyme-linked immunosorbent assay. IFAT = Indirect fluorescent antibody test. IHA = Indirect hemagglutination. LAT = Latex agglutination test. RIA = Radioimmunassay. SAT = Serum or slide agglutination test. CA, CT = Card agglutination, serum card test.

[b] Key to abbreviations: Ba = *B. argentina*, B.big = *B. bigemina*, Bb = *B. bovis*, Bd = *B. divergens*, Bm = *B. major*, Bo = *B. ovata*, Bc = *B. canis*, Bg = *B. gibsoni*, BCb = *B. caballi*, Be = *B. equi*.

B. Serological Tests for Babesiosis

Various serological test systems (Table 1) have been used for the diagnosis of both acute and chronic forms of babesiosis. The antigens utilized are soluble or corpuscular parasite components derived from blood of infected animals or, more recently, from parasites maintained in continuous in vitro cultivation systems.

1. Complement Fixation Test (CFT)

The first serological test system that was developed for the detection of anti-*Babesia*-antibodies is the CFT introduced by Hirato et al.[12] Using lysed, *B. caballi*-parasitized erythrocytes as antigen, they were able to demonstrate early antibodies in horses 11 to 15 days after the onset of parasitemia. Since then, the CFT has found wide application in the serodiagnosis of babesiosis and was found to be rather species specific.[13,14]

A further improvement in the CFT was introduced by Frerichs et al.,[4] who prepared the antigen by passing the hypotonic cell lysates through a refrigerated centrifuge, mixing the corpuscular fraction consisting of parasites and erythrocyte stroma with buffer and filtering the mixture, and then lyophilizing before storage. This CFT based on corpuscular antigen and designed for the detection of anti-*Babesia* antibodies in horses was further improved by

Holbrook et al.[1] Parasites were liberated by passing parasitized RBC through a French pressure cell. The authors claimed that the final antigen (with some erythrocyte stroma and hemoglobin) would change its good antigenicity when stored lyophilized. A test performed according to their protocol was generally specific and was able to differentiate between *B. caballi* and *B. equi* infections. The CFT was subsequently accepted as an official test for equine babesiosis in the U.S. and led finally to the development of the standardized piroplasmosis CFT, published by the U.S. Department of Agriculture.[15] Todorovic and Adams[16] introduced the micro-CFT system for the diagnosis of bovine babesiosis, but stated later[17] that the CFT had some disadvantages over the IFT, mainly in sensitivity and speed of performance. Mahoney[3] developed the most sensitive CFT for cattle babesiosis. The antigen used was a crude extract prepared by distilled water lysis of infected erythrocytes (parasitemia 25 to 30%) followed by centrifugation. Goetz,[18] Merkle,[19] and Aicher[20] completed the extraction (*B. equi*, *B. caballi*) after hypotonic lysis by several freezing and thawing cycles followed by ultra sonication and high-speed centrifugation at 48,000 $\times$ *g* for detection of *Babesia*-antibodies in horses. The antigen used was species specific and no cross reactions were observed between *B. caballi* and *B. equi*.

Fujinaga and Minami[21] evaluated the CFT for the serodiagnosis of experimental and natural infections of cattle with *B. ovata*. Anticomplementary activity was seen in 28 out of 184 bovine sera, 16 of which were positive by the IFAT with titers 1:40. The CFT, however, apparently failed to detect specific antibodies to *B. equi* schizont antigen in horses, but was reliable when merozoite antigen was utilized.[22] A specific increase of serum IgG and IgM antibody could be demonstrated in adult ponies experimentally infected with *B. equi*. The peak of the serum globulin did not show any correlation with the CFT titers. However, Goff et al.[23] noted secondary IgM response in sera of *B. bovis*-infected cattle, indicating that IgM could be responsible for the serological reaction during the first few days when CFT becomes positive.

Cattle vaccinated with nonviable, soluble *B. bovis* cell culture-derived antigens failed to respond serologically in the CFT, even though specific antibodies were detected with the IFAT, and none of the vaccinated animals showed severe clinical signs of the disease after challenge infection. Generally, the CFT for piroplasmosis is considered to be the most specific serologic test for diagnostic purposes; however, it is limited because of its relatively poorer sensitivity compared to the IFAT.[21,24-26] The CFT indicates persistent infections, but tends to produce a relatively high number of false negative results in chronic carriers, probably due to the short half-life of antibodies detectable by the CFT.[1,4,27-29] The test cannot be recommended for the evaluation of the immune status of vaccinated cattle nor can the successful treatment of *B. caballi* infections in horses with Berenil be serologically verified by the CFT.[29] Unfortunately, there is no data on the effect of different antigen preparations (corpuscular-soluble extract) on the sensitivity and reproducibility of the CFT.

2. *Indirect Fluorescent Antibody Test (IFAT)*

Since the original description of the IFAT by Ristic and Sibinovic[30] as a diagnostic tool for the detection of antibodies against *B. caballi* in latent infected horses, it has been extensively applied to the diagnosis of bovine, equine, canine, and even human babesiosis.[3] Based on our survey, it has been the experience of most laboratories that the IFAT is more specific and sensitive than the CFT. Antibodies detectable by the IFAT start to rise shortly after the onset of infection[19,20,22,28,32] and persist over a longer period. The sensitivity and specificity, however, may vary according to the nature of the preparation of antigens used. Until recently, most workers have carried out IFATs on fixed or unfixed blood smears consisting of a variable percentage of parasitized erythrocytes isolated from experimental or naturally infected animals. Since then, the in vitro cultivation of some *Babesia* species became possible, making it possible to employ culture-derived antigens in this test. The sensitivity of the CFT is mainly dependent on the antigenic properties of the isolated parasites and the preservation of the membrane-associated antigens on the fixed or unfixed slides. The influence of various preservation methods on the antigenicity of erythrocytes parasitized with *B. gibsoni* was tested by Tamura et al.[33] Among the fixing solutions tested, acetone gave the best results, probably because acetone removed the erythrocyte associated hemoglobin. The antigenicity of rodent *Babesia* species, however, is not improved by acetone fixation. Acetone-treated and unfixed blood smears maintained their antigenicity over a period of 12 months when stored at −20°C. Anderson et al.[34] used unfixed *B. gibsoni* parasitized RBC as antigen for the detection of canine anti-*Babesia* antibodies and considered the IFAT to be sensitive and reproducible. Heterologous serum titers (*B. gibsoni*, *B. argentina*, *B. equi*, *Plasmodium falciparum*, and *P. brasilianum*) were significantly lower than the homologous values. The use of *B. rodhaini* antigen in the serodiagnosis of canine babesiosis (*B. canis* and *B. gibsoni*) is limited because of the poor sensitivity of the heterologous antigen.[35] Indeed, Goetz[18] did not observe any positive reactions between *B. equi* antibodies and *B. rodhaini* antigen in the IFAT, whereas intraerythrocytic, heat fixed *B. rodhaini* in blood smears were sensitive enough to detect antibodies against *B. divergens* in experimentally infected calves.[36] Fujinaga and Minami[21] inoculated cattle with *B. ovata* and demonstrated IFAT titers lasting until 420 days postinfection. The rate of positivity of the IFAT for *B. ovata* in naturally infected cattle was about 60% higher than that of CFT antibodies.

Naturally infected cattle grazing in *B. ovata* endemic areas showed rises in *B. ovata* IFAT antibody titers during the grazing period, suggesting a subclinical reinfection.[37] Donnelly et al.[28] described a discrepancy between the test results obtained by means of the IFAT and the CFT in sera of horses naturally infected with *B. caballi* and/or *B. equi*. Serological tests performed with the IFAT showed no cross reaction between the two *Babesia* species and a greater sensitivity of the IFAT compared with the CFT.

It is interesting to note that differences in the antigenic properties of two *B. bovis* strains originating from different continents (one from Mozambique, Africa, and the other from Australia), seem to be negligible when tested in the IFAT.

Jones and Conrad[39] were unable to demonstrate any significant cross reactivity between *B. bigemina*, *B. divergens*, and in vitro-derived *B. bovis* antigens using immune sera from cattle experimentally infected with *B. bigemina* and *B. divergens*. Their serological survey revealed that 97% of the bovine sera tested were positive for *B. bovis* antibody in spite of a relatively low incidence of clinical cases of bovine babesiosis in that region. The serological relationship of *B. bovis* and *B. bigemina* isolated in South Africa was tested by Bessenger and Schoeman.[40] Using the IFAT, cross reactions attributable to previous infection with *B. bigemina* at titers of 1:80 to 1:160 were observed in three out of four sera from cattle 8 weeks after infection with *B. bovis*. Titers of 1:80 were regarded as positive for *B. bigemina*. False positive *B. bovis* reactions were seen only during a short postinfective period and at lower titers (1:40). Sera of noninfected control animals were negative to both antigens at a screening dilution of 1:40.

The reactivity of cattle sera from animals infected with *B. bovis* in the IFAT seems to be related to the presence of specific IgM and IgG_1 immunoglobulin, whereas the IgG_2 fraction does not show any specific seroreactivity.[23]

Sera from *B. equi*-infected, parasitologically negative horses reach titers from 1:10 to 1:80 with the homologous antigen. Cross reactions using *B. bovis* antisera against *B. equi* antigen could not be detected in the IFAT, whereas *B. equi* antisera reacted with *B. bovis* antigen with titers from 1:30 to 1:270.

A rather low degree of cross reactivity between *B. equi* and *B. caballi* at titers of 1:40 was reported by Merkle[19] and Aicher,[20] whereas Rehbein and Heidrich-Joswig[22] could not demonstrate significant reactions between *B. caballi* or *B. bigemina* antisera when tested against *B. equi* schizont or *B. equi* merozoite antigen at serum dilutions of 1:20 and above. They also showed that in ponies experimentally infected with *B. equi*, IFAT antibody titers to schizont antigen decreased more rapidly than those detected to merozoite. Sera from various animals containing antibodies against several *Babesia* species were tested for their reactivity with *B. equi* and *B. caballi* antigens in the IFAT. Antibody titers with heterologous antigens ranged from 1:16 to 1:128.

The immunization of cattle with nonviable, culture-derived *B. bovis* antigens leads to the production of antibodies detectable by the IFAT, but not by the CFT when homologous antigen is used.[41] Following immunization of cattle with different fractions derived from *B. divergens* parasitized cattle erythrocytes, Taylor et al.[42] demonstrated that the most pronounced humoral immune response measurable by the IFAT was directed to a glycoprotein-protein component of the immunogen. This data might explain the observations that humoral response measured by the IFAT is predominantly directed against the surface membrane associated parasite antigens. The IFAT provides a good test method for human babesiosis caused mostly by *B. microti* and *B. divergens*. These organisms fortunately do not cross react in the IFAT. Anti-*B. microti* antibodies, however, react with other *Babesia* species (*B. bovis*, *B. bigemina*, *B. equi*, *B. canis*) and *Plasmodium* species.[43] The latter reaction is of major importance in regions where babesiosis and malaria are coendemic. The IFAT allows the clinician to diagnose an acute or chronic *Babesia* infection and also enables him to recognize the *Babesia* species involved, particularly in mixed infections. Because of its relatively high sensitivity, IFAT can also be used to assess the prevalence of *Babesia* infections in endemic areas since parasitemia drops below the detectable level long before IFAT antibody titers begin to decline.[44] Under optimal working conditions, end point titers by one serum dilution have been reported.[45] The finding of significant IFAT antibody titers in asymptomatic cases indicates a carrier state.

3. Enzyme-Linked Immunosorbent Assay (ELISA)

Engvall and Perlmann[46] described the ELISA as a very sensitive and economic assay for the detection of antibodies. With the availability of highly purified immunological reagents, the technique has now been widely adapted to the diagnosis of parasitic infections. The suitability of the ELISA for the demonstration of antibodies against *Babesia* species was first shown by Purnell et al.[47] who used sera of *B. divergens*-infected cattle and a homologous somatic extract antigen. Later Bidwell et al.[48] found good correlation between ELISA and IFAT in cattle infected with either *B. divergens* or *B. major*.

The use of a *B. rodhaini* antigen in the serodiagnosis of experimental infections with *B. canis* and *B. gibsoni* in dogs was described by Weiland and Kratzer.[35] They found the somatic fraction of the extract antigen more reliable in the test than the metabolic one. Reif[36] studied the antibody response to *B. divergens* in experimentally infected cattle using either *B. divergens* or *B. rodhaini* somatic antigens in the ELISA. Positive reactions were demonstrated for a period of 250 days postinfection, and up to 70 days after curative treatment with Berenil.

The serodiagnosis of a *B. divergens* infection by the ELISA using blood samples dried on filter paper was described by Young and Purnell.[49] This technique appears highly suitable for field conditions. Similarly encouraging results with a microplate ELISA for the titration of antibodies to *B. bovis* in cattle sera are reported by Barry et al.[50] According to these authors, the one-step ELISA with a serum working dilution of 1:100 was very specific, showing no cross reactions against *B. bigemina*, *Anaplasma centrale*, and *Theileria* spp. When the ELISA results were compared with those of the IFAT, the ELISA proved to be more sensitive. Cross reactions of equine sera positive for antibodies against *B. equi* with *B. caballi* and *B. rodhaini* antigens were reported by Goetz,[18] who recommended the ELISA as a screening test, and cross reactions in 3 of 14 sera with antibodies against *B. caballi* to

B. equi antigens were also described by Soule et al.[24] When the ELISA was compared with the approved CFT and IFAT, its sensitivity was again comparable to that of the IFAT. Weiland[29] confirmed that ELISA for babesiosis has low specificity for the serological differentiation of *B. equi* and *B. caballi* infections in horses. Furthermore, the test cannot be meaningfully used to monitor the treatment of infections due to *B. caballi*.

O'Donoghue et al.[8] fractionated *B. bigemina* crude extract antigen on Sephadex® G-200 columns and found that only two antigens with molecular weights of 600 and 15 kdaltons were suitable in the ELISA. Using these antigens, they investigated the antibody response within the IgG and IgM classes in experimentally infected calves. Antibody responses to both fractions were first observed on day 7 postinfection. The IgG response persisted for 49 days postinfection, whereas the IgM response had declined to low levels by 28 days postinfection.

Martinod et al.[51] described "light positive reactions" ($OD_{490} = 0.15$) in sera of dogs from a low-risk region in France. The antigen was prepared from in vitro cultured *B. canis* parasites. Sera from dogs with clinical symptoms of babesiosis and from animals living in an endemic area showed strongly positive reactions ($OD_{490} = 0.64 - 0.88$).

Recent studies by Stek et al.[7] have demonstrated the applicability of the ELISA in detecting specific IgM and IgG antibodies in human cases of babesiosis. Employing a *B. microti* purified, soluble antigen, the authors found the test to be highly suitable for the diagnosis of *B. microti* infection with only a low degree of cross reactivity to *P. falciparum*.

The advantage and practicability of a DOT-ELISA for the detection of antibodies in canine babesiosis are currently being investigated by Kakoma et al.[52] Snary and Smith[53] detected one monoclonal antibody raised against *B. rodhaini* in mice that bound to mouse RBC protein (Western blotting) and suggested a common antigenic determinant between mouse erythrocyte and parasitic membrane proteins. When the antibody was tested in the IFAT, there was no reactivity seen with the red cell membrane. According to the authors, a destruction or masking of the determinant or a lower affinity of the antibody might be responsible for this failure. Whether background reactions or false positive values obtained in the ELISA are due to these autoantibodies remains to be determined.

At present, the ELISA for babesiosis is a useful diagnostic aid for the detection of anti-*Babesia* antibodies in recently or chronically infected animals or human beings. Due to cross reactions among antigens extracted from different species used in the test systems, the ELISA cannot identify *Babesia* species in mixed infections. The test should be considered as a screening method for seroepidemiological surveys and needs more comprehensive field studies on specificity, reproducibility, and cost-effectiveness. The preparation of defined antigens and the extended use of purified culture antigens will hopefully improve the diagnostic value of the ELISA in the serodiagnosis of babesiosis.

4. *Agglutination Tests*

The first reported use of an indirect hemagglutination test (IHA) for the diagnosis of *Babesia* infections in animals was by Curnow and Curnow in 1967.[54] These authors used sheep erythrocytes sensitized with a *B. argentina* antigen for the diagnosis of bovine babesiosis. The authors considered the assay to be comparable to the CFT in sensitivity and specificity. Easy to carry out and requiring little technical equipment, this test has been widely applied and modified in the past.[31] Despite the great sensitivity of the IHA, its practical use is limited by the occurrence of nonspecific reactions, probably due to isoagglutinins. Most of the nonspecific reactions occurring with cattle sera tested against *B. argentina*, however, could be eliminated by absorbing sera with pyruvic aldehyde-treated homologous RBC and an extract of normal bovine erythrocytes.[55] The *Babesia* species used for the sensitization and the antigen preparation seem to play an important role on the final sensitivity of the assay. Goodger[56] isolated highly active HA-antigens from *B. argentina* and coupled it to sheep erythrocytes. The resulting system is reported to have raised the specificity of the IHA to 100% in the homologous assay. A specificity of 99.3% is reported with cattle sera diluted 1:100. The use of a crude *B. bigemina* extract antigen coupled to erythrocytes, however, resulted in a lower sensitivity, and heterologous antisera strongly cross reacted with the sensitized cells.

Encouraging results are reported by Goodger and Mahoney,[55] who described the IHA as being more sensitive than IFAT and CFT. Contrary to these findings, Reif[36] observed that the IHA, using erythrocytes sensitized with chromatographically purified extract antigen fractions from *B. rodhaini* and *B. divergens*, was less sensitive than the IFAT and ELISA. Cross reactions of up to 1:32 between *B. equi* and *B. caballi* were reported by Goetz,[18] who did not observe agglutinations of positive equine sera with *B. rodhaini*. Nikoforenko[57] evaluated the IHA for the detection of *B. bigemina* antibodies in infected (n = 138) and noninfected (n = 743) cattle. Isoagglutination was observed in 8% of the serum samples tested. In a comparative study for the rapid detection of anti-*Babesia* antibodies in cattle, Akinboade and Cole[58] concluded that the IHA is more valid than the bentonite agglutination test based on specificity and sensitivity. The bentonite agglutination test (BA) was introduced by Sibinovic and Sibinovic[59] for the diagnosis of equine babesiosis, but has not been widely used or compared to the other established tests. The advantage of this assay is certainly its speed. The capillary tube agglutination test (CTA), developed by Lohr and Ross[60] (see Todorovic et al.[31]) for the detection of *B. bigemina* antibodies in cattle was recently evaluated by Su et al.[61] in Taiwan. Using an unspecified source of *Babesia* antigen, the comparative evaluation of the CTA and the CFT showed that the CTA was apparently less sensitive than the CFT.

Three serological test systems, CTA, passive hemagglutination (PHA), and gel precipitation (GP), were used by Bhoop Singh and Gautam[62] to measure the humoral immune response in horses infected with *B. equi*. In agreement with other workers, the CTA was found to be less sensitive than the CFT, but was recommended for diagnosis under field and laboratory conditions because of its simplicity, specificity, and economy. The use of a latex agglutination test (LAT) as a practical serodiagnostic assay for the rapid diagnosis of *Babesia* infections has been described by Goodger and Mahoney[55] and Lopez and Todorovic.[63] The use of the latter test for seroepidemiological investigations was limited, however, by lack of consistent results. Recently, preliminary results reported by Montenegro et al.[11] with a LA test sensitized with culture-derived, soluble antigens of *B. bovis* demonstrated an increased sensitivity of this assay for the early detection of anti-*Babesia* antibodies compared to the IFAT. Moreover, using LAT culture-derived antigen was found to be more stable and remained antigenic for over 6 months at 4°C. There was a 100% agreement between the IFAT and LAT.

The slide agglutination test developed either with fixed *B. bigemina* parasites[2] or *B. argentina* antigen sensitized latex particles[55] was introduced as a rapid test for diagnosis of *Babesia* infections in cattle herds. Although the studies indicate that serological distinction between infected and noninfected animals is possible with a false positive rate of 3.0% of samples from noninfected animals,[55] the tests have not provided a satisfactory reproducibility and were therefore not pursued any further.

The use of a card agglutination test as a diagnostic aid for the detection of anti-*Babesia* antibodies was reported by various authors.[64,65] The antigen was prepared either from a pure or a crude parasite erythrocytic stroma suspension. Although the test is easy to perform under field conditions, its diagnostic specificity and reliability after prolonged storage remains to be proved.

Most of the rapid agglutination tests are considered to have a low sensitivity. However, the development of a simplified, cost-effective technique should be emphasized to provide the relatively unskilled personnel with reliable and simple tests for seroepidemiological surveys in the field.

5. Radioimmunoassay (RIA)

Kahl et al.[6] stated that the RIA was a suitable test for the detection of anti-*Babesia* antibodies in sera of cattle experimentally and naturally infected with *B. bovis*. The antigen used consisted of a crude preparation of highly concentrated erythrocyte and parasite membrane material. The antigen preparations from homologous or heterologous strains of *B. bovis* used in the assay showed no reproducible differences in the antibody levels and were observed in sera from experimentally infected, splenectomized calves. Considerably greater sensitivity for the detection of anti-*B. bovis* antibodies was reported when the RIA was compared to the IFAT. However, the use of such a sophisticated test is limited to technically well-equipped laboratories.

C. Evaluation of Serological Tests for Babesiosis

1. Serology of Bovine Babesiosis

Piroplasmosis (i.e., babesiosis, theileriosis) of livestock is more of a herd problem than a matter of the infected individual animal. Definitive diagnosis of babesiosis is by the direct identification of the *Babesia* parasites in stained blood smears. The microscopic examination is labor intensive and time consuming. Negative results, however, do not preclude the possibility of infection. An alternative for the direct identification of the parasites is the indirect detection of *Babesia* infections through the demonstration of antibodies by serological methods. Although serological tests have not been optimized and standardized, they are reliable serodiagnostic adjuncts for the detection of *Babesia* infections, particularly for seroepidemiological surveys in enzootic areas and among individuals exposed for the first time. Many attempts have been made to demonstrate antibodies to the different bovine *Babesia* species, employing the complement fixation test (CFA), the indirect hemagglutination (IHA), the indirect immunofluorescent antibody test (IFAT), the capillary (CA), slide (SA), latex agglutination (LA), the babesiosis card agglutination test (BCT), and the gel precipitation; Todorovic et al.[31] reviewed almost 30 papers which had appeared between 1962 and 1978. A summary of the publications between 1977 and 1985 is shown in Table 1. The IFAT, ELISA, CFT, and IHA have been the main tests employed. Antibody response to *B. argentina* and *B. bigemina* has been reported by Rodriguez et al.[66] in 23.5 to 38.0% and in 30.1 to 44.6%, respectively, of 7652 sera from infected cattle in Cuba. The variation in the percentage of positivity depended on the use of either the standard CFT or cold CFT, or the slow agglutination test (SAT). The prevalence of bovine babesiosis in Guyana was serologically determined utilizing the IFAT and CFT. In native cattle, IFAT and CFT antibody activity was detected in 80 and 40%, respectively; antibodies to *B. bovis* were observed in 61 and 16% of the sera.[67] Madruga et al.[68] analyzed the levels of *B. bigemina* and *B. bovis* antibodies in calves of different breeds or crossbreeds by the IFAT. A significant positive correlation could be demonstrated with the anti-*B. bigemina* and anti-*B. bovis* immunoglobulins of the cows. The level of antibodies to *B. bigemina* was generally higher than those to *B. bovis*. A high prevalence of antibodies to *B. bovis* was observed in cattle of 15 dairy farms in Bolivia.[76] Antibody activity was detected by the IFAT in 94 to 100% of the sera examined. The examination of serum samples from 889 cattle of various breeds in Mexico revealed antibodies to *Babesia* in 65.5%.[69] In Africa, Miller et al.[44] performed an IFAT survey with cattle in Mali to determine the prevalence of antibodies to *B. bigemina* and *B. bovis*. Antibody activity to *B. bovis* was very low, but high anti-*B. bigemina* titers were observed. Prevalence of antibodies was twice as much in N'Dama than in Zebu cattle. Ten randomly selected white Fulani trade cattle in Nigeria were tested for anti-*Babesia* antibodies using both the bentonite agglutination (BA) and the IHA. Antibody positivity by the BA was found in eight serum samples and nine were positive to the IHA.[58] Nikoferenko[57] used the IHA to test cattle sera from endemic and nonendemic areas in the U.S.S.R. He found antibody activity in 91% of cattle from farms located in endemic areas. In Southern

Africa, Bessenger and Schoeman[40] investigated the serological response of cattle infected with *B. bigemina* and/or *B. bovis*. The IFAT was suitable to detect antibodies to both species. Cross reactions were observed between the two species, but were most significant when *B. bigemina* antigen was used. Kahl et al.[6] have recently developed a RIA primarily for quantitation of antibodies to *B. bovis*, as reviewed above. A microplate enzyme immunoassay (EIA) is described by Barry et al.[50] for detecting antibodies to *B. bovis* in cattle serum. No significant cross reactions occurred between *B. bovis* antigen and sera from cattle infected with *B. bigemina*, *Anasplasma*, or *Theileria* species. The test correlated with the IFAT at the 95.2% level. The EIA appeared to be more sensitive than the IFAT. Fujinaga and Minami[21] compared the IFAT and the CFT in experimentally *B. ovata* infected cattle. CFT titers were negative on day 270 postinfection, the IFAT tests remained relatively high until day 420 postinfection. In a survey of sera from naturally infected animals, the IFAT was more effective for the serodiagnosis of *Babesia* infections than the CFT. In the U.S., Kuttler et al.[70] compared the CFT and the IFAT for their suitability in detecting antibodies to *B. bigemina* during the course of infection. Within the first 3 months of the infection, the CFT detected 95% of the 60 sera and the IFAT in 100% of 57 samples tested against *B. bigemina* antigen. The CFT titers decreased between 98 to 175 days postinfection to below 60 to 40%, while the IFAT remained positive in 94.6% of 56 sera, thus indicating the higher sensitivity of the IFAT during the latent phase of infection. Montenegro et al.[11] developed a simple LAT test for the diagnosis of *B. bovis* infection in cattle. Compared with the IFAT, the LAT showed a high degree of specificity and sensitivity and possessed appropriate stability and simplicity suitable for field purposes.

In Great Britain, Trees[71] studied the antibody response in splenectomized calves and adult cattle experimentally infected with *B. divergens*. Antibodies measured by the IFAT appeared following parasitemia, and titers peaked 35 days postinfection. The splenectomy did not significantly influence the IFAT response of calves to *B. divergens* infection. Bidwell et al.[48] compared the ELISA, IFAT, and CFT for the detection of antibodies in the serum of cattle experimentally infected with *B. divergens* and *B. major*. Antibodies to both species could be detected using all three tests, but they were detected earlier by the CF test. Titers in the CFT were lower than those obtained by the other tests. No significant differences in titers and sensitivity were observed between ELISA and IFAT, and the authors recommended the ELISA particularly for field use, especially since it is less subjective. Various authors in Germany studied the antibody development after *B. divergens* experimental infection of cattle[36,72] using ELISA, IHA, and IFAT. Field investigations were carried out in southern Germany by Ullmann et al.[73] and Boch.[74] Weiland et al.[75] compared the IFAT, IHA, and ELISA for the detection of antibodies to *B. divergens* in experimentally infected, intact cattle. Parasitemias occurred 2 to 8 days postinfection with 4.7, 5.0, and 5.3% parasitized erythrocytes. Antibodies were detected 6 to 9 days postinfection by means of IFAT and IHA, and 10 days postinfection also by the ELISA. Maximum titers were attained 16 to 33 days postinfection (IFAT), 12 to 28 days postinfection (IHA), and 15 to 34 days postinfection (ELISA) (Figure 2). During the latent course of infection, all three tests remained positive up to 2 1/2 years. Subinoculation of blood into intact or splenectomized jirds (*Meriones* spp.) showed that the cattle were carriers until the end of the experiment.

2. Serology of Canine Babesiosis

Relatively little attention has been paid to serodiagnosis of *Babesia* infections of dogs. Acute babesiosis can easily be diagnosed by the direct demonstration of *B. canis* or *B. gibsoni* in the stained blood smear. The increasing international tourism, however, has increased the demand for diagnostic methods, i.e., indirect procedures have to be developed for the diagnosis of latent infections in the absence of the typical clinical symptomatology and parasitemia. Only serodiagnostic methods are capable of detecting carriers of the disease.

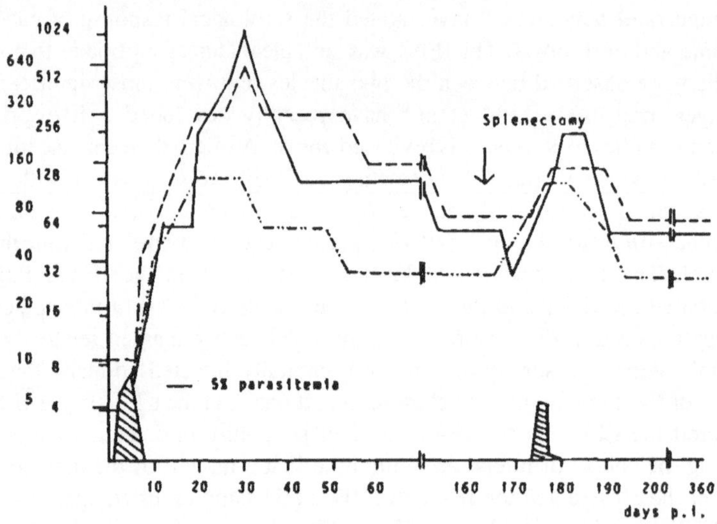

FIGURE 2. Parasitemia and antibody-formation in a calf experimentally infected with *Babesia divergens* (ELISA —; IFAT - - -; IHA ·······).

Various serological methods have been developed and used in the serodiagnosis of *Babesia* infections in domestic animals.[25] Schindler et al.[77-79] employed the CFT and later the IFAT for the assessment of humoral immunity in dogs immunized with soluble or corpuscular antigens. Both of the tests were suitable for the demonstration of antibodies; higher titers were obtained in animals after infection with *B. canis* than in those immunized with soluble antigen. The authors concluded that the serological reactions reflect the multiplication of the parasites in the host. This concept was discussed in relation to the immune status of the dogs studied. During the state of sterile immunity, however, antibody tests in some cases yielded negative results. Popovic and Ristic[80] prepared a soluble *B. canis* antigen by protaminsulfate precipitation for the demonstration of anti-*B. canis* antibodies by means of immunodiffusion. This antigen was suitable for the demonstration of precipitating antibodies in dogs experimentally infected with *B. canis*. The antibodies were demonstrable 17 days postinfection and persisted over a period of 10 months at the same level. No cross reactions were found with sera from dogs with distemper, infectious canine hepatitis, or leptospirosis, thus indicating the specificity of the test. More recently the IFAT has been the major test employed for the demonstration of antibodies to canine piroplasma. Tamura et al.[33] obtained good results with the IFAT for the demonstration of antibodies to *B. gibsoni* in experimentally infected dogs, using acetone-fixed thin blood films as antigen. The IFAT titers of 1:20 to 1:80 were first detected on the 6th day after the experimental infection with *B. gibsoni*. The titers increased rapidly with increasing parasitemia ranging from 1:2560 to 1:5120. These high titers persisted until the 50th day postinfection, the end of the experiment. Also, Anderson et al.[34] describe the IFAT as a sensitive and reproducible test for canine piroplasmosis, using a North American strain of *B. gibsoni* as a source of antigen. In a *B. gibsoni* infected dog, antibody titers developed as the parasitemia increased, i.e., 1:256 after 3.5 weeks when parasitemia was 5.2%, and 1:4096 after parasitemia had peaked to 8% at 5 weeks. At 18 to 24 weeks, the titers fluctuated between 1:4096 and 1:6536. Titers with heterologous *Babesia* and *Plasmodium* antigens were one to three fourfold dilutions lower than those with the homologous antigen. In a limited serological survey, Breitschwerdt et al.,[81] by means of IFAT, detected many infected dogs in various greyhound kennels in different parts of the U.S. Weiland et al.[9,35] investigated the suitability of the enzyme-linked immunosorbent assay (ELISA) for *Babesia* antibodies in comparison to the IFAT as sero-

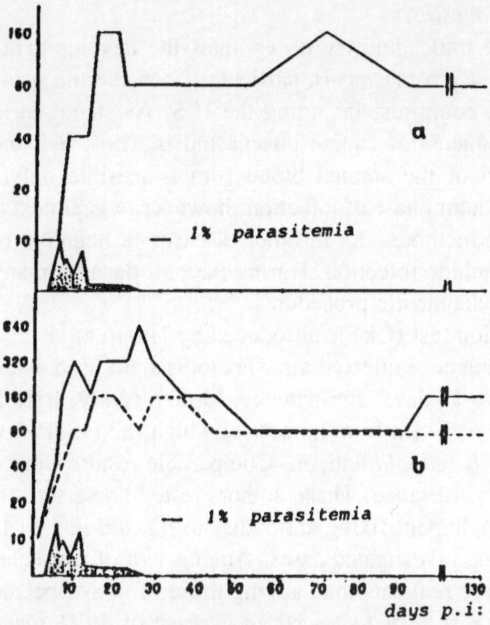

FIGURE 3. Parasitemia and antibody formation in three *B. canis* infected dogs. (a) IFAT; (b) ELISA: — somatic antigen, - - - metabolic antigen.

diagnostic tools for canine *Babesia* infections. The antibody response was studied in three dogs experimentally infected with *B. canis* and two with *B. gibsoni*. As antigen, metabolic products were employed, somatic extracts, and intact parasites of *B. rodhaini*, which strongly cross reacted with antibodies to heterologous *Babesia* species.[82] Antibodies could be demonstrated with both ELISA and IFAT during the fresh and latent phases of infection. Antibody titers of 1:160 and 1:320 to *B. canis* were observed with the ELISA with the somatic extract antigen 5 to 7 days postinfection, and maximum titers (1:640 to 1:1280) were reached 19 days postinfection. During the latent phase, the titers declined to between 1:80 and 1:160. With the metabolic antigen, antibodies occurred a few days later, and the titers remained higher than those obtained with somatic extracts. Antibodies detected by the IFAT occurred 7 days postinfection with titers of 1:160 and persisted during the latent phase with values of 1:80 (Figure 3). Dogs infected with *B. gibsoni* showed a rapid increase in titers 6 to 17 days postinfection up to 1:320. These values persisted during the late stage, while titers obtained with metabolic antigens declined to between 1:80 and 1:160. By means of IFAT, antibodies could be demonstrated at 10 days postinfection; they peaked 30 days postinfection at 1:160. During the latent phase, IFAT titers were between 1:40 and 1:80. The authors recommended both ELISA and IFAT as suitable for the detection of *B. canis* and *B. gibsoni* infections in dogs if antigens of the rodent *B. rodhaini* were employed. These workers, supported by Denning et al.,[83] concluded that dogs remain carriers for over a period of several months or even years since splenectomology of these dogs precipitates babesiosis. Nontreated animals from endemic areas therefore have to be considered as carriers. Epidemiological studies by Martinod et al.,[51] in France, in a territory without any risk of *B. canis* infections of dogs and in a hyperendemic area by means of micro-ELISA using *B. canis* antigen, showed that most of the 70 dogs tested from the endemic area possessed antibodies to *B. canis*. Boch[74] detected 46 dogs positive for *B. canis* out of 116 suspected animals in southern Germany. In both studies, a number of the dogs tested were asymptomatic.

3. Serology of Equine Babesiosis

The increasing equine trade industry necessitates the development of a reliable immunodiagnostic test for equine piroplasmosis, particularly considering stringent restriction measures imposed by various countries, including the U.S. As stated earlier, the direct identification of the etiologic agents of equine piroplasmosis, (i.e., *B. caballi* and *B. equi*), by microscopic examination of the stained blood film is feasible only during the phase of parasitemia. During the latent phase of infection, however, it is almost impossible to identify the piroplasms by direct methods. As in other diseases, a negative result by microscopic examination does not preclude infection. During the past decade, many attempts have been made to standardize serodiagnostic procedures.[25,31]

The complement fixation test (CFT), introduced by Hirato et al.,[12] remains the serologic test of choice. Whole parasites collected after hemolysis are used as antigen. Antibodies to *B. caballi* first appear 4 to 15 days after patency. More recently, Bertoldini et al.,[84] in Italy, using the CFT examined 617 sera from horses, of which, 82 (13.1%) were positive with *B. equi* and 16 (2.6%) with *B. caballi* antigen. Comparable results were obtained by Chevrier et al.[85] and Soule et al.[86] in France. These authors tested horse sera from different regions of France and found complement-fixing antibodies to *B. equi* in 1 to 12%, antibodies to *B. caballi* in 0.2 to 2% of the investigated cases. Among sera of suspected animals were three to four times more positive reagents than among those of nonsuspected horses. The results of Soule et al.[86] on 950 sera tested showed an average of 4.2% reactivity to *B. equi* and 1% to *B. caballi* in the CFT.

Investigations for the prevalence of anti-piroplasm antibodies were also carried out by Joyner et al.[87] on 3887 horses in Great Britain and Ireland using the CFT. None of 3241 horses, which were born and permanently resided in the British Isles, had antibodies to *B. equi* or *B. caballi*. However, 42% (262) of 618 horses of foreign origin were positive for either or both species. Seropositive reactors came from Arabian Gulf countries (15 of 19), Argentina (1 of 19), Belgium (2 of 15), France (1 of 5), Federal Republic of Germany (13 of 148), Poland (2 of 24), Portugal (2 of 3), Spain (222 of 346), U.S.S.R. (3 of 3), and some other countries. Tests repeatedly performed on 95 horses showed that some animals lost their CFT titers within a few months of the first test, while others retained titers to *B. caballi* for up to 13 months and to *B. equi* for 28 months. Also, Chevrier et al.[85] found higher percentages of positive reactors in sera from Belgium (10%), Morocco (17%), and Spain (21%). Donnelly et al.[88] found complement fixing antibodies to *B. equi* and/or to *B. caballi* in 4 of 14 English-born foals of Portuguese-born seropositive mares. While the titers in the mares remained constant throughout a period of 3 months observation, the titers in the foals declined steadily to baseline level by about 4 months from birth. This suggests passively acquired maternal antibodies rather than a congenital infection of the foals. Friedhoff[27] examined 765 horse sera from Germany and the Netherlands and found 42 (5.5%) reactive with *B. equi*, and 5 (0.7%) with *B. caballi* antigen. In the case of five foals, he could not exclude congenital infection. The overnight CFT (Kolmer technique, 4°C) was recommended by Petrovskii et al.[89] for the detection of antibodies to *B. equi*. Positive reactions were obtained by this technique in 36% of more than 3000 samples examined. Tests performed during the 1980 Olympiad gave some positive reactions to *B. equi* in sera of horses from Poland, Romania, Italy, and India. In Cuba, 25.3% of 525 horse sera were positive for *B. equi* by means of the CFT,[90] and in Mexico, 28 (9.3%) sera from 300 healthy horses were positive for *B. equi* by both tube and microplate CFT, the latter proving to be more sensitive.[91]

Heidrich-Joswig[92] inoculated two ponies with *B. equi* from ticks. The CFT became positive about the same time as the parasites appeared in the blood, i.e., 14 to 18 days postinfection. The parasites disappeared from peripheral blood, but after day 28, the CFT titer remained high until day 35 and was still elevated 144 days postinfection.

Amerault et al.[65] studied the CFT in comparison with the card test (CT). Out of 394 sera

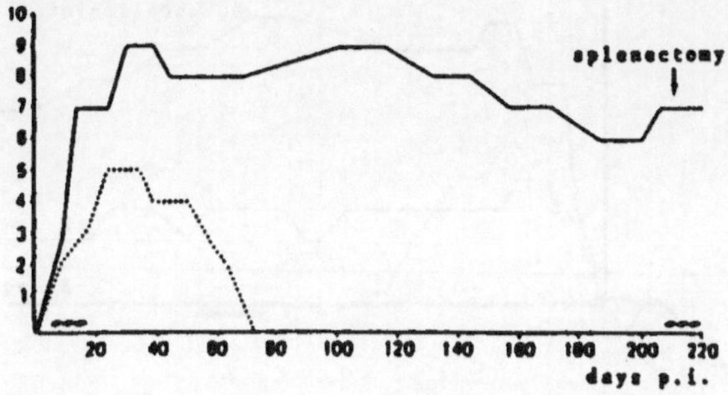

FIGURE 4. Antibody formation after experimental *B. caballi* infection. IIF (—), CFT (····), parasitemia (-●-●-).

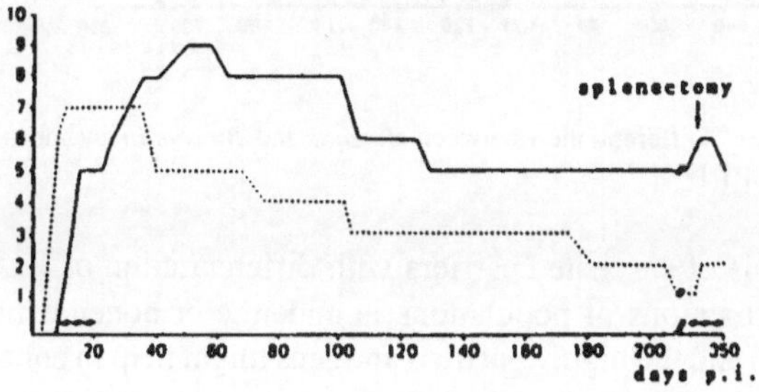

FIGURE 5. Antibody formation after experimental *B. equi* infection. IIF (—), CFT (····), parasitemia (-●-●-). (Titer steps are twofold dilutions, stating 1:4 [IIF] or 1:5 [CFT].

tested, 194 (49.2%) were positive to *B. caballi*. The authors found a good correlation between both tests. The typical serologic response in experimental equine babesiosis is depicted in Figures 4 and 5 (splenectomy) and intact animals (Figure 6).

During the last 10 years, more emphasis has been put on the combined use of the CFT and IFAT. Accordingly, many experts strongly recommend that these two tests be performed to complement each other.[24,28,93,94]

III. CONCLUSIONS AND RECOMMENDATIONS

This review has attempted to compile the most currently used techniques in the serology of babesiosis (Table 1). One must keep in mind that the definitive diagnosis of babesiosis continues to be the direct demonstration of the parasites. If direct demonstration fails during the preparasitemic stage of the infection as well as during its latent course, indirect methods for the demonstration of specific antibodies to the causative agent may help to ascertain the diagnosis as babesiosis. Now that many immunodiagnostic techniques have become available for the demonstration of antibodies to the various *Babesia* species, they complement the parasitological examinations. The diagnostic value of serologic methods for the demonstration of antibodies is, however, limited by (1) the impossibility of diagnosing prepatent infections; (2) the persistence of titers after natural recovery or chemotherapy, and (3) the difficulty to interpret low grade and/or cut-off titer. The exact state of infection may, therefore, pass unrecognized. Highly positive serological results provide a definite diagnosis; negative serological results do not necessarily preclude infection, and antibody titers bear no direct relationship to parasitemia.

The application of serological methods is governed by the diagnostic requirements for,

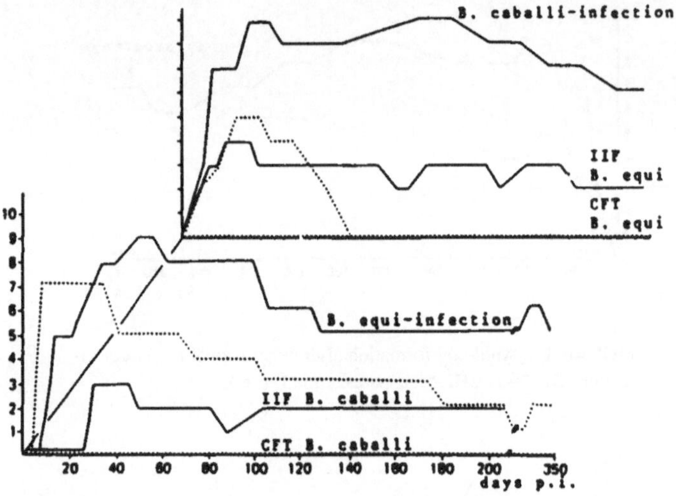

FIGURE 6. Differentiation between *B. equi* and *B. caballi* by means of IFF (—) and CFT (····).

(1) individual diagnosis of suspected carriers with differentiation of *Babesia* species or, (2) epidemiological investigations of populations in endemic or nonendemic areas. The use of monoclonal antibodies and the highly purified antigens might help in enhancing the sensitivity and specificity of immunodiagnostic techniques.

More emphasis should be placed on the development of diagnostic techniques for the quantitation of parasite-specific circulating antigens and immune complexes in body fluids.

The intensive international cooperation of institutions and laboratories concerned with babesial serology could lead to the standardization of various immunodiagnostic procedures.

REFERENCES

1. **Holbrook, A. A., Frerichs, W. M., and Allen, P. C.,** Laboratory diagnosis of equine piroplasmosis, *Proc. 3rd Int. Conf. Equine Infectious Diseases, Paris,* Vol. 3, 1972, 467.
2. **Curnow, J. A.,** The use of a slide agglutination test to demonstrate antigenic differences between *Babesia bigemina* parasites, *Aust. Vet. J.*, 49, 290, 1973.
3. **Mahoney, D. F.,** Bovine babesiosis: preparation and assessment of complement fixing antigen, *Exp. Parasitol.*, 20, 232, 1967.
4. **Frerichs, W. M., Holbrook, A. A., and Johnson, A. J.,** Equine piroplasmosis: complement fixation titers of horses injected with *Babesia caballi, Am. J. Vet. Res.*, 30, 697, 1969.
5. **Goodger, B. V.,** *Babesia bovis* (= *argentina*): changes in erythrophylic and associated proteins during acute infection of splenectomized and intact calves, *Z. Parasitenkd.*, 55, 1, 1978.
6. **Kahl, L. P., Anders, R. F., Callow, L. L., Rodwell, B. J., and Mitchell, G. F.,** Development of a solid phase radioimmunoassay for antibody to antigens of *Babesia bovis* infected bovine erythrocytes, *Int. J. Parasitol.*, 12, 103, 1982.
7. **Stek, M., Ruebush, M. J., Benach, J. L., Duncan, J. F., and Deisanti, C.,** Enzyme immune assays for specific IgM and IgG antibodies in experimental and human babesiosis, 2nd Int. Conf. Malaria and Babesiosis, Annecy, France, September 19 to 22, 1983.
8. **O'Donoghue, P. J., Friedhoff, K. T., Vizcaino, O. G., and Weyreter, H.,** The detection of IgM and IgG antibodies against *Babesia bigemina* in bovine sera using semi-defined antigens in enzyme immunoassay, *Vet. Parasitol.*, 18, 1, 1985.
9. **Kratzer, I.,** Die Brauchbarkeit der *Babesia rodhaini*-Antigene zum Nachweis von Babesien-Infektionen im ELISA, Vet. Med. thesis, University of Munich, 1979.

10. **Montenegro-James, S., James, M. A., and Ristic, M.,** Localization of culture-derived soluble *Babesia bovis* antigens in the infected erythrocyte, *Vet. Parasitol.*, 13, 311, 1983.
11. **Montenegro, S., James, M. A., Levy, M. G., Preston, M. D., Esparza, H., and Ristic, M.,** Utilization of culture-derived soluble antigen in the latex agglutination test for bovine babesiosis and anaplasmosis, *Vet. Parasitol.*, 8, 291, 1981.
12. **Hirato, K., Ninomiya, N., Uwano, Y., and Kutil, T.,** Studies on the complement fixation reaction for equine piroplasmosis, *Jpn. J. Vet. Sci.*, 1, 204, 1945.
13. **Mahoney, D. F.,** Bovine babesiosis: diagnosis of infection by a complement fixation test, *Aust. Vet. J.*, 38, 48, 1962.
14. **Mahoney, D. F.,** Bovine babesiosis: an assessment of the significance of complement fixing antibodies based upon experimental infection, *Aust. Vet. J.*, 40, 369, 1964.
15. U.S. Department of Agriculture, 1. Equine Piroplasmosis Complement Fixation Test Antigen Production. 2. Piroplasmosis Complement Fixation Test — Micro Method, Animal and Plant Health Inspection Service, Natl. Vet. Serv. Lab., Diagnostic Reagents Laboratory, U.S. Department of Agriculture, Ames, Iowa, 1985.
16. **Todorovic, R. A. and Adams, L. G.,** Serodiagnosis of babesiosis, Proc. 19th World Veterinary Congr., Mexico City, Vol. 3, 1971, 1114.
17. **Todorovic, R. A. and Long, R. F.,** Comparison of indirect fluorescent antibody (IFA) with complement fixation (CF) tests for diagnosis of *Babesia* spp. infection in Colombian cattle, *Tropenmed. Parasitol.*, 27, 169, 1976.
18. **Goetz, F.,** Untersuchungen über die Brauchbarkeit von ELISA, IFAT, IHA und KBR zum Nachweis von *Babesia equi* Infektionen, Vet. Med. thesis, University of Munich, 1982.
19. **Merkle, F. J.,** Untersuchungen über die Brauchbarkeit von KBR, IFAT und ELISA zum Nachweis der *Babesia caballi* Infektion des Pferdes, Vet. Med. thesis, University of Munich, 1983.
20. **Aicher, B. M.,** Möglichkeiten der serologischen Differenzierung der *Babesia caballi*-Infektion des Pferdes mit Hilfe von KBR, IFAT und ELISA, Vet. Med. thesis, University of Munich, 1984.
21. **Fujinaga, T. and Minami, T.,** Indirect fluorescent antibody and complement fixation tests in the diagnosis of bovine theileriosis and babesiosis in Japan, *Vet. Parasitol.*, 8, 115, 1981.
22. **Rehbein, G. and Heidrich-Joswig, S.,** Use of schizont and piroplasm antigens of *Babesia equi* in the indirect fluorescent antibody and complement fixation tests, *Vet. Parasitol.*, 12, 135, 1983.
23. **Goff, W. L., Wagner, G. G., Craig, T. M., and Long, R. F.,** The role of specific immunoglobulins in antibody-dependent cell mediated cytotoxicity assays during *Babesia bovis* infection, *Vet. Parasitol.*, 14, 117, 1984.
24. **Soule, C., Perret, C., and Dorchies, P.,** Babesiose equine a *Babesia equi*: comparison des techniques de fixation du complement d'immunofluorescence indirecte et ELISA, *Rev. Med. Vet.*, 135, 419, 1984.
25. **Todorovic, R. A.,** Serological diagnosis of babesiosis: a review, *Trop. Anim. Health Prod.*, 7, 1, 1975.
26. **Tenter, A. M. and Friedhoff, K. T.,** Serodiagnosis of experimental and natural *B. caballi* infections, *Vet. Parasitol.*, 20, 43, 1986.
27. **Friedhoff, K. T.,** Die Piroplasmen der Equiden — Bedeutung für den internationalen Pferdeverkehr, *Berl. Muench. Tieraerztl. Wochenschr.*, 95, 368, 1982.
28. **Donelly, J., Joyner, L. P., Graham-Jones, O., and Ellis, C. P.,** A comparison of complement fixation and immunofluorescent antibody tests in a survey of the prevalence of *Babesia equi* and *Babesia caballi* in horses in the Sultanate of Oman, *Trop. Anim. Health Prod.*, 12, 50, 1980.
29. **Weiland, G.,** Species specific serodiagnosis of equine piroplasma infections by means of complement fixation test (CFT), immunofluorescence (IIF), and enzyme-linked immunosorbent assay (ELISA), *Vet. Parasitol.*, 20, 43, 1986.
30. **Ristic, M. and Sibinovic, S.,** Serologic diagnosis of *Babesia bigemina* infection in cattle by the indirect fluorescent antibody test, *Res. Vet. Sci.*, 9, 557, 1964.
31. **Todorovic, R. A. and Carson, C. A.,** Methods for measuring the immunological response to *Babesia* in *Babesiosis*, Ristic, M. and Kreier, J. P., Eds., Academic Press, New York, 1981, 381.
32. **Weiland, G., Aicher, B. M., and Boch, J.,** Serodiagnostik und Therapiekontrolle der Pferdepiroplasmose mit KBR und IFAT, *Berl. Muench. Tieraerztl. Wochenschr.*, 97, 341, 1984.
33. **Tamura, T., Takahashi, K., Sonada, M., and Koiwa, M.,** The indirect fluorescent antibody test as a method for detecting antibody in dogs infected with *Babesia gibsoni*, *J. Coll. Dairy*, 8, 249, 1980.
34. **Anderson, J. F., Magnarelli, L. A., and Sulzer, A. J.,** Canine babesiosis: indirect fluorescent antibody test for a North American isolate of *Babesia gibsoni*, *Am. J. Vet. Res.*, 41, 2102, 1980.
35. **Weiland, G. and Kratzer, I.,** Fluoreszenz- und enzymserologische Untersuchungen zum Nachweis der latenten Hundebabesiose, *Berl. Muench. Tieraerztl. Wochenschr.*, 92, 398, 1979.
36. **Reif, L.,** Der serologische Nachweis der *Babesia divergens* Infektion des Rindes mit der Immunfluoreszenz, der indirekten Hämagglutination und dem Immunoenzyme test, Vet. Med. thesis, University of Munich, 1980.

37. **Fujinaga, T. et al.**, Changes of immunofluorescent antibody titers against *Babesia ovata* in grazing cattle in an enzootic area and its sub-detectable infection, *J. Jpn. Vet. Med. Assoc.*, 38, 162, 1985.
38. **Callow, L. L., Kanahai, G. K., and Vandenberghe, A.**, Serological comparison of strains of *Babesia bovis* occurring in Australia and Mozambique, *Trop. Anim. Health Prod.*, 13, 79, 1981.
39. **Jones, T. W. and Conrad, P. A.**, Detection of antibody to *Babesia bovis* using a fluorescent antibody staining technique with *in vitro* and *in vivo*-derived antigens, 2nd Int. Conf. Malaria and Babesiosis, Annecy, France, September 19 to 22, 1983.
40. **Bessenger, R. and Schoeman, J. H.**, Serological response of cattle to infection with *Babesia bigemina* and *Babesia bovis* in Southern Africa, *Onderstepoort J. Vet. Res.*, 50, 115, 1983.
41. **Plössl, J.**, Untersuchungen zur Spezifität des Indirekten Immunfluoreszenzantikörpertests (IFAT) zum Nachweis von Antikorpern gegen verschiedene *Babesia* Arten, Vet. Med. thesis, University of Munich, 1984.
42. **Taylor, S. M., Elliott, C. T., and Kenny, J.**, Isolation of antigenic proteins from erythrocytes parasitized with *Babesia divergens* and a comparison of their immunizing potential, *Vet. Parasitol.*, 21, 99, 1986.
43. **Chisholm, E., Ruebush, T. K., Sulzer, A. J., and Healy, G. R.**, *Babesia microti* infection in man: evaluation of an indirect immunofluorescent antibody test, *Am. J. Trop. Med. Hyg.*, 27, 14, 1978.
44. **Miller, D. K., Diall, O., Craig, T. M., and Wagner, G. G.**, Serological prevalence of bovine babesiosis in Mali, *Trop. Anim. Health Prod.*, 16, 71, 1984.
45. **Tenter, A. M.**, Serodiagnose experimenteller und natürlicher Piroplasmeninfektionen der Pferde, Vet. Med. thesis, University of Munich, 1984.
46. **Engvall, E. and Perlmann, P.**, Enzyme linked immunosorbent assay, ELISA. III. Quantitation of specific antibodies by enzyme labelled anti-immunoglobulin in antigen coated tubes, *J. Immunol.*, 109, 129, 1972.
47. **Purnell, R. E., Hendry, D. J., Bidwell, D. E., and Turp, P.**, Microplate enzyme linked immunosorbent assay for antibody to *Babesia divergens* in cattle, *Vet. Rec.*, 99, 102, 1976.
48. **Bidwell, D. E., Turp, P., Joyner, L. P., Payne, R. C., and Purnell, R. E.**, Comparison of serological tests for babesiosis in British cattle, *Vet. Rec.*, 103, 446, 1978.
49. **Young, E. R. and Purnell, R. E.**, Evaluation of dried blood samples as a source of antibody in the micro-ELISA test for *Babesia divergens*, *Vet. Rec.*, 106, 60, 1980.
50. **Barry, D. N., Rodwell, B. J., Timms, P., and McGregor, W.**, A microplate enzyme immunoassay for detecting and measuring antibodies to *Babesia bovis* in cattle serum, *Aust. Vet. J.*, 59, 136, 1982.
51. **Martinod, S., Brossard, M., and Moreau, Y.**, Immunity of dogs against *B. canis*, its vector tick *Dermacentor reticulatus*, and *Ixodes ricinus* in endemic area, *J. Parasitol.*, 71, 269, 1985.
52. **Kakoma, I.**, personal communication, 1986.
53. **Snary, D. and Smith, M. A.**, An antigenic determinant common to both mouse red blood cells and several membrane proteins of the parasitic protozoa *Babesia rodhaini*, *Mol. Biochem. Parasitol.*, 20, 101, 1986.
54. **Curnow, J. A. and Curnow, B. A.**, An indirect haemagglutination test for the diagnosis of *Babesia argentina* infection in cattle, *Aust. Vet. J.*, 49, 286, 1967.
55. **Goodger, B. V. and Mahoney, D. F.**, A rapid slide agglutination test for the diagnosis of *Babesia argentina* infections, *Aust. Vet. J.*, 50, 250, 1974.
56. **Goodger, B. V.**, Preparation and assessment of purified antigens in the passive haemagglutination test for bovine babesiosis. *Aust. Vet. J.*, 47, 251, 1971.
57. **Nikiforenko, V. I.**, Passive (indirect) haemagglutination test for surveying the prevalance of piroplasmosis in cattle, *Tr. Vses. Inst. Eksp. Vet.*, 56, 58, 1982.
58. **Akinboade, O. A. and Cole, A.**, The application of an agglutination system in the rapid detection of *Babesia* antibodies, *Anim. Technol.*, 36, 185, 1985.
59. **Sibinovic, S., Sibinovic, K. H., and Ristic, M.**, Equine babesiosis: diagnosis by bentonite agglutination and passive haemagglutination tests, *Am. J. Vet. Res.*, 30, 691, 1969.
60. **Lohr, K. F. and Ross, I. P. I.**, Capillary tube agglutination test for the detection of *Babesia bigemina* antibody, *Z. Tropenmed. Parasitol.*, 20, 287, 1969.
61. **Su, J. F., Cheng, J. S., Liaw, S. J., and Lin, P.C.**, Studies on the serological diagnosis of bovine babesiosis in Taiwan, *Taiwan J. Vet. Med. Anim. Husb.*, 41, 45, 1983.
62. **Bhoop Singh, and Gautam, O. P.**, A comparison of serological tests for the diagnosis of equine babesiosis, *Indian J. Vet. Med.*, 1, 44, 1981.
63. **Lopez, G. and Todorovic, R. A.**, Rapid latex agglutination (RLA) test for the diagnosis of *Babesia argentina*, *Vet. Parasitol.*, 4, 1, 1978.
64. **Todorovic, R. A. and Kuttler, K. L.**, Babesiosis card agglutination test, *Am. J. Vet. Res.*, 35, 1347, 1974.
65. **Amerault, T. E., Frerichs, W. M., and Stiller, D.**, Comparative serologic study of equine piroplasmosis, with card and complement fixation tests, *Am. J. Vet. Res.*, 40, 529, 1979.
66. **Rodriguez, O. N., Espaine, L., Rodriguez, P., and Rivas, A.**, Applicacion de un sistema de metodos parael serodiagnostico de la anaplasmosis y babesiosis en bovinos, *Rev. Cubana Cienc. Vet.*, 10, 1, 1979.

67. **Applewhaite, L. M., Craig, T. M., and Wagner, G. G.,** Serological prevalence of bovine babesiosis in Guyana, *Trop. Anim. Health Prod.*, 13, 13, 1981.
68. **Madruga, C. R., Aycardi, E., Kessler, R. H., Schenk, M. A. M., De Figueiredo, G. R., and Curvo, J. B. E.,** Niveis de anticorpos anti-*Babesia bigemina* c *Babesia bovis*, cm bezerros da raca Nelore, Ibage e cruzamentos de Nelore, *Pesq. Agropec. Bras.*, 19, 1163, 1984.
69. **Sanchez, F. L., Fajardo, J., and Alarcon, G. J. C.,** Prevalencia de anticorpos contra anaplasmosis y babesiosis e incidencia de infeccion diaria de babesiosis en bovinos del municipio de Playa Vincente, Veracruz, *Tec. Pecu. Mex.*, 44, 82, 1983.
70. **Kuttler, K. L., Adams, L. G., and Todorovic, R. A.,** Comparisons of the complement fixation and indirect fluorescent antibody reactions in the detection of bovine babesiosis, *Am. J. Vet. Res.*, 38, 153, 1977.
71. **Trees, A. I.,** Indirect fluorescent antibody levels in experimental *Babesia divergens* infections of cattle, *Res. Vet. Sci.*, 24, 126, 1978.
72. **Rauch, J.,** Untersuchungen zum Nachweis von Antikörpern und von zirkulierenden Antigenen an experimentall mit *Babesia divergens* infizierten und an babesioseverdächtigen Rindern mit dem Enzyme-linked-Immunosorbent-Assay (ELISA), Vet. Med. thesis, University of Munich, 1981.
73. **Ullmann, B., Centaurier, C., Weiland, G., and Boch, J.,** Ein Beitrag zur Rinderbabesiose im westlichen, *Berl. Muench. Tieraerztl. Wochenschr.*, 97, 265, 1984.
74. **Boch, J.,** Babesieninfektionen bei Pferd, Rind und Hund in Süddeutschland, *Tieraerztl. Prax. Suppl.*, 1, 3, 1985.
75. **Weiland, G., Reif, L., Schmidt, R., and Boch, J.,** Serologische Untersuchungen zum Nachweis der *Babesia divergens*-Infektion des Rindes, *Berl. Muench. Tieraerztl. Wochenschr.*, 93, 261, 1980.
76. **Nicholls, M. J., Ibata, G., and Rodas, F. V.,** Prevalence of antibodies to *Babesia bovis* and *Anaplasma marginale* in dairy cattle in Bolivia, *Trop. Anim. Health Prod.*, 12, 48, 1980.
77. **Schindler, R. and Dennig, H. K.,** Versuche zum Nachweis von Antikörpern gegen *Babesien* mit der Komplementbindungsreaktion, *Z. Tropenmed. Parasitol.*, 13, 480, 1962.
78. **Schindler, R., Wokatsch, R., and Schröder, G.,** Immunität und serologische Reaktionen gegen *Babesia canis* bei Hunden nach Infektion mit lebenden Parasiten und nach Immunisierung mit einem löslichen Antigen, *Z. Tropenmed. Parasitol.*, 17, 226, 1966.
79. **Schindler, R., Schröder, G., Stieger, R., Wirahadiredja, S., and Kessler, W.,** Weitere Untersuchungen über Immunitat und serologische Reaktion bei Hunden nach einer Infektion mit *Babesia canis*, *Z. Tropenmed. Parasitol.*, 21, 182, 1970.
80. **Popovic, N. A. and Ristic, M.,** Diagnosis of canine babesiosis by a gel precipitation test, *Am. J. Vet. Res.*, 31, 2201, 1970.
81. **Breitschwerdt, E. B., Malone, J. B., MacWilliams, P., Levy, M. G., Wualls, C. W., Jr., and Prudich, M. J.,** Babesiosis in the Greyhound, *J. Am. Vet. Med. Assoc.*, 182, 978, 1983.
82. **Weiland, G. and Dohna, H.,** Die Brauchbarkeit des Enzyme-linked Immunosorbent Assay (ELISA) zum Nachweis von Babesien-Antikörpern in Hochimmunseren mit Hilfe von Ektoantigenen, *Berl. Muench. Tieraerztl. Wochenschr.*, 91, 33, 1978.
83. **Dennig, H. K., Centurier, C., Göbel, E., and Weiland, G.,** Ein Beitrag zur Babesiose des Hundes und ihrer Bedeutung in der Bundesrepublik Deutschland und Berlin West, *Berl. Muench. Tieraerztl. Wochenschr.*, 93, 373, 1980.
84. **Bertoldini, G., Codazza, D., and Redaelli, G.,** Serological diagnosis of *Babesia* infection of horses by complement fixation, *Clin. Vet.*, 101, 490, 1978.
85. **Chevrier, L., Soule, C., and Dorchies, Ph.,** Les Piroplasmoses inapparentes, *Bull. Acad. Vet. France*, 52, 37, 1979.
86. **Soule, C., Chevrier, L., and Dorchies, P.,** Diagnostic serologique des prioplasmoses equines. Utilisation d'une microtechnique de fixation du complement, *Rev. Med. Vet.*, 130, 1523, 1979.
87. **Joyner, L. P., Donnelly, J., and Huck, R. A.,** Complement fixation tests for equine piroplasmosis (*Babesia equi* and *B. caballi*) performed in the UK during 1976 to 1979, *Equine Vet. J.*, 13, 103, 1981.
88. **Donnelly, J., Phipps, L. P., and Watkins, K. L.,** Evidence of maternal antibodies to *Babesia equi* and *B. caballi* in foals of seropositive mares, *Equine Vet. J.*, 14, 126, 1982.
89. **Petrovskii, V. V., Malyshev, S. N., and Semina, L. Yu.,** Immunodiagnosis of equine piroplasmosis, *Tr. Vses. Inst. Eksp. Vet.*, 57, 86, 1983.
90. **Salabarria, F. F., Gonzales, M., and Jimenez, T.,** La RFC en el diagnostico serologico *Nuttalia equi*, *Rev. Cubana Cienc. Vet.*, 13, 81, 1982.
91. **Duran Diaz, A.,** Evaluacion de la prueba de fijacion de complemento en tubo y en microplaca para el diagnostico de la piroplasmosis equina, *Vet. (Mexico City)*, 14, 202, 1983.
92. **Heidrich-Joswig, S.,** Vergleichende Untersuchung von Antikörpertitern sowie der Immunglobuline IgM und IgG während einer *Babesia equi* Infektion, *Berl. Muench. Tieraerztl. Wochenschr.*, 95, 439, 1982.

93. **Callow, L. L., McGregor, W., Rodwell, B. J., Rogers, R. J., Fraser, G. G., Mahoney, D. F., and Robertson, G. M.,** Evaluation of an indirect fluorescent antibody test to diagnose *Babesia equi* infection in horses, *Aust. Vet. J.*, 55, 555, 1979.
94. **Donnelly, J., Joyner, L. P., and Frank, C.,** Quantitative epidemiological studies on the prevalance of babesiosis in horses in Kuwait, *Trop. Anim. Health Prod.*, 12, 253, 1980.
95. **Herrman, M., Baumann, D., Weiland, G., and von Salis, B.,** Das Auftreten von equiner Piroplasmose als Bestands problem in der Schweiz, *Pferdeheilkunde*, 3, 17, 1987.
96. **Malhotra, D. V., Banerjee, D. P., and Gautam, O. P.,** Prevalance of latent cases of *Babesia equi* infection in some parts of North West India as measured by the capillary agglutination test, *Equine Vet. J.*, 10, 24, 1978.
97. **Miranda, A. Q. and Colman, O. L. R.,** Babesiosis y anemia infecciosa equina: actualizacion an el empleo de papeles absorbentes en la remision de muestras para el diagnostico serologico, *Vet. Argent.*, 1, 250, 1984.

Chapter 10

IMMUNIZATION AGAINST *BABESIA*

Miodrag Ristic and Sonia Montenegro-James

TABLE OF CONTENTS

I. INTRODUCTION

A. General Background

Babesiosis is a tick-transmitted disease of animals that is manifested by fever, anemia, occasional hemoglobinuria, and the appearance of infecting protozoa in the host erythrocytes.[1] Currently, 17 distinct species of *Babesia* are known to infect various vertebrate hosts. Depending on the species of *Babesia,* the transmission in the tick can be transovarial or transtadial.[2] Losses due to babesiosis may be as high as 50% when susceptible cattle are introduced into endemic areas of the disease.

The term *Babesia* was adopted to honor a Rumanian scientist, Babes, who in 1888 first described the *Babesia* parasite in the blood of African cattle showing hemoglobinuria.[3] The classic work of Smith and Kilbourne, which was underway in the U.S. at approximately the same time, signified the beginning of the scientific history of babesiosis.[4] As a result of this study, the tick was demonstrated for the first time as a vector of a protozoan agent, *Babesia bigemina.* For several decades during the early part of this century, babesiosis, caused by *B. bigemina,* was responsible for severe losses of cattle in the U.S. These losses, estimated to be in millions of dollars, prompted national tick eradication programs which started in 1906 and were successfully completed during a 4-decade period.

The success of the U.S. in eliminating babesiosis by eradication of the vector remains an isolated example. Babesiosis persists as one of the most important diseases of cattle in all tropical and semitropical regions of the world. Most of the developing countries are situated in these regions. There are approximately 1.2×10^9 cattle in the world, and the majority of these animals are potentially threatened by one or more *Babesia* spp.[5] Pressed by a need for economic improvements, new methods of animal husbandry are being gradually instituted in countries affected by babesiosis. Among the several methods used to increase the productivity of commercial operations are the importation and local adaptation of more productive breeds of cattle from countries with temperate climates. Also, the newly imported cattle are more economically raised on farm lands located at lower altitudes which are operationally more effective and commercially more accessible. Tick-borne diseases, babesiosis in particular, constitute one of the major constraints for the expected performance of these animals, thus limiting the availability of food for human consumption and the potential supply of export products on international markets.

The example of the U.S. in controlling tick-borne diseases by eradication of the tick vector has met with serious difficulties in countries located in tropical and semitropical regions. Some of the obstacles to this approach are the development of tick resistance to acaricides,[6] excessive cost of modern acaricides, and the ecological features of the tick, which often feeds on various domestic and wild animals.[2] Another method of controlling babesiosis is the use of chemoprophylactic and chemotherapeutic compounds. In small herds and barn-raised cattle, there is a good chance of detecting animals during the initial phase of the disease and promptly administering chemotherapy which frequently results in a favorable outcome. The situation with range-raised cattle spread over vast pasture areas is, however, more difficult and less practical.[7] Consequently, as with other infectious diseases, the only effective means for controlling the disease is by use of immunoprophylactic measures, namely, safe and effective vaccines.

B. Protective Immunity

For many decades, the consensus of opinion was that induction of effective protective immunity against bovine babesiosis required prior exposure of the host to the live, preferably replicating, causative agents of this disease. A procedure for induction of protective immunity by infection and treatment, known as premunization, is the oldest known method. Safer immunization procedures have been developed subsequently by altering the virulence of the immunizing organism by rapid serial passages in splenectomized calves (*B. bovis*) and exposure of the infective blood to irradiation (*B. bovis* and *B. bigemina*). Cattle immunization studies utilizing blood-derived nonliving *Babesia* antigens and protection to clinical infection after elimination of the agent from carrier animals by chemotherapy indicated that the sterile immunization approach to vaccine development against babesiosis is possible.[8]

A major breakthrough in babesiosis vaccine research was achieved by the development of methods for continuous in vitro propagation of *B. bovis*.[9] This finding was followed by the development of methods for short-term cultivation of *B. canis* and *B. bigemina*. These culture procedures provided abundant quantities of protective soluble exoantigens in the supernatant medium that were found to be effective anti-*Babesia* immunogens.[10]

Immunochemical analysis of culture supernatants has demonstrated that at least three parasite antigens are released in vitro.[11] These antigens were found to be effective immunogens useful for the induction of protective humoral immunity against infections caused by *B. bovis*, *B. bigemina*, and *B. canis*. In addition, the ability of exoantigens to induce cell-mediated immune (CMI) responses was demonstrated by means of the in vitro interaction between soluble antigens and mononuclear phagocytes derived from normal, *Babesia*-free cattle. This procedure resulted in the production of soluble fractions by treated phagocytes which inhibited growth and development of *Babesia* parasites.[12] Furthermore, inoculation of cattle with *B. bovis* exoantigens fortified by the use of a suitable adjuvant has revealed CMI responses in inoculated cattle by means of the lymphocyte transformation technique.[13]

Various laboratory and field vaccination studies conducted in several regions of the world using culture-derived exoantigens of *B. bovis*, in combination with appropriate adjuvants, yielded satisfactory results.[8]

A joint research program between the University of Illinois and the Venezuelan Veterinary Research Institute, now in its 6th year of existence, has led to the development of a combined *B. bovis* and *B. bigemina* vaccine. After a series of controlled laboratory and field studies, the combined culture-derived vaccine is now officially produced by Venezuelan government scientists and widely used for the control of bovine babesiosis in that country.

The exoantigen culture system has also been successfully applied to the development of an immunogen against *B. canis*.[14] Further refinements of the system have led to the development of the first commercial vaccine for canine babesiosis.[15]

II. IMMUNIZATION OF CATTLE WITH BLOOD AND TICK-DERIVED ANTIGENS

Various vaccination methods have been developed and studied under laboratory and field conditions as immunoprophylactic means against bovine babesiosis. In most instances, the vaccine antigens were derived from animals in the acute phase of infection, usually splenectomized, and were introduced into recipient animals in live or inactivated forms. Other research efforts explored the use of tick-derived preinfectious *Babesia* forms as a means of inducing protective immunity against bovine babesiosis.

A. Live Vaccines

A limited degree of protection was produced by cattle injected with irradiated *B. bovis*, *B. bigemina*, or *B. major* infected erythrocytes. The cells were irradiated with a dose sufficient to prevent a parasitemia from developing after injection of the irradiated parasites.[16] Where the irradiation dose only "inactivated" or "killed" the majority of the parasites, the calves were later strongly immune to challenge with viable parasites. The immunizing efficiency of irradiated parasites may be related to the fact that even after irradiation there is some residual metabolic activity retained by the organism.[17]

Immunization of cattle against babesiosis caused by *B. bovis* using an "avirulent" vaccine strain has successfully been practiced in Australia since 1965. A vaccinal strain is produced by rapid passage of *B. bovis*-infected blood through splenectomized calves.[18] The organism appears safe for inoculation into animals with spleen *in situ*; however, subsequent inoculation of blood from vaccinated animals into susceptible cattle may induce clinical babesiosis. The mechanism involved in change of the virulence of the organism by passage through splenectomized calves has not yet been fully analyzed.

In order to study the underlying mechanism of the attenuation process, we subjected tick-derived *B. bovis* to 20 consecutive blood passages in splenectomized calves.[8] Studies of the parasite in infected animals and in vitro cultures revealed that with each passage there was an increase in blood parasitemia level and that each increase was followed by a gradual shift of the parasite morphology from the pyriform type to predominantly ring forms. Subsequent studies revealed that the selection from pyriform to ring form parasites was due to the ability of the latter to multiply at nearly twice the rate of the pyriform type. Finally, an electron microscopic examination of the parasite population at various passages showed that the majority of erythrocytes hosting low passage parasites (passages 1 to 4) possessed membrane knobs and that these structures were rarely, if at all, observed in parasitized erythrocytes beyond the eighth consecutive passage in splenectomized calves. It could, therefore, be concluded that rapid passages in splenectomized calves abrogated the ability of the organism to escape splenic filtration, thereby rendering it less virulent. When a knobless, "attenuated" parasite population was passed two times through intact (spleen *in situ*) cattle, a reversion to the knobbed virulent form occurred in a rather rapid fashion. Thus, the knobs appear to be a reversible specific "marker" for the "wild-type" *B. bovis*. Infection with the knobless type, however, confers complete protection to cattle against the virulent knobbed type. Consequently, the attenuation process of *B. bovis* developed by the Australian workers must be considered as a selection process of more rapidly multiplying *B. bovis* forms which lack the ability to sequester in blood capillaries and thus are unable to avoid destruction by splenic filtration.

From a practical point of view, the continuous use of an "attenuated" *Babesia* vaccine obviously results in an establishment and maintenance of bovine reservoirs of infection for the tick vector. In addition, adverse reactions have been occasionally observed in vaccinated cattle. This situation prompted many workers to investigate the usefulness of inactivated vaccines and the sterile immunity induced by such vaccines.

B. Vaccines Using Blood Plasma-Derived Exoantigens

Our knowledge of blood plasma exoantigens dates some 2 decades back when we first observed their presence in the serum of dogs, horses, and rats acutely infected with *B. canis*, *B. caballi*, *B. equi*, and *B. rodhaini*, respectively.[8,19] During these studies, an interrelationship between parasitemia levels, concentration of plasma antigens, and the level of antibodies to these antigens was established. There was a direct relationship between the levels of parasitemia and plasma antigens while an inverse relationship existed between these elements and serum antibody levels. Recognizing the immunogenic potential of these antigens, we initiated vaccination studies of dogs against infections caused by *B. canis* utilizing soluble exoantigens as immunogens. It is in the course of these studies that we, for the first time, recognized the protective nature of these antigens. However, while such potential was clearly in evidence, it was also recognized that the immunogenicity of soluble plasma-derived antigens was low, probably because of immune complex formation and enzymatic degradation.

The antigens showed serologic cross reaction between species, and the antibodies synthesized after immunization killed the parasites in neutralization tests performed in vitro and in vivo. Direct fluorescent antibody tests showed that these antibodies also combined with antigens on both erythrocyte stroma and on the organism itself.[20]

Vaccination studies with *B. bovis* and *B. divergens* serum soluble antigens conferred partial protection in susceptible cattle against homologous challenge.[21]

Biologically intact exoantigens derived from the supernatant medium of babesial and plasmodial cultures and used in crude and purified forms induced protective immunity in inoculated animals.[19,22]

C. Vaccines Using Corpuscular Erythrocyte-Derived Antigens

Mahoney[23] and Phillips[24] first showed that protection could be induced against *B. bovis* and *B. rodhaini*, respectively, by the inoculation of washed parasite suspensions emulsified in Freund's complete adjuvant (FCA). For *B. bovis*, infected erythrocytes were first concentrated to a 95% suspension and then lysed in distilled water, and parasites mixed with erythrocyte stroma were recovered by centrifugation.

Immunization failed to prevent the establishment of infection, but multiplication of the organism was suppressed. The vaccinated animals did not become ill and experienced only mild temperatures and transient parasitemias.

Kuttler and Johnson[25] have demonstrated protective immunity in a group of adult intact cattle and splenectomized calves following vaccination with a particulate *B. bigemina* antigen. The immunogen was prepared from parasitized erythrocytes lysed in a 0.35% NaCl solution and subsequently washed with veronal buffer pH 7.2. The antigen was administered subcutaneously in an equal volume of FCA.

Among several disadvantages, there are two major difficulties with the use of infected cattle-derived blood for the preparation of anti-*Babesia* vaccines. The first obvious difficulty is that the method is too laborious and costly for mass vaccine production. The second difficulty is the development of antibodies to isoerythrocytic antigens in vaccinated animals. The latter constitutes a serious obstacle for wider use of the method.

D. Solubilized Erythrocyte-Derived Vaccines

The soluble material from infected erythrocytes disintegrated by sonic oscillation was fractionated by precipitation of the fibrinogen-like (FL) proteins and by absorption of hemoglobin on a CM-Sephadex® column.[20] Three fractions — FL proteins, the supernatant after precipitation of FL proteins, and the hemoglobin-free fraction of the crude soluble material — were used to immunize groups of five splenectomized calves. All three fractions induced protective immunity in vaccinated calves. It was also found that FL proteins contained all of the specific babesial antigens located on the stroma of infected erythrocytes, but little of that located on the parasite itself.[20] In other similar studies utilizing solubilized *B. bovis* antigens, at least two fractions were found to be immunogenic in splenectomized calves. One of these fractions appeared to have greater potential because of its lower contamination with native bovine erythrocyte antigens.[26]

More recently, a number of similar antigens were isolated using monoclonal antibodies.[27] One of these antigens, fortified with FCA, induced protective immunity in splenectomized calves against challenge 2 weeks following the second vaccinal dose.[27]

E. Vaccines from Infected Ticks

Prior to the development of cultures for in vitro propagation of *Babesia* species and in absence of a laboratory animal that is susceptible to infection with these organisms (*B. divergens* in mice is an exception), there has been a continuous effort to use tick stages of *Babesia* as potential immunogens.

The final tick *Babesia* form derived from repeated asexual division of kinetes has been known as vermicule.[28] The current terminology for this process is sporogony, and the resulting *Babesia* form is termed sporokinea.

It was reported that sera from animals that had recovered from *B. bigemina* infections cross reacted serologically with tick-derived vermicules of that organism.[29] In order to test the possible use of tick-derived vermicules as immunogens against bovine blood-phase infections, we conducted several vaccination and challenge experiments. The vermicules derived from engorged *Boophilus microplus* ticks were purified on a discontinuous Ficoll density gradient and then emulsified with FCA and used to vaccinate four calves. The calves were inoculated with 1.9×10^7 vermicules divided into two doses given at 2-week intervals. Autofluorescence of vermicule preparations interfered with the interpretation of IFA tests when vermicule antigens were used. No significant cross reactivity with blood-phase parasites was observed.

This negative finding is consistent with the negative IFA test results obtained when serum from naturally infected calves was tested against vermicular antigen. To further elucidate a possible relationship between tick-derived vermicules and larvae-derived infectious *B. bovis* forms, 5 calves were each inoculated with 1.7×10^7 vermicules and challenged with 1000 larvae 8 weeks later. There was no detectable difference in the resulting infection in the animals in the control and immunized groups. The IFA test with blood-phase antigen became positive approximately 10 days postchallenge. When vermicules were used as antigens in the IFA test, no difference could be demonstrated between sera from control and immunized calves, even after homologous challenge. Despite the failure to detect antibodies crossreacting with blood and tick-stage parasites, 38% less ($p < 0.05$) adult females dropped from the immunized calves than from the control calves. This reduction was probably mediated by an immune response to tick antigens present in the vermicule preparations injected into these calves.

Even if an immune response and protection against tick-borne challenge could have been induced with one or more injections of vermicules, the large number of female ticks needed to produce that many vermicules under the conditions employed would make this approach to vaccination impractical.

III. IMMUNIZATION OF CATTLE WITH CULTURE-DERIVED *BABESIA* EXOANTIGENS

A. Introduction

Control of vector ticks and chemotherapeutic measures cannot be expected to provide the complete resolution of babesiosis. Furthermore, serious disadvantages have been attributed to the use of living *Babesia* organisms as vaccines against the disease.[20] Thus, development of a soluble, exoantigen-containing *Babesia* immunogen, free of erythrocytic antigens, has been proposed as a practical and realistic solution for the immunoprophylactic control of babesiosis.[10]

Exoantigens are soluble proteinaceous moieties naturally released into the blood plasma of animals infected with *Babesia* or into the supernatant medium of in vitro cultures of these organisms. Because the immunogenic potential of exoantigens derived from infected animals is minimized due to complex in vivo interactions,[30] in vitro culture systems have been optimized for antigen production. Consequently, a safe and effective babesiosis vaccine, consisting of culture-derived exoantigens and a potent adjuvant, is already in the final phase of development.

The continuous in vitro cultivation system[9] for *Babesia* organisms offers at least two sources of vaccine material: (1) a whole merozoite immunogen, and (2) parasite-free, soluble exoantigens derived from the merozoite surface coat.[31] Intensive research has been devoted to the merozoite vaccine; however, it is extremely difficult to purify parasites and parasitic antigens from erythrocytic components.[32] Recently, several target host-protective antigens associated with *B. bovis*-infected erythrocytes have been shown to be of host rather than parasite origin.[33,34] Culture-derived exoantigens are essentially free of erythrocytic stromal antigens, thus there is no induction of isoantibodies associated with neonatal isoerythrolysis.[13]

The merozoite surface coat elicits the production of antibodies that react directly with the parasite, causing aggregation and lysis of merozoites.[31] Consequently, antibodies produced against exoantigens are important for protection against disease.

Exoantigens from in vitro cultures of *B. bovis* and *B. canis* have been shown to be highly immunogenic in cattle[35-38] and dogs,[14,39] respectively. Immunogenicity has been confirmed in independent trials conducted in Mexico,[31,35,40] the U.S.,[36,37] Colombia,[41] Israel,[42] Venezuela,[38,44] France,[14] and Australia.[13] In vitro cultivation of *B. bigemina*[43,44] and *B. divergens*[45,46] also has made possible the incorporation of exoantigens from these *Babesia* species in vaccine preparations.

Culture-derived vaccines are advantageous in that in vitro systems may be scaled up to accomodate production demands, and the storage and shipment of these vaccines may be easily facilitated by preparation in lyophilized form. Development of semiautomated culture systems will optimize exoantigen output (see Chapter 4).

Culture-derived *B. bovis* and *B. bigemina* exoantigens have been isolated and identified by high performance liquid chromatography, gel electrophoresis, and immunoblotting,[47,48] These studies have allowed the analytical characterization of these antigens. Additionally, *Babesia* exoantigen production and purification has been monitored and quantitatively assessed by a two-site immunoassay.[48,49] This assay can be effectively utilized to control the dosage of antigenically defined immunogens.

The protective capacity of such immunogens has to be exhaustively tested before mass application. Factors such as the immunogenicity of different strains, duration of immunity, minimal protective dose, the protective efficacy of a combined *B. bovis-B. bigemina* immunogen, vaccine stability, and immunopotentiation with various adjuvants must be investigated. During the last 6 years, most of these aspects have been studied as a result of a cooperative research program between the University of Illinois and the Veterinary Research Institute in Maracay, Venezuela. A critical appraisal of field vaccination trials began 3 years ago on private- and government-owned farms selected in areas where large scale dairy and beef production is crucial for the Venezuelan livestock industry. Much progress has been made in the analysis of protective immunity conferred by culture-derived *Babesia* exoantigens. Advances in this area have been made possible by close collaboration between university scientists, government institutes in tropical countries, and industrial laboratories.

At the present time, exoantigen-containing vaccines are being used against *B. bovis, B. bigemina,* and *B. canis* parasites. With each specific vaccine, extensive laboratory and field immunization trials have been conducted. The characteristics of an effective inactivated vaccine for bovine babesiosis have been described,[50] and it is believed that *Babesia* exoantigens meet these requirements: prevent clinical disease, effective against different parasite strains, two doses when combined with an acceptable adjuvant induce protective immunity for at least a 12-month period, no concurrent immunization against host blood groups, antigenically stable for at least 2 years when stored at 4°C and supplied in large quantities. Continued optimization of currently available culture-derived immunogens will best guarantee the successful production of food-producing animals in the tropics.

B. Vaccines from *Babesia* Cell Cultures

1. Preparation of Culture-Derived B. bovis *and* B. bigemina *Immunogens*

The MASP[9] erythrocyte culture system is initiated utilizing either *B. bovis* or *B. bigemina* organisms selected as seed material. All parasite strains are maintained by transmission in *Boophilus microplus* ticks under laboratory conditions and/or by no more than four syringe-passages in splenectomized calves. Initiation of cultures of *B. bovis, B. bigemina,* and *B. canis* requires fresh, infected blood because suboptimal growth results from cryopreserved *Babesia.*[51] Reactivation of each strain is established separately by inoculation of infected blood stabilities into splenectomized animals. Infections are monitored by examining thin

Preparation of Culture-Derived *Babesia bovis* and *Babesia bigemina* Immunogens

MASP culture 3-day cycle

↓

pool supernatant fluids collected daily

↓

centrifuge, filter (0.45 μm) and lyophilize

↓

add 1.5 mg saponin (QUIL-A)/dose and store at 4°C

Each ml-eq contains approximately 650 EIA units of *Babesia bovis* and 500 EIA units of *Babesia bigemina*

FIGURE 1. Flow diagram for the preparation of culture-derived *B. bovis* and *B. bigemina* immunogens.

peripheral blood smears and by determining body temperatures and packed cell volumes (PCV). When a certain preestablished parasitemia level (1 to 3%) is reached, infected blood is aseptically collected to begin in vitro cultivation. Healthy, susceptible calves (or dogs for *B. canis*) are maintained in tick-free isolation units and serve as blood and/or serum donors. Cultures are set up at 10% hematocrit in a medium consisting of 60 or 80% Medium 199 and 40 or 20% adult bovine serum (*B. bovis* and *B. bigemina*, respectively) supplemented with HEPES and antibiotics. Erythrocyte cultures are incubated at 37°C in an atmosphere of 3% CO_2 in humidified air. Average parasitemia for a 72-hr culture cycle is 8% for *B. bovis* and 3% for *B. bigemina*. Figure 1 depicts a flow diagram of the procedure followed in the preparation of culture-derived *B. bovis* and *B. bigemina* immunogens.

2. Quantitation of Culture-Derived Babesia Exoantigens

A two-site immunoassay (EIA) was developed for detection of culture-derived *B. bovis*[49] and *B. bigemina*[48] exoantigens. Principally, the assay consisted of four major steps: (1) bind purified specific antibody to microtiter wells, (2) add culture supernatants containing exoantigens, (3) add the enzyme-labeled specific antibody, and (4) add a substrate and record the subsequent amount of color development. Bovine IgG was isolated from hyperimmune serum obtained from cattle that had been previously immunized with either culture-derived *B. bovis* or *B. bigemina* exoantigens and later needle challenged with virulent *B. bovis* or *B. bigemina*-infected erythrocytes (IFA titers = 1:163,840 for *B. bovis* and 1:40,960 for *B. bigemina*). Immune IgG was fractionated by caprylic acid[52] or ammonium sulfate precipitation[53] and further purified by anion-exchange chromatography. Anti-*Babesia* IgG was then conjugated to horseradish peroxidase (HRPO)[54] or alkaline phosphatase (AP)[55] for *B. bovis* and *B. bigemina*, respectively. The basic test procedure was adapted from a two-site EIA method previously described.[56] Optimal concentrations of capture antibody were 10 μg/mℓ and 100 μg/mℓ for antibodies to *B. bovis* and *B. bigemina*, respectively.[44] Results were expressed in EIA units calculated as follows:

$$\text{EIA unit or P/N ratio} = \frac{\text{OD 414 infected culture supernatant fluid}}{\text{OD 414 uninfected culture supernatant fluid}}$$

Figure 2 depicts the typical reactivity pattern observed with the respective titration of 3-day culture supernatant pools containing *B. bovis* or *B. bigemina* exoantigens.[44] The values determined at an antigen dilution of 1:80 averaged 8.5 EIA units or 680 EIA units/milliliter equivalent (mℓeq) for *B. bovis*, and 6.5 EIA units or 520 EIA units/mℓeq for *B. bigemina*. The two-site EIA has also proved useful for monitoring antigenic activity during purification

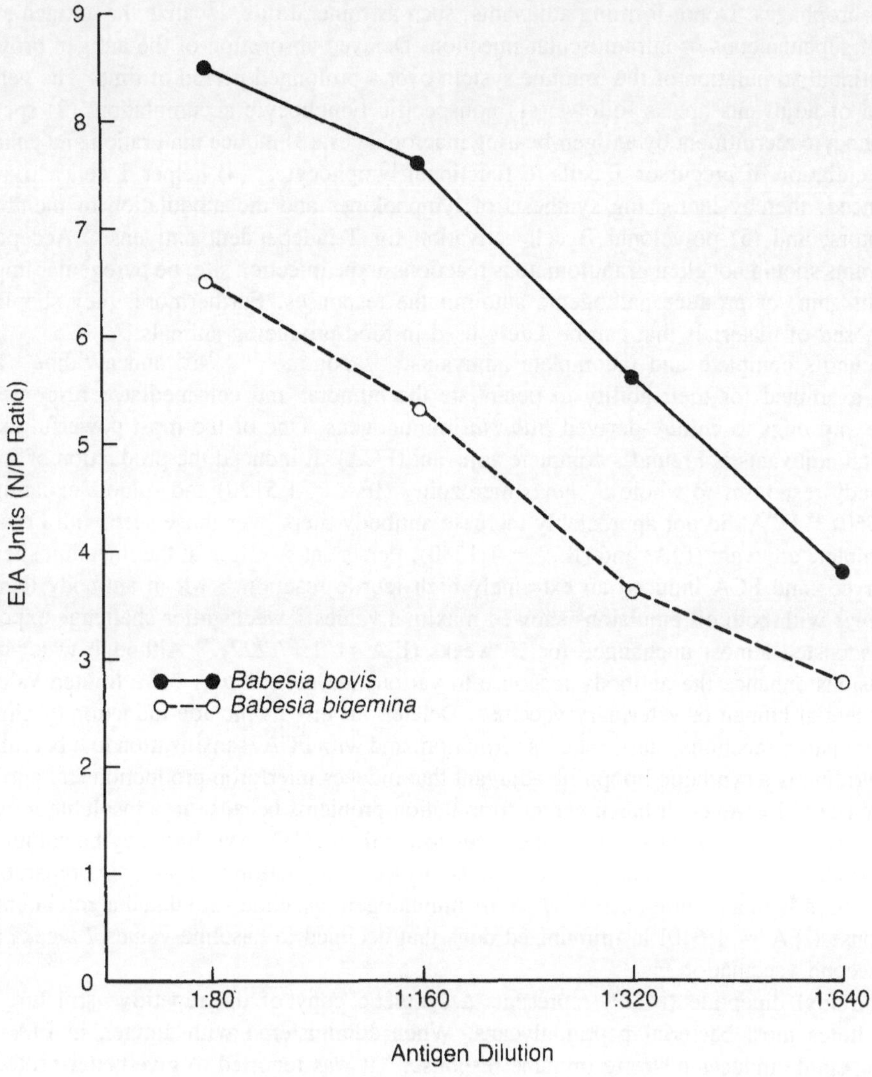

FIGURE 2. Titration of *B. bovis* and *B. bigemina* exoantigens using a two-site enzyme immunoassay.

procedures.[47,48] Furthermore, with this assay, the kinetics of antigen released into supernatants can be measured. Exoantigen production correlated with the rate of early in vitro growth as 24-hr cultures showed the greatest antigenic activity.[49] Using a two-site EIA to detect malarial antigens released in vivo during *Plasmodium yoelii* infection in mice, Taylor et al.[57] observed similar kinetics of exoantigen release.

3. *Adjuvant Selection and Immunization Procedure*

Coincident with the application of selected antigens, rather than intact *Babesia* organisms, comes the need for adjuvants and/or appropriate delivery systems to present an immunogen to the immune system of the host in its most effective form. An immunologic adjuvant can be defined as any agent that acts nonspecifically to enhance specific immune responses to an antigen. Mechanisms of action are complex due to the heterogeneity of many adjuvant preparations, the multiple effects of various adjuvant materials, and the complexity of the immune system.[58] Adsorption of antigens onto carriers, such as aluminum hydroxide, latex particles, or liposomes, increases the size of the antigenic complex, facilitating its uptake by macrophages. Depot-forming adjuvants, such as mineral oils, localize the antigen at the site of subcutaneous or intramuscular injection. Delayed absorption of the antigen provides a continual stimulation of the immune system over a prolonged period of time. The general action of adjuvants are as follows: (1) nonspecific lymphocyte accumulation; (2) specific lymphocyte recruitment by antigen-bearing macrophages; (3) induce maturation and enhance differentiation of precursor T cells to functional lymphocytes; (4) helper T cell activity is enhanced, thereby increasing synthesis of lymphokines and the modulation of membrane receptors; and (5) polyclonal B cell activation for T-independent antigens.[58] Acceptable adjuvants should not elicit granulomatous reactions at the injection site, be pyrogenic, impede weight gain, or produce pathogenic autoimmune responses. Furthermore, they should be composed of materials that can be safely used in food-producing animals.[59]

Freund's complete and incomplete adjuvants,[35] saponin,[13,31,36-38,44] and avridine[61] have been examined for their ability to potentiate the humoral and cell-mediated responses of cattle and dogs to culture-derived *Babesia* immunogens. One of the most powerful experimental adjuvants is Freund's complete adjuvant (FCA). It induced the production of strong antibody responses to whole *B. bovis* merozoites (IFA = 1:5120) and soluble exoantigens (1:2560).[35] FCA did not appreciably increase antibody titers over those seen with Freund's incomplete adjuvant (FIA) and (IFA = 1:1280). Persistent swelling at the injection site was observed, and FCA induced an extremely high febrile reaction.[35] Mean antibody titers to *B. bovis* with both oil emulsions showed maximal values 3 weeks after challenge exposure and persisted almost unchanged for 10 weeks (IFA = 1:17,222).[35] Although water-in-oil emulsions enhance the antibody response to various antigens,[60] they have limited value in commercial human or veterinary vaccines. Deleterious effects include induction of chronic inflammatory reactions, sterile abcess formation, and with FCA, sensitization to tuberculin.[60]

Avridine is a synthetic lipophilic adjuvant that induces interferon production and antiviral resistance.[61] However, it has inherent formulation problems because it is insoluble in water and exists as waxy particles at temperatures lower than 39°C. Avridine may be solubilized in absolute ethanol and mixed with oil emulsions for combination with antigen preparations. When used with a culture-derived *B. canis* immunogen, avridine stimulated a good humoral response (IFA = 1:640) in immunized dogs that declined to baseline values 7 weeks after the second vaccination.[62]

Muramyl dipeptide (MDP) represents a synthetic copy of the repetitive structure that constitutes most bacterial peptidoglycans. When administered with antigen in FIA, this glycopeptide induces a strong immune response.[63] It was reported to give better protection than saponin when mixed with an antigenic fraction derived from *B. divergens*-infected erythrocytes.[64] However, MDP is pyrogenic, and mineral oil and some emulsifiers are potentially oncogenic.[63]

Before vaccines are used in commercial livestock, it is essential to determine the efficacy of adjuvants and their acceptability with regard to potential effects on carcass quality. Thus, relatively innocuous adjuvants, such as aluminum hydroxide and saponin, are often administered to increase antibody production against common bacterial toxins and inactivated viruses.[65] Saponin is a glycoside of plant origin (*Quillaia saponaria*) that functions as a surface-active adjuvant, associating with cell membranes to facilitate cell interaction and synthesis of regulatory molecules.[65] Antibody responses to *Babesia* are augmented when exoantigens are injected in combination with saponin (Quil-A).[36-38,44] High antibody levels are obtained 1 week (IFA = 1:2229) and 2 weeks (IFA = 1:5583) after the second vaccinal dose, followed by a transient decline in antibody titers during a 3-month period.[38] CMI, as assessed by specific lymphocyte transformation in vitro is also enhanced by saponin. Peripheral blood lymphocytes may be significantly stimulated up to 6 months after vaccination with *B. bovis* exoantigen.[13] Saponin probably stimulates a subpopulation of T lymphocytes reactive to *Babesia* antigens, thus explaining the characteristic anamnestic response seen after challenge exposure.[13,31,36-38,44] Another report has confirmed the CMI-enhancing capacity of saponin as delayed hypersensitivity reactions were observed in cattle immunized with a foot-and-mouth vaccine containing the adjuvant.[66]

Wells et al.[67] used nine different adjuvants to examine their ability to potentiate the humoral and cell-mediated immune responses of cattle to a soluble glycoprotein antigen prepared from *Trypanosoma brucei*. Saponin was the only nonoil based adjuvant that induced a protective immune response without concomitant side effects observed in animals that received oil-containing adjuvants. A wide range of immunological adjuvants have been employed to potentiate the protective immunity elicited by inactivated vaccines prepared from various *Plasmodium* species in different animal hosts.[68] Saponin has been shown to be a powerful adjuvant for amplifying the immunity conferred by killed vaccines against *Plasmodium* spp.,[69-71] *T. congolense*,[72] and *T. brucei*.[67] Apparently, saponin is particularly efficacious with protozoan vaccines as it potentiates the humoral response to cell surface parasitic antigens.[68] Results of the above studies illustrate the overall effectiveness of saponin as an adjuvant for vaccines in cattle. Last, its use is practical and acceptable because of its ease in formulation due to high water solubility and because no tissue inflammation is evident at the inoculation site.

Qualitative and quantitative differences in immunogenicity are also subject to the influence of the immunization procedure. Antigen composition, immunogen preparation, antigen dosage, and immunization schedule are all factors that relate to immunogenicity and the type of immune response.[58] Closely spaced inoculations produce a rapid decline in antibody levels and a poor booster effect. Vaccination at 2-week intervals with *B. bovis* exoantigens gave suboptimal results.[13,35] Intervals of 3 to 4 weeks have produced consistently better results in terms of the persistence of high antibody levels after two vaccinal doses.[36-38,44] Similarly, intervals of 15 to 32 weeks used for a foot-and-mouth disease virus polypeptide immunogen have also given promising results as an antibody response was maintained for at least 8 months.[73] Ongoing babesiosis vaccination trials are attempting to establish the duration of immunity conferred by *Babesia* exoantigens beyond the 6-month period,[37] considering the practicability and convenience of greater vaccination intervals.

C. Efficacy of *Babesia* Exoantigens

1. Cross-Protective Capacities of Culture-Derived B. bovis Exoantigens

An important consideration for the use of culture-derived *B. bovis* immunogens is the cross-protective capacity between different geographical isolates. Many species of blood protozoa represent a mosaic of different antigenic types, and in every infection, the parasite circumvents the immune response of the host by continually changing important surface antigens.[74] Interstrain antigenic differences exist within the *B. bovis* species and have been well documented.[75-78] However, this antigenic variation appears to be of minor importance in the control of clinical disease. Strong cross immunity has been observed in animals immunized with one strain and challenged with another.[77] Moreover, most antigens are found to be conserved among different isolates of *B. bovis*.[79] Attenuated Australian *B. bovis* vaccine strains have been well characterized and shown to be effective in inducing heterologous strain immunity.[80-82]

The logistics of vaccine development and its subsequent application would probably require heterologous immunity to geographically isolated *B. bovis* strains. Consequently, a study was designed to assess the degree of heterologous strain immunity induced by culture-derived soluble immunogens against *B. bovis* strains from different geographical locations of Latin America (Argentina, Colombia, Ecuador, Mexico, and Venezuela).[38] Susceptible, yearling, crossbred cattle were vaccinated with immunogens derived from Venezuelan or Mexican strains and challenged with virulent *B. bovis* organisms of the strains specified above. Table 1 summarizes some of the most relevant findings in this study. Vaccinated animals were

Table 1
EFFICACY[a] OF A VENEZUELAN CULTURE-DERIVED *BABESIA BOVIS* IMMUNOGEN AGAINST HETEROLOGOUS CHALLENGE

Babesia bovis challenge strain	Group	Duration of fever (days)	Duration of parasitemia (days)	Max % PVC reduction	Max % platelet reduction
Venezuela	Vaccinated, n = 6	0.6 ± 0.1[b]	1.2 ± 0.3[b]	25.7 ± 1.3[b]	51.8 ± 1.3[b]
	Controls, n = 6	3.6 ± 0.4	3.0 ± 0.5	41.2 ± 3.6	82.0 ± 3.2
Argentina	Vaccinated, n = 6	1.3 ± 0.2[b]	2.4 ± 0.5[b]	31.2 ± 1.0[b]	67.4 ± 2.4[b]
	Controls, n = 6	3.5 ± 1.1	4.7 ± 0.4	46.2 ± 3.0	81.4 ± 4.8
Mexico	Vaccinated, n = 6	5.0 ± 0.8	6.0 ± 1.0	50.0 ± 4.7	78.0 ± 3.8
	Controls, n = 6	6.0 ± 0	8.0 ± 0.5	56.3 ± 2.3	86.0 ± 2.8

[a] Group means ± standard error.
[b] Significantly different from control values, $p < 0.01$.

fully protected against challenge with the homologous strain, as their clinical condition remained unaltered after challenge exposure and continued to gain weight throughout the experimental period. Significant clinical differences were in the duration of fever (vaccinates = 0.6 days; controls = 3.6 days) and maximum percent PCV reduction (vaccinates = 25.7%; controls = 41.2%). These findings confirmed previous reports where strong homologous immunity was induced with culture-derived *B. bovis* exoantigens of the Mexican strain.[31,36,37] The Venezuelan immunogen also demonstrated the ability to induce heterologous strain immunity against challenge with the Argentinian and Ecuadorian strains (data not shown). This good degree of cross protection indicates that common protective antigens exist among strains from various geographical regions. An interesting finding in this study was the difference in cross immunity between the Mexican isolate and all the other strains utilized. Cattle vaccinated with the Venezuelan immunogen failed to protect against challenge with the Mexican strain. Similarly, there was a low degree of cross protection when animals vaccinated with the Mexican strain were challenged with any of the heterologous isolates (data not shown). One possible explanation for this uniqueness may be the long-standing prohibition of cattle importation into Mexico for the purpose of controlling foot-and-mouth disease.

A comparison of mean antibody titers between homologous (Venezuela) and heterologous strains, indicative of the presence and absence of cross protection (Argentina and Mexico, respectively), is shown in Figure 3. Maximal homologous antibody responses were reached 1 week after the second vaccination: IFA titer for homologous Venezuela strain = 1:2229; heterologous Argentinian strain = 1:1152, and heterologous Mexican strain = 1:448. The presence of strain-specific antigens among various Latin American isolates was evident as heterologous antibody levels were invariably lower for all strains, with the Mexican isolate demonstrating the greatest difference during the vaccination period. Fourfold differences in postvaccination titers have been reported between a *B. bovis* strain from Israel and the Mexican isolate.[42] A strong anamnestic response began soon after challenge with titers peaking approximately 4 weeks postchallenge. Higher levels of antibody to all *B. bovis* strains were eventually obtained. It is possible that elevated acute phase proteins observed

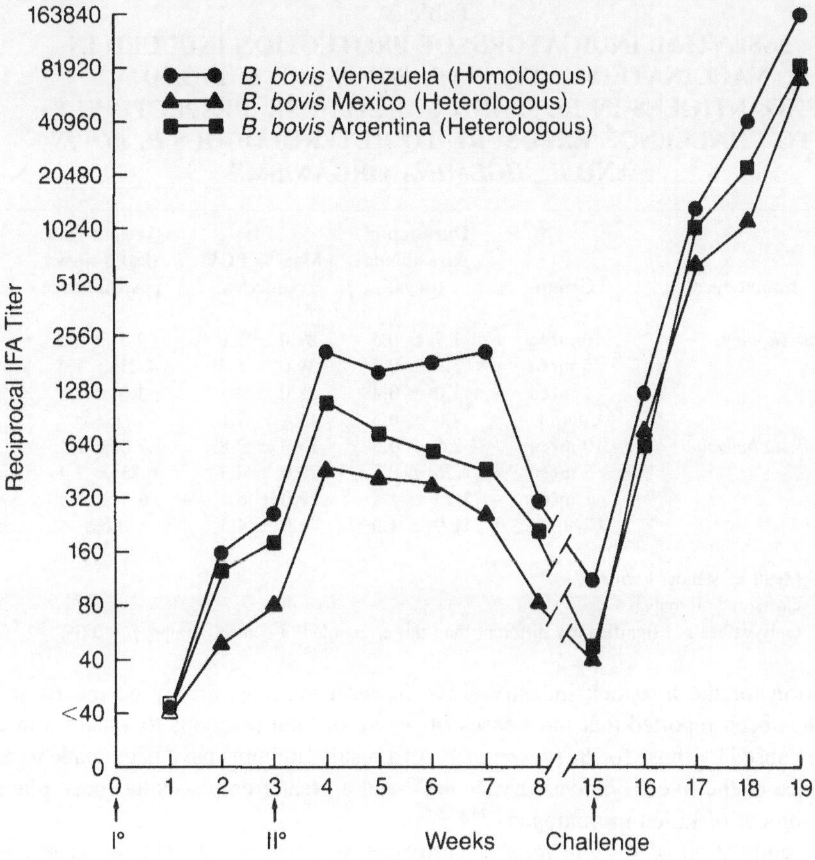

FIGURE 3. Mean homologous and heterologous anti-*B. bovis* IFA titers following vaccination with the Venezuelan immunogen.

during *B. bovis* infections[83] could also lead to the production of high amounts of nonprotective antibody (IFA-reactive) directed against fibrinogen-related products.[84]

Immunogens derived from the Venezuelan *B. bovis* isolate appear to be efficacious in protecting against different parasite strains. To prevent disease, the host must be primed for a secondary response against the unrelated epitopes of the heterologous strain. Therefore, complete purification and characterization of these protective antigens are relevant and worthwhile goals. In the final analysis, culture-derived *Babesia* exoantigens, even when containing bovine serum components are nevertheless safe and able to protect cattle against disease.

2. Efficacy of Culture-Derived B. bigemina Exoantigens: Dose-Response Evaluation of B. bovis and B. bigemina Immunogens

In the tropics and subtropics, *B. bovis* and *B. bigemina* coexist, resulting in mixed infections.[5] Host susceptibility, management practices, and the virulence of local strains influence the relative importance of each species within a given endemic area.[18] Thus, cattle in Australia are usually vaccinated only against *B. bovis*, which is responsible for more than 90% of babesiosis outbreaks.[18] In Israel,[7] India,[85] and South Africa,[86] cattle are also vaccinated against *B. bigemina* because of its pathogenicity.

Studies in Bolivia,[81] Colombia,[87] Mexico,[88] and Venezuela[89] have shown that *B. bigemina* infections are highly prevalent. Since dairy herds commonly graze in tick-infested pastures throughout these endemic areas, vaccination against *B. bigemina* may be an important

Table 2
ESSENTIAL INDICATORS[a] OF PROTECTION INDUCED IN VACCINATED CATTLE[b] (*B. BOVIS* AND *B. BIGEMINA* EXOANTIGENS IN 10, 5, AND 1 MℓEQ/DOSE, RESPECTIVELY) TO CHALLENGE EXPOSURE TO HETEROLOGOUS *B. BOVIS* AND *B. BIGEMINA* ORGANISMS

Immunogen	Group	Duration of parasitemia (days)	Max % PCV reduction	Weight gain (kg) 3 weeks postchallenge
Babesia bovis	10 mℓeq	1.7 ± 0.5	29.0 ± 2.0	9.5 ± 2.0
	5 mℓeq	3.5 ± 0.5	39.0 ± 6.3[c]	4.25 ± 1.6
	1 mℓeq	3.0 ± 0.4	38.0 ± 4.1[c]	4.0 ± 1.4
	Controls	7.0 ± 0.5	43.4 ± 5.1	Loss
Babesia bigemina	10 mℓeq	2.5 ± 0.5	25.0 ± 2.8	7.0 ± 1.4
	5 mℓeq	2.25 ± 0.8	29.0 ± 4.3	6.25 ± 2.4
	1 mℓeq	2.75 ± 1.1	22.0 ± 6.1	6.5 ± 2.3
	Controls	11.0 ± 1.0	46.0 ± 4.3	Loss

[a] Mean ± standard error.
[b] Crossbred animals.
[c] Only values not significantly different than those recorded in control groups, $p < 0.05$.

consideration for the livestock industry. Live *Babesia* vaccines are by no means safe for cattle. It has been reported that most cases of severe or fatal reactions to vaccines in South Africa are caused by those for *B. bigemina*.[86] As a result, attempts have been made to reduce the virulence of the live *B. bigemina* vaccine,[90] and a greater emphasis has been placed on the development of killed immunogens.[25,85,91]

In vitro cultivation of *B. bigemina*[43,48] provides an excellent system for generation of culture-derived *B. bigemina* exoantigens. Recently, the identification, quantitation, and physicochemical characterization of these *B. bigemina* exoantigens was studied.[48] Thereafter, two pilot vaccination experiments demonstrated the immunogenic potential of *B. bigemina* exoantigens[48] and provided the basis for two subsequent immunization studies that, (1) determined the comparative efficacy of culture-derived *B. bovis* and *B. bigemina* exoantigens to heterologous challenge, and established the antigenic threshold for induction of protective immunity, and (2) assessed the efficacy of a combined *B. bovis*-*B. bigemina* immunogen with a dose-response analysis in highly susceptible, purebred cattle.[44]

Studies of the efficacy of *B. bigemina* and *B. bovis* exoantigens applied as single immunogens were conducted on six vaccinated groups of four animals each (yearling crossbred cattle). They received either 10, 5, or 1 mℓeq of culture-derived *B. bigemina* or *B. bovis* immunogens, respectively. The immunizations were administered subcutaneously on days 0 and 21. Eight nonvaccinated cattle were included as negative controls. Three months after the second vaccination, all animals were challenged with 1×10^9 virulent *B. bovis* or *B. bigemina* organisms of heterologous strains. The immune response was assessed by comparing parasitemia levels, PCV reductions, and the weight gains in vaccinated and control groups (Table 2). Strong protection against heterologous challenge was observed in animals vaccinated with 10 mℓeq of either *B. bovis* or *B. bigemina* exoantigens. Most characteristic was the rapid clearance of parasites by vaccinated animals (duration of parasitemia: *B. bovis* vaccinates = 1.7 days, controls = 7.0 days; *B. bigemina* vaccinates = 2.5 days, controls = 11.0 days). However, at 5- and 1-mℓeq doses, *B. bovis* exoantigen was less efficient as determined by the maximum percent PCV reduction (10 mℓeq = 29%, 5 mℓeq = 39%, 1 mℓeq = 38%). Nevertheless, the return to normal PCV levels was rapid, suggesting vascular congestion rather than erythrocyte destruction. Vaccinated cattle also eliminated parasites at a faster rate and gained weight after challenge as compared with control animals (Table 2). It is interesting to note that *B. bigemina* exoantigens induced protection at all dosage levels, as indicated by the clinical responses analyzed. These data showed that *B. bigemina*-vaccinated cattle were more resistant to infection with heterologous strains.

Persistence of infection and parasitemic relapses are often a result of antigenic variation. Relapses have been observed with *B. bovis* for at least 4 years, whereas relapses to *B. bigemina* cease within 2 months.[92] Thus, it is possible that *B. bovis* parasites have a greater capacity for manifesting antigenic changes than *B. bigemina*. The results obtained in the above study reveal that with *B. bovis* strains, the specificity of the protective response is more strain-dependent than for that of *B. bigemina* parasites.

The kinetics of IFA responses observed during the vaccination and challenge periods are shown in Figure 4. After vaccination, *B. bigemina* exoantigens elicited comparatively lower antibody titers than did *B. bovis* antigens. Maximal antibody levels were achieved with the 10-mℓeq dosage, illustrating a dose-dependent response pattern (*B. bovis* = 10 mℓeq = 1:2560, 5 mℓeq = 1:1122, 1 mℓeq = 1:501; *B. bigemina* = 10 mℓeq = 1:1071, 5 mℓeq = 1:501, 1 mℓeq = 1:562). *B. bovis* antibody levels induced by the 5-mℓeq dosage were intermediary between those produced in response to the 10- and 1-mℓeq doses. *B. bigemina* exoantigens administered in 5- and 1-mℓeq dosages elicited similar antibody responses throughout the vaccination period. In summary, these experiments indicated that vaccination with *B. bovis* and *B. bigemina* exoantigens can effectively protect animals against a virulent challenge exposure. Important characteristics of host-protective immunity are rapid elimination of parasites and the ability to withstand clinical manifestations of infection as demonstrated by gains in body weight. The 10-mℓeq dosage appears to be optimal for ensuring complete immunity to heterologous *B. bovis* challenge. Finally, knowledge of the best cross-protective isolates would be advantageous in the selection of strains for vaccine production.

3. Dose-Response Evaluation of a Combined B. bovis-B. bigemina Immunogen

Simultaneous vaccination against *B. bovis* and *B. bigemina* with a combined immunogen is recommended in the tropics where susceptible cattle are concurrently affected by these parasites. Use of live, multistrain vaccines for the control of bovine babesiosis has been reported in Australia[93,94] and South Africa.[95]

In order to optimize results obtained in future babesiosis vaccination regimes utilizing culture-derived exoantigens, an evaluation of simultaneous vaccination with a combined *B. bovis-B. bigemina* immunogen was undertaken.[44] Animals used in this study were highly susceptible, purebred Holstein cattle imported from the U.S. to Venezuela. Productive breeds are commonly imported to improve the quality of native livestock. Susceptible, imported cattle are the principal targets for babesiosis and anaplasmosis, as they experience severe morbidity and mortality upon introduction into endemic areas.[80] A total of 23 Holstein cattle approximately 15 months of age were used in this study. Cattle were divided into three groups of five animals each and received the combined *B. bovis-B. bigemina* immunogen containing either 10, 5, or 1 mℓeq for each *Babesia* species. The vaccine was administered subcutaneously on days 0 and 21. Three months after the second immunization, all vaccinated animals were challenged intravenously with 1×10^9 virulent parasites of each *Babesia* species. Eight nonvaccinated animals received the same number of either *B. bovis* (n = 4) or *B. bigemina* (n = 4) organisms. Body temperature, PCV levels, platelet counts, parasitemia, IFA titers, and selected erythrophilic proteins were monitored throughout the experiment. Body weights were determined monthly prior to challenge and then biweekly postinoculation. A comparison of several essential indicators of the protection afforded by three dosages of the combined *B. bovis-B. bigemina* immunogen is shown in Table 3. Immunization with the *B. bovis-B. bigemina* exoantigens protected cattle against simultaneous, homologous *B. bovis* and *B. bigemina* challenge exposure as all vaccinated animals

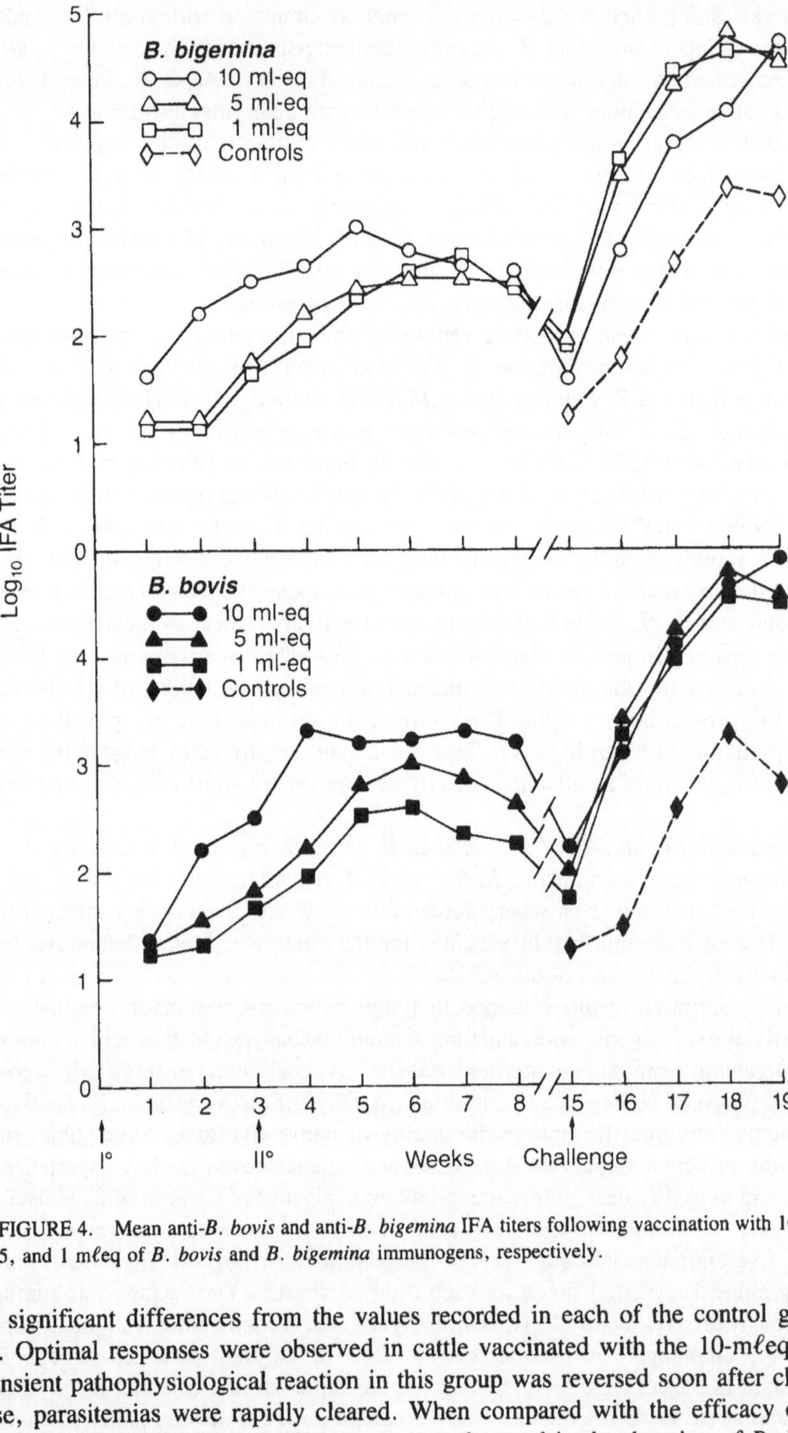

FIGURE 4. Mean anti-*B. bovis* and anti-*B. bigemina* IFA titers following vaccination with 10, 5, and 1 mℓeq of *B. bovis* and *B. bigemina* immunogens, respectively.

showed significant differences from the values recorded in each of the control groups ($p < 0.05$). Optimal responses were observed in cattle vaccinated with the 10-mℓeq dosage. Any transient pathophysiological reaction in this group was reversed soon after challenge. Likewise, parasitemias were rapidly cleared. When compared with the efficacy of the 5- and 1-mℓeq dosages, significant differences were observed in the duration of *B. bigemina* parasitemia (10 mℓeq = 2.5 days, 5 mℓeq = 5.3 days, 1 mℓeq = 4.0 days), maximum percent PCV reduction (10 mℓeq = 27.7%, 5 mℓeq and 1 mℓeq = 36.0%) and weight gain 3 weeks postchallenge (10 mℓeq = 4.5 kg, 5 mℓeq = 1.7 kg, 1 mℓeq = 1.4 kg). No marked differences were observed between the effects induced by either 5- or 1-mℓeq dosages. Nonvaccinated control animals showed acute clinical disease following challenge

Table 3
ESSENTIAL INDICATORS[a] OF PROTECTION INDUCED IN VACCINATED CATTLE[b] (COMBINED *BABESIA* EXOANTIGENS IN 10, 5, AND 1 MℓEQ/DOSE) TO SIMULTANEOUS CHALLENGE EXPOSURE TO HOMOLOGOUS *B. BOVIS* AND *B. BIGEMINA* ORGANISMS

Immunogen	Vaccinated[c]	Duration of parasitemia (days)	Max % PCV reduction	Weight gain (kg) 3 weeks postchallenge
	10 mℓeq	0.5 ± 0.5 (*B. bovis*) 2.5 ± 0.5 (*B. bigemina*)	27.7 ± 3.5	4.5 ± 0.6
Combined *B. bovis* and *B. bigemina* exoantigens	5 mℓeq	0.8 ± 0.3 (*B. bovis*) 5.3 ± 0.9[d] (*B. bigemina*)	36.0 ± 2.0[d]	1.7 ± 0.6[d]
	1 mℓeq	1.4 ± 0.4 (*B. bovis*) 4.0 ± 0.2[d] (*B. bigemina*)	36.0 ± 3.0[d]	1.4 ± 0.5[d]
Controls	*B. bovis*	7.0 ± 0.6	53.0 ± 5.0	Loss
	B. bigemina	10.0 ± 1.3	50.0 ± 4.0	Loss

[a] Mean ± standard error.
[b] Purebred Holstein animals.
[c] The vaccinated group showed significant differences from the values recorded in each of the control groups.
[d] Significantly different from standard 10 mℓeq dose, $p < 0.05$.

infection with either *B. bovis* or *B. bigemina* organisms, characterized by ataxia, labored respiration, profuse salivation, anorexia, anemia, and hemoglobinuria (*B. bigemina*). Antibody levels produced in response to vaccination with the combined *B. bovis*-*B. bigemina* immunogen at 10, 5, and 1 mℓeq/dose are depicted in Figure 5. Anti-*B. bovis* IFA titers after vaccination were similar for all dosages (10 mℓeq = 1:640, 5 mℓeq = 1:480, 1 mℓeq = 1:416), although antibody levels were lower than those elicited by the single *B. bovis* immunogen (10 mℓeq = 1:2560, 5 mℓeq = 1:1122, 1 mℓeq = 1:501). The anti-*B. bigemina* responses to the combined *Babesia* immunogen also did not show marked differences in titers between the various dosages (10 mℓeq = 1:2432, 5 mℓeq = 1:1408, 1 mℓeq = 1:1280). However, all responses were greater than those elicited by the single *B. bigemina* immunogen (10 mℓeq = 1:1071, 5 mℓeq = 1:501, 1 mℓeq = 1:562). Serological cross reactivity between *B. bovis* and *B. bigemina* has been reported with complement fixation,[96,98] indirect hemagglutination,[97] IFA,[88,98] and latex agglutination[99] tests. It is believed that one or more antigens may be shared by *B. bovis* and *B. bigemina*.[100] It has been suggested that common antigens are either highly immunogenic in *B. bovis* infections or present in greater amounts.[101] Higher and more persistent titers to *B. bigemina* antigen with *B. bovis* immune serum have been reported,[88] and our results seem to support these findings. Additionally, quantitative estimation of cross-reacting antigens has further confirmed the assumption that the common antigen(s) was present in larger amounts in *B. bovis* parasites (Figure 6). Serologic cross reactivity between *Babesia* parasites has not been conclusively correlated with significant protection against challenge exposure,[88] suggesting that monospecific vaccines against *B. bovis* or *B. bigemina* cannot be expected to provide adequate protection from disease caused by the heterologous species. Therefore, a combined *B. bovis*-*B. bigemina* immunogen is deemed necessary.

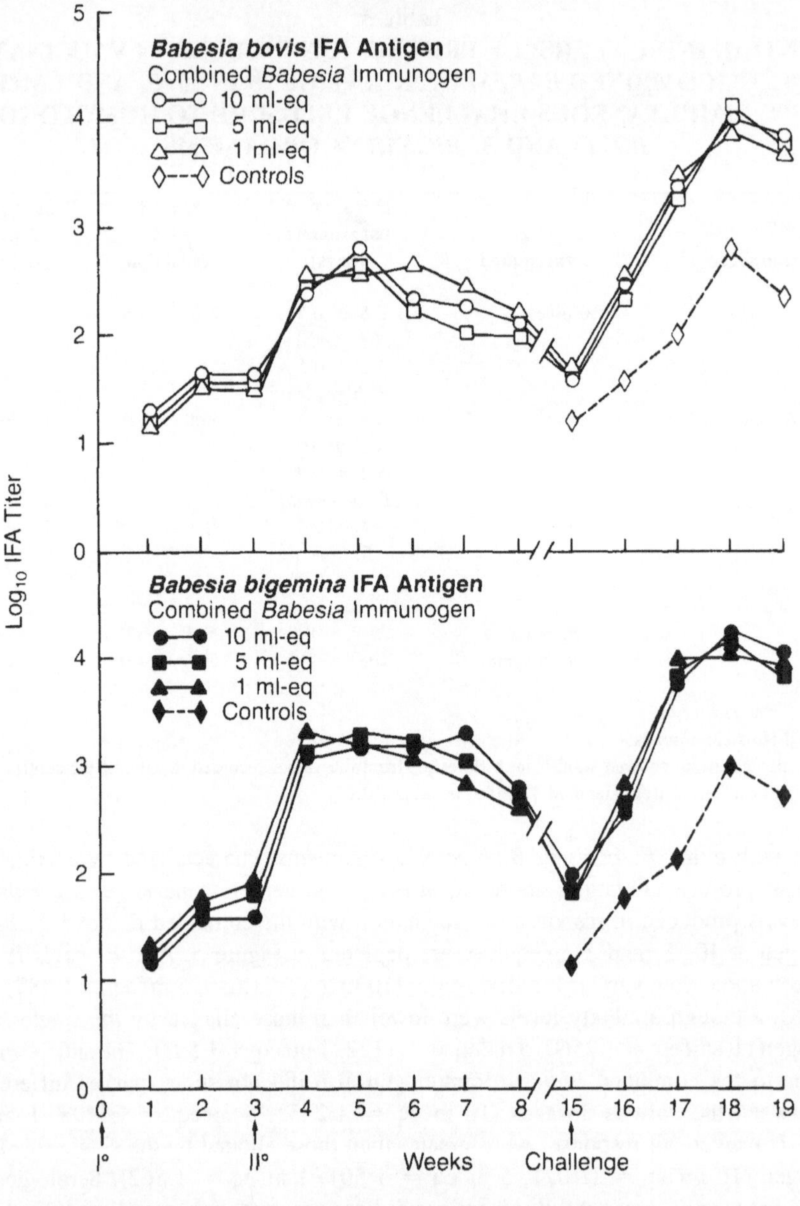

FIGURE 5. Mean anti-*B. bovis* and anti-*B. bigemina* IFA titers following vaccination with a combined *Babesia* immunogen at 10, 5, and 1 mℓeq dosages.

D. Influence of Cattle Breed on Vaccine Performance in the Control of Bovine Babesiosis

Nonspecific mechanisms of host immunity to babesial infections need to be analyzed in order to determine the relative importance of such factors during active immunization against babesiosis. Innate immunity contributes significantly to the overall immunoresponsiveness of the mammalian host, and has been shown to play an important role in bovine immunity.[102] This natural resistance affects pathogenicity, a measure of the interaction between a particular parasite and its host.[80] In cattle, it is believed that a long evolutionary association between tropical breeds of Zebu or buffalo, *B. bovis* and *B. bigemina* parasites, and the tick vector *Boophilus microplus* has resulted in a relatively high resistance to pathogenic effects in these hosts.[103,104] Disease occurs when susceptible cattle, not well-adapted to a *Babesia* species, are subjected to fluctuating levels of infection by vector ticks.[105] Animal hosts also show

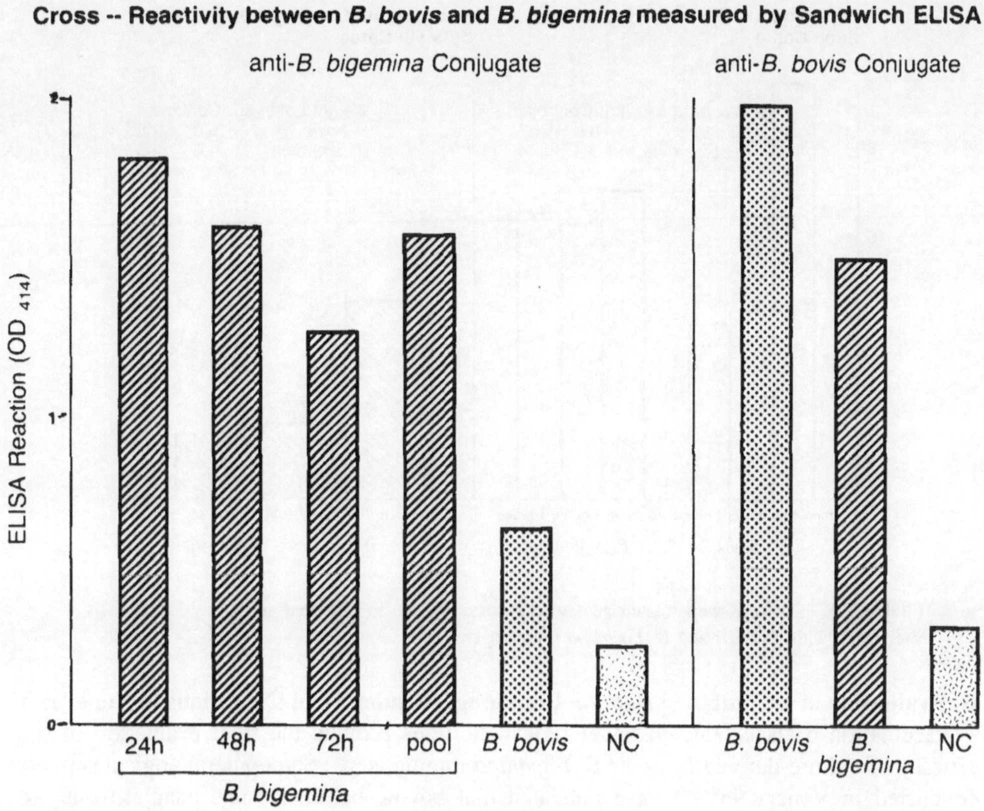

FIGURE 6. Cross reactivity between *B. bovis* and *B. bigemina* exoantigens as measured by a two-site enzyme immunoassay.

variable susceptibility to the vector, with *Bos indicus* cattle demonstrating greater resistance to infestation with *Boophilus microplus* than European breeds (*Bos taurus*).[105] Crossbreeding with Zebu animals helps to conserve this resistance, although occasionally *Babesia* may cause serious disease in previously unexposed hybrid cattle.[106]

In our immunization studies, comparable groups of crossbred (*Bos indicus* × *Bos taurus*) and purebred Holstein (*Bos taurus*) cattle have been utilized during trials with the culture-derived *Babesia* immunogen.[38,44] The most consistent indicator of protection against clinical disease is the gain in body weight after challenge exposure. Figure 7 depicts the influence of cattle breeds on the overall efficiency of culture-derived *Babesia* exoantigens. Vaccinated crossbred cattle showed greater weight gains after virulent heterologous challenge than did vaccinated purebred cattle after homologous challenge. The pathogenicity of *Babesia* infections was more accentuated in unvaccinated Holstein animals than in crossbred controls, as demonstrated by the greater weight loss in the former (Holstein unvaccinated cattle — weight loss = *B. bovis* = 15.0 kg, *B. bigemina* = 9.0 kg; crossbred unvaccinated cattle = weight loss = *B. bovis* = 3.5 kg, *B. bigemina* = 3.3 kg).

Culture-derived *Babesia* exoantigens have proved useful and effective in protecting highly susceptible cattle against babesiosis. Vaccination of crossbred cattle which exhibit a degree of innate resistance is still beneficial and recommended because active immunization will further enhance the protective immune response against this economically important disease.

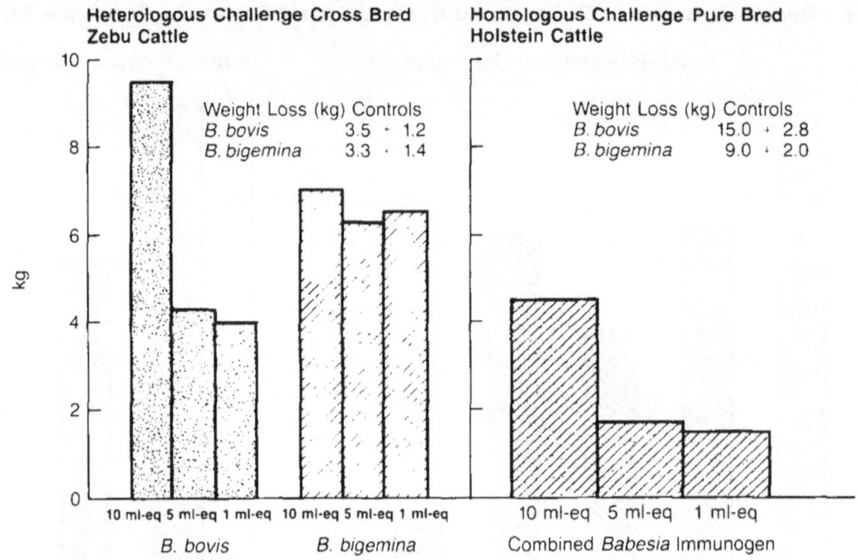

FIGURE 7. Average weight gain at 3 weeks postchallenge in purebred and crossbred animals vaccinated with *B. bovis* and *B. bigemina* immunogens.

E. Evaluation of a Combined *Babesia* Immunogen Under Field Conditions in Venezuela

Vaccination trials conducted under field conditions provide the final evaluation of the efficacy of culture-derived *B. bovis-B. bigemina* immunogens. Seroepidemiological surveys conducted in Venezuela[89,107] have indicated that bovine babesiosis and anaplasmosis are endemic hemoparasitic diseases with a prevalence rate of approximately 50%. Venezuela represents an ideal epidemiological setting for immunization studies because (1) highly susceptible, purebred cattle are regularly imported from the U.S. and Canada, and (2) native crossbred animals affect varying degrees of enzootic stability for these infections.

Since the initiation of field vaccination experiments in July 1984, three such trials have been completed. As of July 1986, eight trials were being conducted in six states of Venezuela (Figure 8). Dairy cattle were selected for evaluation of the exoantigen vaccine because most importation involved this type of animal. Furthermore, diversity in management practices, geographic characteristics, and farm size offered the opportunity for analysis within a wide range of epidemiologic conditions. Private- and government-owned sites were selected in areas where large-scale dairy and beef production is crucial for the livestock industry. These ranches encompass a total cattle population of 10,000 head, of which 2000 were vaccinated with the combined *B. bovis-B. bigemina* immunogen in a regime of two subcutaneous inoculations at a 4-week interval. With the collaboration of local veterinarians, serological, parasitological, and hematological data are being collected on a monthly basis from 10% of the animals. Data are also collected from a similar number of unvaccinated control cattle.

Farms corresponding to field trials Numbers 1 to 4 and 10 are in the category of enzootic instability (susceptible imported, purebred animals in tick-infested areas), the remaining farms (Numbers 5 to 8 and 11) fall into the enzootically stable category (purebred or crossbred cattle with early exposure to *Babesia* parasites, demonstrating specific antibody titers). Herd size, breed, and age at the time of vaccination are indicated in Figure 8. Most animals also received other standard vaccinations for dairy cattle in Venezuela, e.g., immunizations against foot-and-mouth disease, brucellosis, and a bacterin containing a combination of *Clostridium* spp. and *Pasteurella* spp.

Results obtained in the first three trials during a 1-year monitoring period following vaccination and natural tick exposure are presented in Table 4. Preliminary indications are

Location of Controlled Field Vaccination Trials in Venezuela with a Combined *Babesia bovis* - *Babesia bigemina* Vaccine (1984-1986)

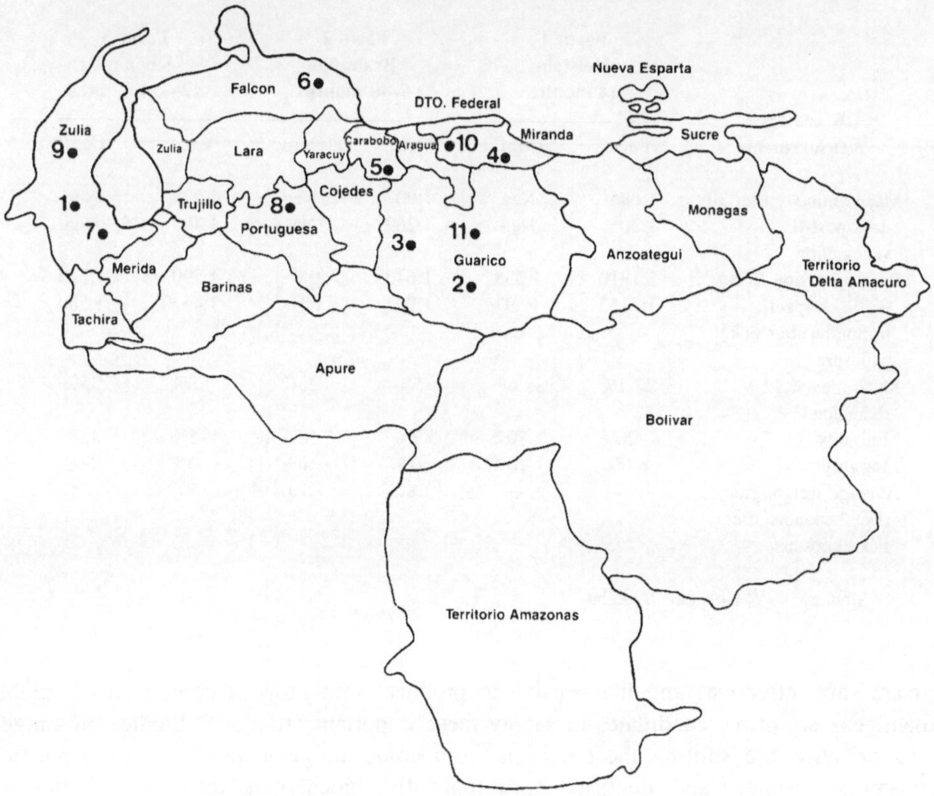

Farm	Herd Size	Breed	Age (months)	Epidemiologic Status
1	2000	Holstein	24	Enzoot. Unstable
2	1500	Brahman	6–8	Enzoot. Unstable
3	500	¾ Holstein	12–14	Enzoot. Unstable
4	300	Brown Swiss	6–12	Enzoot. Unstable
5	120	Cross Bred	24	Enzoot. Stable
6	150	Cross Bred	18	Enzoot. Stable
7	3000	Cross Bred	24	Enzoot. Stable
8	120	Holstein	18	Enzoot. Stable
9	500	Holstein	6–8	Enzoot. Unstable
10	600	Holstein	10–18	Enzoot. Unstable
11	500	Cross Bred	5–6	Enzoot. Stable

FIGURE 8. Geographic distribution of field vaccination trials in Venezuela.

encouraging and suggest that control of babesiosis can be accomplished. Early data show a reduced prevalance of babesiosis on these farms and indicate that the use of the culture-derived *Babesia* vaccine has effectively controlled the clinical manifestation of infection. With these promising results, it is anticipated that within the next year, the Venezuelan Institute for Veterinary Research, under the auspices of the Ministry of Agriculture, will begin production and distribution of the vaccine. The first priority is for immunization of valuable livestock imported from the U.S. and Canada.

IV. FUTURE CONSIDERATIONS FOR IMMUNIZATION AGAINST BABESIOSIS: MOLECULAR VACCINES AGAINST PARASITES

The knowledge and technology required to develop vaccines that can reliably protect man and animals against parasitic infections are rapidly becoming available. In search of vaccines

Table 4
DATA FROM FIELD VACCINATION TRIALS IN VENEZUELA

Breed age measurements	Farm 1 Holstein 24 months		Farm 2 Brahman 6—8 months		Farm 3 Siboney[a] 12—14 months	
	Vacc.	**Cont.**	**Vacc.**	**Cont.**	**Vacc.**	**Cont.**
Mean antibody titer 30 days post-II vaccination	1:686 1:905	Neg Neg	1:90 1:269	Neg Neg	1:278 1:403	Neg Neg
Seroconversion *B. bo.*, *B. bi.*, respect., 6 months after tick exposure	1:1810 1:2152	1:335 1:597	1:640 1:907	1:80 1:160	1:2903 1:4530	1:320 1:580
Mean max % PCV reduction	22.1%	38.6%	5%	23%	8%	35%
Morbidity	20%	70%	3%	18%	15%	33%
Mortality	0%	10%	0%	0%	0%	5%
Average weight gain (kg) 3 months after tick exposure	—	—	18.3	10	50	24

[a] Siboney = 3/4 Holstein, 1/4 Zebu.

that are safe, effective, and inexpensive to produce, immunogens comprised of soluble exoantigens are prime candidates to satisfy these important criteria.[108] Studies on antigen characterization are shifting the emphasis from crude antigenic preparations to purified polypeptides, isolated and identified using innovative biochemical techniques. Future research efforts will be directed toward isolation of protective antigens and the production of synthetic vaccines. Subunit vaccines can be produced in large amounts as the desired protein can be synthesized in bacteria, yeast, or animal cells using recombinant DNA technology.[109] Moreover, experimental animals infected with vaccinia virus recombinants containing genes coding for a variety of immunogenic proteins may be protected against challenge infection with the corresponding agent.[110] Undoubtedly, we have entered a new era in vaccine development, with recombinant DNA technology and other molecular and biochemical methodologies being applied to the synthesis of new types of immunizing agents for man and animals. While most of these vaccines are still in the developmental stages, it is expected that they may eventually supplement or even replace the use of conventional vaccines.[111] Five strategies are currently being applied for the generation of new vaccines: (1) recombinant DNA cloning of immunogenic surface proteins, (2) chemical synthesis of peptide vaccines, (3) construction of recombinant viruses containing gene inserts which code for foreign surface proteins, (4) genetic engineering of nonpathogenic mutant agents, and (5) production of monoclonal antiidiotype antibodies that mimic the major immunogenic epitopes of surface proteins of infectious agents.[111]

Priorities for research support in genetic engineering for livestock disease prevention were formulated at a workshop sponsored in 1982 by the U.S. National Research Council, U.S. Agency for International Development, the World Bank, and the Rockefeller Foundation.[112] Diseases selected for funding were (1) neonatal diarrheas, (2) bovine respiratory disease complex, (3) African swine fever, (4) hemotropic parasites — anaplasmosis, babesiosis, theileriosis, trypanosomiasis, and (5) animal tuberculosis.

Eventual replacement of conventional vaccines can be expected to proceed slowly and with caution. Considerable basic work still needs to be conducted with the molecular vaccines, not only in the definition of inherent immunogenic properties, but also on immunopotentiation and construction of effective delivery systems universally acceptable for use. Nevertheless, certain distinct advantages of molecular vaccines are already well recognized. They include increased safety, greater stability, improved potency, simplified production, and reduced allergenicity.[113] The new products must be economically feasible to produce, e.g., cattle immunization, and highly efficacious in order to compete with the use of established vaccines.[114,115]

Although potentially safe and immunogenic when presented with an appropriate adjuvant, a disadvantage of molecular vaccines may be derived from their relatively restricted antigenic composition as contrasted with parasitic extracts or secretions that present a multiplicity of antigens to the immune system. The assumption with the molecular vaccine approach is that a critical subset of antigens or epitopes will be identified, and the correspondingly limited immune response will be sufficient to protect the majority of individuals in a genetically diverse host population on subsequent contact with genetically diverse parasites.[116]

The identification and production of *Babesia* parasite antigens relevant to host-protective immunity is the necessary next step in the search for molecular vaccines against babesiosis. However, even after host-protective molecules are identified and produced, an extensive period of research and development will be required before field application can be accomplished. Until optimal molecular vaccines are developed, the culture-derived soluble *Babesia* immunogens offer the best combination of potency, efficacy, and safety in fulfilling the critical need for effective control measures against babesiosis.

REFERENCES

1. **Purnell, R. E.,** Babesiosis in various hosts, in *Babesiosis*, Ristic, M. and Kreier, J. P., Eds., Academic Press, New York, 1981, 25.
2. **Young, A. S. and Morzaria, S. P.,** Biology of Babesia, *Immunol. Today*, 2, 211, 1986.
3. **Babes, V.,** Sur L'Hemoglobinuria Bacteriene Du Boeuf, *C. R. Acad. Sci.*, 107, 692, 1888.
4. **Smith, T. and Kilbourne, F. L.,** Investigations into the nature, causation and prevention of Texas or southern cattle fever, *U.S. Dept. Agric. Bur. Anim. Ind. Bull.*, 1, 1, 1893.
5. **McCosker, P. J.,** The global importance of babesiosis, in *Babesiosis*, Ristic, M. and Kreier, J. P., Eds., Academic Press, New York, 1981, 1.
6. **Wharton, R. H. and Roulston, W. J.,** Resistance of ticks to pesticides, *Annu. Rev. Entomol.*, 15, 381, 1970.
7. **Pipano, E. and Hadani, A.,** Control of bovine babesiosis, in *Malaria and Babesiosis*, Ristic, M., Ambroise-Thomas, P., and Kreier, J. P., Eds., Martinus Nijhoff, Dordrecht, Netherlands, 1984, 263.
8. **Ristic, M.,** Research on babesiosis vaccines, in *Malaria and Babesiosis*, Ristic, M., Ambroise-Thomas, P., and Kreier, J. P., Eds., Martinus Nijhoff, Dordrecht, Netherlands, 1984, 103.
9. **Levy, M. G. and Ristic, M.,** Babesia bovis: continuous cultivation in microaerophilous stationary phase culture, *Science*, 207, 1218, 1980.
10. **Ristic, M. and Levy, M. G.,** A new era of research toward solution of bovine babesiosis, in *Babesiosis*, Ristic, M. and Kreier, J. P., Eds., Academic Press, New York, 1981, 509.
11. **Montenegro-James, S., James, M. A., and Ristic, M.,** Localization of culture-derived soluble *Babesia bovis* antigens in the infected erythrocyte, *Vet. Parasitol.*, 13, 311, 1983.
12. **Montealegre, F., Levy, M. G., Ristic, M., and James, M. A.,** Growth inhibition of *Babesia bovis* by secretions from bovine mononuclear phagocytes, *Infect. Immun.*, 50, 523, 1985.
13. **Timms, P., Dalgliesh, R. J., Barry, D. N., Dimmock, C. K., and Rodwell, B. J.,** *Babesia bovis*: comparison of culture-derived vaccine, non-living antigen and conventional vaccine in the protection of cattle against heterologous challenge, *Aust. Vet. J.*, 60, 75, 1983.
14. **Moreau, Y. and Laurent, N.,** Antibabesial vaccination using antigens from cell cultures: industrial requirements, in *Malaria and Babesiosis*, Ristic, M., Ambroise-Thomas, P., and Kreier, J. P., Eds., Martinus Nijhoff, Dordrecht, Netherlands, 1984, 129.

15. **Moreau, Y., Laurent, N., Martinod, S., Mackowiak, A., and Dubrent, N.,** Immunology, immunopathology and immunoprevention of canine piroplasmosis, *Prat. Med. Chirarqicalde Anim. Cie,* 21, 85, 1986.
16. **Wright, I. G., Mahoney, D. F., Mirre, G. B., Goodger, B. V., and Kerr, J. D.,** The irradiation of *Babesia bovis.* II. The immunogenicity of irradiated blood parasites for intact cattle and splenectomized calves, *Vet. Immunol. Immunopathol.,* 3, 591, 1982.
17. **Purnell, R. E., Brocklesby, D. W., and Stark, A. J.,** Protection of cattle against *Babesia major* by the inoculation of irradiated piroplasms, *Res. Vet. Sci.,* 25, 388, 1978.
18. **Callow, L. L.,** Vaccination against bovine babesiosis, *Adv. Exp. Med. Biol.,* 93, 121, 1977.
19. **Ristic, M. and Kreier, J. P.,** Malaria and Babesiosis: similarities and differences, in *Malaria and Babesiosis,* Ristic, M., Ambroise-Thomas, P., and Kreier, J. P., Eds., Martinus Nijhoff, Dordrecht, Netherlands, 1984, 3.
20. **Mahoney, D. F.,** Immunization against blood-derived antigens of *Babesia,* in *Babesiosis,* Ristic, M. and Kreier, J. P., Eds., Academic Press, New York, 1981, 475.
21. **Purnell, R. E. and Brocklesby, D. W.,** *Babesia divergens* in splenectomized calves: immunogenicity of lyophilized plasma from an infected animal, *Res. Vet. Sci.,* 23, 255, 1977.
22. **James, M. A., Kakoma, I., Ristic, M., and Cagnard, M.,** Induction of protective immunity to *Plasmodium falciparum* in *Saimiri sciureus* monkeys with partially purified exoantigens, *Infect. Immun.,* 49, 476, 1985.
23. **Mahoney, D. F.,** Bovine babesiosis: the immunization of cattle with killed *Babesia argentina, Exp. Parasitol.,* 20, 125, 1967.
24. **Phillips, R. S.,** Active immunization of rats against *Babesia (Nuttallia) rodhaini* using killed vaccine, *Parasitology,* 57, 11, 1967.
25. **Kuttler, K. L. and Johnson, L. W.,** Immunization of cattle with *Babesia bigemina* antigen in Freund's complete adjuvant, *Am. J. Vet. Res.,* 41, 536, 1980.
26. **Mahoney, D. F., Wright, I. G., and Goodger, B. V.,** Bovine babesiosis: the immunization of cattle with fractions of erythrocytes infected with *Babesia bovis* (Syn. *B. argentina*), *Vet. Immunol. Immunopathol.,* 2, 145, 1981.
27. **Wright, I. G., White, M., Tracey-Patte, P. D., Donaldson, R. A., Goodger, B. V., Waltisbuhl, D. J., and Mahoney, D. F.,** *Babesia bovis*: isolation of protective antigen by using monoclonal antibodies, *Infect. Immun.,* 41, 244, 1983.
28. **Ristic, M.,** Babesiosis, in *Diseases of Cattle in the Tropics,* Ristic, M. and McIntyre, I., Eds., Martinus Nijhoff, Dordrecht, Netherland, 1981, 443.
29. **Morzaria, S. P. and Young, A. S.,** Identification of *Babesia bigemina* in the tick *Boophilus decoloratus* by the indirect fluorescent antibody technique, *Res. Vet. Sci.,* 23, 55, 1977.
30. **Mahoney, D. F.,** *Babesia argentina*: immunogenicity of plasma from infected animals, *Exp. Parasitol.,* 32, 71, 1972.
31. **Smith, R. D., James, M. A., Ristic, M., Aikawa, M., and Vega, C. A.,** Bovine babesiosis: protection of cattle with culture-derived soluble *Babesia bovis* antigen, *Science,* 212, 335, 1981.
32. **Ristic, M., Smith, R. D., and Kakoma, I.,** Characterization of *Babesia* antigens derived from cell cultures and ticks, in *Babesiosis,* Ristic, M. and Kreier, J. P., Eds., Academic Press, New York, 1981, 337.
33. **Cox, F. E. G.,** Spurious *Babesia* antigens, *Nature (London),* 321, 384, 1986.
34. **Goodger, B. V., Wright, I. G., and Waltisbuhl, D. J.,** *Babesia bovis*: the effect of acute inflammation and isoantibody production in the detection of babesial antigens, *Experientia,* 41, 1577, 1985.
35. **Smith, R. D., Carpenter, J., Cabrera, A., Gravely, S. D., Erp, E. E., Osorno, M., and Ristic, M.,** Bovine babesiosis: vaccination against tick-borne challenge exposure with culture-derived *Babesia bovis* immunogens, *Am. J. Vet. Res.,* 40, 1078, 1979.
36. **Kuttler, K. L., Levy, M. G., James, M. A., and Ristic, M.,** Efficacy of a non-viable culture-derived *Babesia bovis* vaccine, *Am. J. Vet. Res.,* 43, 281, 1982.
37. **Kuttler, K. L., Levy, M. G., and Ristic, M.,** Cell culture-derived *Babesia bovis* vaccine. Sequential challenge exposure of protective immunity during a 6-month post-vaccination period, *Am. J. Vet. Res.,* 44, 1456, 1983.
38. **Montenegro-James, S., Toro Benitez, M., Leon, E., Lopez, R., and Ristic, M.,** Heterologous strain immunity in bovine babesiosis using a culture-derived soluble *Babesia bovis* immunogen, *Vet. Parasitol.,* 18, 321, 1985.
39. **Molinar, E., James, M. A., Kakoma, I., Holland, C. J., and Ristic, M.,** Antigenic and immunogenic status on cell culture-derived *Babesia canis, Vet. Parasitol.,* 10, 29, 1982.
40. **Osorno, M.,** Immunologic Characterization of Soluble Antigen Derived from Cell Culture of *Babesia bovis,* Ph.D. thesis, University of Illinois, Urbana, 1980.
41. **Todorovic, R.,** Personal communication, 1982.
42. **Pipano, E.,** Personal communication, 1985.

43. **Vega, C. A., Buening, G. M., Green, T. J., and Carson, C. A.,** *In vitro* cultivation of *Babesia bigemina, Am. J. Vet. Res.*, 46, 416, 1985.
44. **Montenegro-James, S., Toro Benitez, M., Leon, A., Lopez, R., and Ristic, M.,** Bovine babesiosis: induction of protective immunity with culture-derived Babesia bovis and Babesia bigemina immunogens, *Parasitol. Res.*, 74, 142, 1987.
45. **Vaayrynen, R. and Tuomi, J.,** Continuous *in vitro* cultivation of *Babesia divergens, Acta Vet. Scand.*, 23, 471, 1982.
46. **Moreau, Y.,** Personal communication, 1986.
47. **Montealegre, F.,** *Babesia bovis:* Detection and Characterization of Culture-Derived Exoantigens, Ph.D. thesis, University of Illinois, Urbana, 1986.
48. **Wanduragala, L.,** 1988, in preparation.
49. **Montealegre, F., Montenegro-James, S., Kakoma, I., and Ristic, M.,** Detection of culture-derived *Babesia bovis* exoantigen using a two-site immunoassay, *J. Clin. Microbiol.*, 25, 1648, 1987.
50. **Mahoney, D. F., Wright, I. G., and Goodger, B. V.,** Immunization against babesiosis: current studies and future outlook, *Prev. Vet. Med.*, 2, 401, 1984.
51. **Palmer, D. A., Buening, G. M., and Carson, C. A.,** Cryopreservation of *Babesia bovis* for *in vitro* cultivation, *Parasitology,* 62, 221, 1982.
52. **Steinbuch, M. and Audran, R.,** The isolation of IgG from mammalian sera with the aid of caprylic acid, *Arch. Biochem. Biophys.*, 134, 279, 1969.
53. **Harboe, N. M. G. and Ingild, A.,** Immunization, isolation of immunoglobulins and antibody titre determination, *Scand. J. Immunol.*, 17 (Suppl. 10), 345, 1983.
54. **Wilson, M. B. and Nakane, K. P.,** Recent developments in the periodate method of conjugating horseradish peroxidase (HPPO) to antibodies, in *Immunofluorescence and Related Staining Techniques,* Knapp, W., Holubar, K., and Wick, G., Eds., Elsevier/North Holland, Amsterdam, 1978, 215.
55. **Avrameas, S.,** Coupling of enzymes to proteins with glutaraldehyde. Use of the conjugate for the detection of antigens and antibodies, *Immunochemistry,* 6, 43, 1969.
56. **Burkot, T. R., Williams, J. L., and Schneider, I.,** Identification of *Plasmodium falciparum*-infected mosquitoes by a double antibody enzyme-linked immunosorbent assay, *Am. J. Trop. Med. Hyg.*, 33, 783, 1984.
57. **Taylor, D. W., Evans, C. B., Hennessy, G. W., and Aley, S. B.,** Use of two-sited monoclonal antibody assay to detect a heat-stable malarial antigen in the sera of mice infected with *Plasmodium yoelii, Infect. Immun.*, 51, 884, 1986.
58. **Kaeberle, M. L.,** Function of carriers and adjuvants in induction of immune responses, in *Advances in Carriers and Adjuvants for Veterinary Biologics,* Nervig, R. M., Gough, P. M., Kaeberle, M. L., and Whetstone, C. A., Eds., Iowa State University Press, Ames, 1986, 11.
59. **Chedid, L., Carelli, C., and Audibert, F.,** Use of adjuvants, antigens and carriers in synthetic veterinary vaccines, in *Advances in Carriers and Adjuvants for Veterinary Biologics,* Nervig, R. M., Gough, P. M., Kaeberle, M. L., and Whetstone, C. A., Eds., Iowa State University Press, Ames, 1986, 51.
60. **Murray, R., Cohen, P., and Hardegree, M. C.,** Mineral oil adjuvants: biological and chemical studies, *Ann. Allergy.*, 30, 146, 1972.
61. **Jensen, E. K.,** Synthetic adjuvants: avridine and other interferon inducers, in *Advances in Carriers and Adjuvants for Veterinary Biologics,* Nervig, R. M., Gough, P. M., Kaeberle, M. L., and Whetstone, C. A., Eds., Iowa State University Press, Ames, 1986, 79.
62. **Kakoma, I.,** Unpublished data, 1985.
63. **Chedid, L., Audibert, F., and Johnson, A. G.,** Biological activities of muramyl dipeptide, a synthetic glycopeptide analogous to bacterial immunoregulating agents, *Prog. Allerg.*, 25, 63, 1980.
64. **Taylor, S. M., Kenny, J., Mallow, T., and Elliott, C. T.,** The immunization of cattle against *Babesia divergens* with fractions of parasitized erythrocytes, *Vet. Parasitol.*, 16, 235, 1984.
65. **Allison, A. C.,** Mode of action of immunological adjuvants, *J. Reticuloendothel. Soc.*, 26, (Suppl.), 619, 1979.
66. **Morvat, G. M.,** A comparison of the adjuvant effects of saponin and oil emulsions in foot-and-mouth vaccines for cattle, *Bull. Off. Int. Epizool.*, 81, 1319, 1974.
67. **Wells, P. W., Emery, D. L., Henson, C. A., Morrison, W. I., and Murray, M.,** Immunization of cattle with a variant-specific surface antigen of *Trypanosoma brucei*: influence of different adjuvants, *Infect. Immun.*, 36, 1, 1982.
68. **McColm, A. A., Bomford, R., and Dalton, L.,** A comparison of saponin with other adjuvants for the potentiation of protective immunity by a killed *Plasmodium yoelii* vaccine in the mouse, *Parasite Immunol.*, 4, 337, 1982.
69. **Saul, K. W. and Kreier, J. P.,** *Plasmodium berghei*: immunization of rats with antigens from a population of free parasites rich in merozoites, *Tropenmed. Parasitol.*, 28, 302, 1977.
70. **Mitchell, G. H., Richard, W. H. G., Voller, A., Dietrich, F. M., and Dukor, P.,** Non-MDP, saponin, corynebacteria, and pertussis organisms as immunological adjuvants in experimental malaria vaccination of macaques, *Bull. WHO,* 57 (Suppl. 1), 189, 1979.

71. **McDonald, V. and Sherman, I. W.,** *Plasmodium chabaudi*: humoral and cell-mediated responses of immunized mice, *Exp. Parasitol.*, 49, 442, 1980.
72. **Johnson, P. R. A., Neal, A., and Gall, D.,** Protective effect of killed trypanosome vaccine with incorporated adjuvants, *Nature (London)*, 200, 83, 1963.
73. **McKercher, P. D., Moore, D. M., Morgan, D. O., Robertson, B. H., Callis, J. J., Kleid, D. G., Shire, S. J., Yansura, D. G., Dowbenko, D., and Small, B.,** Dose-response evaluation of a genetically engineered foot-and-mouth disease virus polypeptide immunogen in cattle, *Am. J. Vet. Res.*, 46, 587, 1985.
74. **Callow, L. L.,** Strain immunity in babesiosis, *Nature (London)*, 204, 1213, 1964.
75. **Callow, L. L.,** A note on homologous strain immunity in *Babesia argentina* infections, *Aust. Vet. J.*, 44, 208, 1968.
76. **Curnow, J. A.,** *In vitro* agglutination of bovine erythrocytes infected with *Babesia argentina, Nature (London*, 217, 207, 1968.
77. **Curnow, J. A.,** Studies on antigenic changes and strain differences in *Babesia argentina* infections, *Aust. Vet. J.*, 49, 279, 1973.
78. **Brown, I. N., Brown, K. N., and Hills, L. A.,** Immunity to malaria: the antibody response to antigenic variation by *Plasmodium knowlesi, Immunology*, 14, 127, 1968.
79. **Kahl, L. P., Anders, R. F., Rodwell, B. J., Timms, P., and Mitchell, G. F.,** Variable and common antigens of *Babesia bovis* parasites differing in strain and virulence, *J. Immunol.*, 129, 1700, 1982.
80. **Callow, L. L. and Dalgliesh, R. J.,** Immunity and immunopathology in babesiosis, in *Immunology of Parasitic Infections*, 2nd ed., Cohen, S. and Warren, K. S., Eds., Blackwell Scientific, Oxford, 1982, 307.
81. **McCosker, P. J.,** Control of piroplasmosis and anaplasmosis in cattle, *A Practical Manual*, Food and Agriculture Organization, Santa Cruz, Bolivia, 1975.
82. **Callow, L. L., Quioga, O. E., and McCosker, P. J.,** Serological comparisons of Australian and South American strains of *Babesia argentina* and *Anaplasma marginale, Int. J. Parasitol.*, 6, 307, 1976.
83. **Goodger, B. W., Wright, I. G., and Mahoney, D. F.,** The use of pathophysiological reactions to assess the efficacy of the immune response to *Babesia bovis* in cattle, *Z. Parasitenkd.*, 66, 41, 1981.
84. **Thoongsuwan, S., Cox, H. W., and Rickman, W. J.,** Antibody to fibrinogen/fibrin products (anti-F) in malaria, babesiosis, and trypanosomiasis of rodents, *J. Parasitol.*, 65, 426, 1979.
85. **Banerjee, D. P. and Prasad, K. D.,** The use of killed vaccine in immunization of cattle against *Babesia bigemina* infection, *Indian J. Vet. Med.*, 5, 1, 1985.
86. **DeVos, A. J., Combrink, M. P., and Bessenger, R.,** *Babesia bigemina* vaccine: comparison of the efficacy and safety of Australian and South African strains under experimental conditions in South Africa, *Onderstepoort J. Vet. Res.*, 49, 155, 1982.
87. **Corrier, D. E., Gonzalez, E. F., and Betancourt, A.,** Current information on the epidemiology of bovine anaplasmosis and babesiosis in Colombia, in *Tick-Borne Diseases and Their Vectors*, Wilde, J. K. H., Ed., University of Edinburgh, 1978, 114.
88. **Smith, R. D., Molinar, E., Larios, F., Monroy, J., Trigo, F., and Ristic, M.,** Bovine babesiosis: pathogenicity and heterologous species immunity of tick-borne *Babesia bovis* and *B. bigemina* infections, *Am. J. Vet. Res.*, 41, 1957, 1980.
89. **James, M. A., Coronado, A., Lopez, W., Melendez, R., and Ristic, M.,** Seroepidemiology of bovine anaplasmosis and babesiosis in Venezuela, *Trop. Anim. Health Prod.*, 17, 9, 1985.
90. **Dalgliesh, R. J., Callow, L. L., Mellors, L. T., and McGregor, W.,** Development of a highly infective *Babesia bigemina* vaccine of reduced virulence, *Aust. Vet. J.*, 57, 8, 1981.
91. **Todorovic, R. A., Gonzalez, E. F., and Adams, L. G.,** Bovine babesiosis: sterile immunity to *Babesia bigemina* and *Babesia argentina* infections, *Trop. Anim. Health Prod.*, 5, 234, 1973.
92. **Mahoney, D. F., Wright, I. G., and Mirre, G. B.,** Bovine babesiosis: the persistence of immunity to *Babesia argentina* and *B. bigemina* in calves (*Bos taurus*) after naturally acquired infection, *Ann. Trop. Med. Parasitol.*, 67, 197, 1973.
93. **Emmerson, F. R., Knott, S. G., and Callow, L. L.,** Vaccination with *Babesia argentina* in 5 beef herds in South Eastern Queensland, *Aust. Vet. J.*, 52, 451, 1976.
94. **Mellors, L. T., Dalgliesh, R. T., Timms, P., Rodwell, B. J., and Callow, L. L.,** Preparation and laboratory testing of a frozen vaccine containing *Babesia bovis, B. bigemina, and Anaplasma centrale, Res. Vet. Sci.*, 32, 194, 1982.
95. **DeVos, A. J.,** Immunogenicity and pathogenicity of three South African strains of *Babesia bovis* in *Bos indicus* cattle, *Onderstepoort J. Vet. Res.*, 45, 119, 1978.
96. **Mahoney, D. F.,** Bovine babesiosis: diagnosis of infections by a complement fixation test, *Aust. Vet. J.*, 38, 48, 1962.
97. **Goodger, B. V.,** Preparation and preliminary assessment of purified antigens in the passive hemagglutination test for bovine baoesiosis, *Aust. Vet. J.*, 47, 251, 1971.

98. **Todorovic, R. A. and Long, R. F.,** Comparison of indirect fluorescent antibody (IFA) with complement fixation (CF) tests for diagnosis of *Babesia* spp. infections in Colombian cattle, *Tropenmed. Parasitol.*, 27, 169, 1976.
99. **Montenegro, S., James, M. A., Levy, M. G., Preston, M. D., Esparza, H., and Ristic, M.,** Utilization of culture-derived soluble antigen in the latex agglutination test for bovine babesiosis and anaplasmosis, *Vet. Parasitol.*, 8, 291, 1981.
100. **Mahoney, D. F.,** Circulating antigens in cattle infected with *Babesia bigemina* or *Babesia argentina*, *Nature (London)*, 211, 422, 1966.
101. **Goodger, B. V.,** Further studies of hemagglutinating antigens of *Babesia bigemina*, *Aust. Vet. J.*, 49, 81, 1973.
102. **Zwart, D. and Brockesby, D. W.,** Babesiosis: nonspecific resistance, immunological factors and pathogenesis, *Adv. Parasitol.*, 17, 49, 1979.
103. **Daly, G. D. and Hall, W. T. K.,** Note on the susceptibility of British and some Zebu-type cattle to tick fever (babesiosis), *Aust. Vet. J.*, 31, 152, 1955.
104. **Johnston, L. A. Y. and Sinclair, D. F.,** Differences in response to experimental primary infection with *Babesia bovis* in Hereford, Droughtmaster and Brahman cattle, in *Proc. Symp. 56th Annu. Conf. Australian Veterinary Assoc.*, Johnston, L. A. Y. and Cooper, M. G., Eds., Australian Veterinary Association, Sydney, 1980, 14.
105. **Joyner, L. P. and Donnelly, J.,** The epidemiology of babesial infections, *Adv. Parasitol.*, 17, 115, 1979.
106. **Davidson, K. B.,** An usual outbreak of babesiosis, *Vet. Rec.*, 85, 391, 1969.
107. **Toro, M., Lopez, R., Leon, E., Ruiz, A., and Garcia, J. A.,** Epidemiologic and critical features of bovine babesiosis in Venezuela, *Vet. Trop.*, 2, 91, 1982.
108. **Cox, F. E. G.,** Vaccines for parasitic disease, in *Parasites — Their World and Ours*, Mettrick, D. F. and Desser, S. S., Eds., Elsevier Biomedical, 1982, 395.
109. **Morein, B. and Simons, K.,** Subunit vaccines against enveloped viruses, microsomes, micelles and other protein complexes, *Vaccine*, 3, 83, 1985.
110. **Brown, F., Schild, G. C., and Ada, G. L.,** Recombinant vaccinia viruses as vaccines, *Nature (London)*, 319, 549, 1986.
111. **Bachrach, H. L.,** New approaches to vaccines, *Adv. Vet. Sci. Comp. Med.*, 30, 1, 1985.
112. Board on Science and Technology for International Development, *Priorities in Biotechnology Research for International Development*, National Research Council, National Academy Press, Washington, D.C., 1982.
113. **Cavanagh, D.,** *New Biotechnology for Animal Health and Production*, Ross, D. B. and Sellers, K., Eds., Association of Veterinarians in Industry, London, 1984, 67.
114. **Murphy, F. A., Brown, F., and Obijeski, J. F.,** Genetic engineering technology in vaccine production and control of animal virus diseases, in *Applied Virology*, Academic Press, New York, 1984, 17.
115. **Pond, W. G., Merkel, R. A., McGilliard, L. D., and Rhodes, V. J.,** *Animal Agriculture — Research to Meet Human Needs in the 21st Century*, Westview Press, Boulder, Colo., 1980, chap. 2.
116. **Mitchell, G. F.,** Towards molecular vaccines against parasites, *Parasite Immunol.*, 6, 493, 1984.

Chapter 11

EPIDEMIOLOGIC AND IMMUNOPROPHYLACTIC ASPECTS OF CANINE BABESIOSIS IN FRANCE

Yves Moreau, Serge Martinod, and Georg Fayet

TABLE OF CONTENTS

I. INTRODUCTION

Canine babesiosis is a tick-borne disease of dogs and other Canidae characterized by fever and varying degrees of anemia which often develops rapidly.[1] Puppies are highly susceptible, untreated cases are invariably fatal,[2] and transplacental transmission has been suggested since high parasitemias have been observed in neonates.[2,3] The most important etiologic agent for canine babesiosis is *Babesia canis*, a large *Babesia* transmitted by a number of ticks, but mostly by *Rhipicephalus sanguineus*.[3-6] More recently, *Dermacentor reticulatus* has been shown to be the main vector of this disease in France.[7]

Canine babesiosis is no longer considered a strictly tropical disease in that about 50 years ago, cases were confirmed in the U.S., and the disease appears to be increasing on a worldwide basis.[8-10] These data have been based on epidemiologic studies in Europe and the U.S.

Tremendous efforts, described previously[11,12] and discussed elsewhere in this book, have been directed toward the control of bovine babesiosis. Similar efforts involving our team in collaborative work with the University of Illinois have culminated in the development of an organism-free vaccine against *B. canis*,[13] which has been evaluated under laboratory and field conditions. Last year, a commercial product, Pirodog®, was introduced onto the French market, thus representing the first antiprotozoan vaccine ever produced commercially. In this review, we will describe some major epidemiologic data on canine babesiosis in France and also highlight pertinent aspects of the field trials evaluating safety and efficacy of Pirodog®.

II. EPIDEMIOLOGY OF CANINE BABESIOSIS IN FRANCE

A. Tick Vectors

Only adult *D. reticulatus* ticks feed on dogs and are epidemiologically important (larvae and nymphs parasitize other small mammals). In a recent study, Martinod[14] has demonstrated that in endemic areas, most dogs show serologic evidence of exposure to both *B. canis* and its tick vector (Figure 1). Using an enzyme immunoassay (EIA) test, the authors showed a good correlation between *B. canis* and *D. reticulatus* titers in sera of dogs showing clinical symptoms ($r = 0.76$ df 20 $p < 0.001$), thereby confirming that *D. reticulatus* is the main vector of *B. canis* infection.

Other ticks, such as *D. marginatus, Ixodes ricinus, I. hexagonus,* and *R. sanguineus* parasitize dogs in France. Some of them can transmit diseases other than babesiosis, for instance, *R. sanguineus* is the vector for *Ehrlichia canis*. There is evidence, however, that *D. marginatus,* some *Hyaloma,* and *Haemaphysalis* species can also transmit babesiosis.[14]

B. Incidence in Relation to Geographic and Individual Host Factors

A recent survey[15] in contrast to the previous one[16] has shown a very important increase in the incidence and prevalence of canine babesiosis. The current epidemiological map of the disease in France (Figure 2) shows that *B. canis* infection is still prevalent in both regions where it has been previously recorded (Rhone Alps and Aquitaine) and that the disease is now present in several zones north of the river Loire. Another interesting observation is that numerous cases have now been confirmed near Paris, where the disease was totally absent 10 years ago.

This trend toward the northern and eastern regions is also confirmed by Belgian, German, and Swiss veterinary authorities who are now observing autochthonous cases of this disease. Hitherto, it was generally believed that the disease was imported from neighboring countries such as Yugoslavia. The explanation for this upsurge in the number of cases of the disease is mostly speculative, based on factors such as changes in climatic conditions favoring the tick vector and increased traffic of holiday-makers with their dogs traveling into and out of

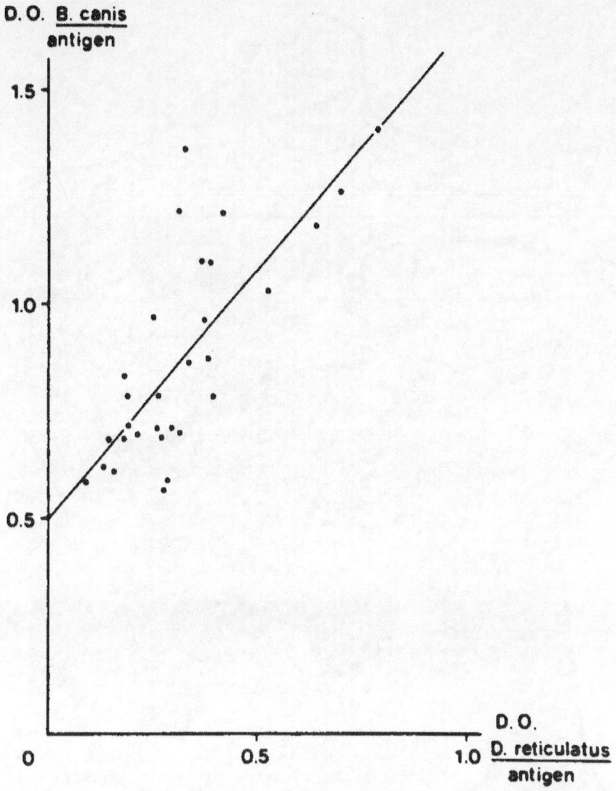

FIGURE 1. Relation between optical densities detected by ELISA with *B. canis* and *D. reticulatus* antigens in sera of dogs showing clinical symptoms. Equation of linear regression line: y = 1.17 × +0.50 (r = 0.76 df 28 p <0.001).

France. In addition, changes in urban and rural planning whereby more favorable tick breeding conditions are created might contribute to the increased prevalence rate. Most likely, however, all these factors combined with genetic traits determine the present epidemiologic trends.

We have previously evaluated the influence of breed, sex, and age on the susceptibility of dogs to *B. canis*.[17] We demonstrated that some breeds, such as beagles and fox terriers, are relatively resistant to *B. canis* infection, whereas breeds like Doberman, Pekingese, and Yorkshire terriers showed a higher degree of susceptibility. In the same study, we found no correlation between susceptibility with sex, size, or age of the dogs.

Other possible factors influencing susceptibility to *B. canis* are immunologically mediated.[18,19] In various publications, evidence has been produced demonstrating generalized immunosuppression as shown by reduced lymphocyte blastransformation indexes.[19] These data indicated that peripheral T lymphocytes were the major target of the immunosuppressive effects of canine babesiosis. Further, the immunosuppression state coincided with high prevalence and recrudescence rates in hyperendemic regions.[18]

III. IMMUNOPROPHYLAXIS WITH CULTURE-DERIVED *BABESIA CANIS* VACCINE

A. Development of the Vaccine

The successful development of short-term cultures for *B. canis*,[20] and its subsequent improvement,[13] facilitated the production of highly immunogenic exoantigens.[21] A master-

FIGURE 2. Epidemiological survey of *Babesia canis* infection in France (1985).

seed parasite strain (MSP) was established as previously described.[13] In addition, a seed lot derived from the MSP was initiated and thus a commercial production scheme for the manufacture of the first protozoal vaccine (Pirodog®) was accomplished.

B. Laboratory and Field Application of the Vaccine

Over the last 4 years, several trials have been accomplished utilizing military or civilian-owned dogs. The results are summarized in Table 1. The degree of protection after 3 years of regular annual vaccination ranged between 70 to 100%. The few clinical cases observed in vaccinated dogs were generally mild. The beneficial aspects of the vaccine were particularly evident in immunosuppressed dogs, which are recognized as a high risk group.[18,19]

IV. CONCLUSIONS

Canine babesiosis continues to be of major importance in France where more than 400,000 cases are recorded every year. In the results of the clinical trials reported herein, protection

Table 1
RESULTS OF VACCINATION TRIALS WITH PIRODOG® IN MILITARY (M) AND CIVILIAN (C) POPULATIONS IN ENDEMIC AREAS OF FRANCE

Center	Location	Babesiosis endemicity	No. of dogs	Use	Estimated percent protection
Suippes	NE	H	500(M)	Breeding, training	−70
Gramat	SW	H	100(M)	Breeding, training	90
La Tremblade	W	H	120(C)	Hunting	100[a]
Belley	E	H	20(C)	Hunting	100[b]

[a] After 3 years of observation, no cases of babesiosis were observed in vaccinated dogs compared with 4% after the 2nd year and 16% prior to initiation of the program.

[b] Prior to vaccination program, there was an incidence of 50% of recrudescent babesiosis. After 3 years of annual vaccination, no cases were reported.

was accomplished with the culture-derived exoantigen[12] at the level of 70 to 100%, and the disease occasionally seen in vaccinates was generally mild. This first-generation vaccine against a protozoal agent marks the era of immunoprophylaxis against apicomplexa and will have a major impact, particularly in southern France. The high efficacy and safety make this vaccination approach attractive, opens great opportunities for practical means of control of babesiosis, and provides a foundation for immunoprophylaxis against other protozoal diseases.

REFERENCES

1. **Bansal, S. R. B. and Gautam, O. P.,** Reports of clinical cases of *Babesia canis* disease among the exotic breeds of dogs, *Indian J. Vet. Med.*, 1, 87, 1981.
2. **Breitschwerdt, E. B., Malone, J. B., MacWilliams, P., Levy, M. G., Qualls, W., and Prudich, M. J.,** Babesiosis in the greyhound, *J. Am. Vet. Med. Assoc.*, 182, 978, 1983.
3. **Farwell, G. E., LeGrand, E. K., and Cobb, C. C.,** Clinical observations on *Babesia gibsoni* and *Babesia canis* infection in dogs, *J. Am. Vet. Med. Assoc*, 180, 507, 1982.
4. **Riek, R. F.,** Babesiosis, in *Infectious Blood Diseases of Man and Animals*, Vol. 2, Weinman, D. and Ristic, M., Eds., Academic Press, New York, 1968, 219.
5. **Liebisch, V. A. and Gillani, S.,** Experimentelle Ubertragung der hundebabesiose (*Babesia canis*) durch Einheimische deutsche zeckenarten. I. Die braune Hundezecke *(Rhipicephalus sanguineus)*, Dtsch. Tieraerztl. Wochenschr., 86, 149, 1979.
6. **Grogan, J. W.,** Piroplasmosis in a dog, *J. Am. Vet. Med. Assoc.*, 123, 234, 1953.
7. **Moreau, Y.,** Immunologie des Babesioses, in *Le Point Veterinaire*, Les Entretiens de Bourgelat, P. V. Maisons, Alfort, France, 1981, 133.
8. **Anderson, J. F. and Magnarelli, L. A.,** Canine *Babesia* new to North America, *Science*, 204, 1431, 1979.
9. **Klinefelter, M. R.,** Cause, diagnosis and treatment of canine piroplasmosis, *Vet. Med. Small Anim. Clin.*, October, 1505, 1982.
10. **Uilenberg, G., Top, P. D. J., Arends, P. J., Kool, P. J., Van Dijk, J. E., van Schieveen, P. B. E., and Zwart, D.,** Indigenous canine babesiosis in the Netherlands, *Tijdschr. Diergeneeskd.*, 110, 93, 1985.
11. **Mahoney, D. F.,** Prospects for an antibabesial vaccine, in *Babesiosis*, Ristic, M. and Kreier, J. P., Eds., Academic Press, New York, 1981.
12. **Ristic, M.,** Comments on prospects for antibabesial vaccine, in *Babesiosis*, Ristic, M. and Kreier, J. P., Eds., Academic Press, New York, 1981, 563.
13. **Moreau, Y. and Laurent, N.,** Antibabesial vaccination using antigens from cell culture fluids: industrial requirements, in *Malaria and Babesiosis*, Ristic, M., Ambroise-Thomas, P., and Kreier, J. P., Eds., Martinus Nijhoff, Dordrecht, Netherlands, 1984, 129.

14. **Martinod, S.,** Contribution aux Recherches de Cartographie Epidemiologique: les Babesioses Canine et Bovine dans le Jura Meridional, Ph.D. thesis, University of Grenoble, France, 1983.
15. **Fayet, G., Pages, J. P., and Trouillet, J. L.,** Babesiose canine: resultats d'une enquete realisee supres de 700 veterinaires practiciens, *Prat. Med. Chirurgicale Anim. Comp.*, 21, 69, 1986.
16. **Fournier, J.,** Compte rendu d'une enquete epidemiologique sur l'importance et la repartition de la piroplasmose canine en France, *Anim. Comp.*, 2, 95, 1974.
17. **Martinod, S., Brossard, M., and Moreau, Y.,** Immunity of dogs against *Babesia canis*, its vector tick *Dermacentor reticulatus* and *Ixodes ricinus* in endemic areas, *J. Parisitol*, 71, 269, 1985.
18. **Moreau, Y., Laurent, N., Martinod, S., Mackowiak, M., and Dubreuil, N.,** Immunologie, immunopathologie et essais d'immunoprevention de la piroplasmose canine, *Prat. Med. Chirurgicale Anim. Comp.*, 21, 85, 1986.
19. **Pages, J. P. and Trouillet, J. L.,** La Babesiose du chien: mode d'action de *Babesia canis* et consequences cliniques, *Prat. Med. Chirurgicale Anim. Comp.*, 21, 97, 1986.
20. **Molinar, E.,** *In Vitro* Propagation of *Babesia canis:* Morphological Ultrastructural and Immunologic Studies on Antigens Derived from Primary Cell Cultures, M.S. thesis, University of Illinois, Urbana, 1979.
21. **Molinar, E., James, M. A., Kakoma, I., Holland, C. J., and Ristic, M.,** Antigenic and immunogenic studies on cell culture-derived *Babesia canis*, *Vet. Parasitol.*, 10, 29, 1982.

Chapter 12

EQUINE BABESIOSIS

Eberhard Schein

TABLE OF CONTENTS

I. INTRODUCTION

Two different protozoa of the genus *Babesia* infect horses: *Babesia caballi*[1] and *B. equi*.[2] Laveran in 1901[2] was the first to describe an intraerythrocytic parasite, which he discovered in peripheral blood smears taken from South African horses. He accordingly named it *Piroplasma equi*.

Nuttall and Strickland[1] noticed another pathogenic organism, *B. caballi*, which was morphologically distinct from *B. equi*, but likewise caused piroplasmosis in the horse.[3] They named the smaller species *Nuttallia equi* (Laveran) and the larger species *Piroplasma caballi* (Nuttall). The genus "*Nuttallia*" had already been assigned to a North American bivale shell species (*Nomen occupatum*) making the designated classification "*Nuttallia equi*" no longer tenable.[4] Consequently, this organism was also classified under the genus *Babesia*.[5,6] The taxonomic position of *B. equi* to the genus *Babesia* was questioned again in the investigations of Schein et al.[7] concerning the developmental cycle in horses. *Babesia equi* exhibits characteristics of the *Theileria* in their morphology and development in the equine host as well as in the tick vector. However, until a new classification is agreed upon, the name *B. equi* will be used in this review.

The current importance of equine babesiosis is due to the fact that these pathogenic organisms can readily be transmitted by carrier horses or by infected ticks introduced into originally *Babesia*-free countries, hence causing a spread of the disease in these areas. The U.S. and Australia can be cited as prime examples.

This review attempts to highlight the current knowledge on the taxanomic, epidemiologic, pathophysiologic, diagnostic, and preventive aspects of equine babesiosis and demonstrate the increasing international importance of this disease.

II. MORPHOLOGY AND LIFE CYCLE OF *BABESIA CABALLI*

A. In Equidae

Babesia caballi exhibits characteristic features of the genus *Babesia* only in the erythrocytes. In Giemsa-stained peripheral blood smears of *B. caballi*, either two pear-shaped 3 to 5 μm merozoites or a single spheric 3 to 4 μm trophozoite can be observed (Figure 1). The merozoites are located centrally in the erythrocytes and form a common pole from a sharp angle. The blueish-colored cytoplasm contains one or two red-stained nuclear chromatin bodies and generally one vacuole. Usually, there are not more than two babesia in a single erythrocyte. The trophozoites appear as circular structures with a blueish anular zone which contain at one location a red-stained chromatin body and a large central vacuole.[8]

B. In the Tick Vector

The developmental cycle of *B. caballi* in the vector, *Dermacentor nitens*, was examined and extensively elucidated by Holbrook et al.[9] The authors discovered that after the intake of blood, the small spherical bodies of 4 to 6 μm in diameter developed during the initial stage in the gut contents of the tick. The characteristic sexual stages as described by Koch[10] and Mehlhorn et al.[11] for *B. canis* as well as Friedhoff and Buscher[12] for *B. bigemina* could not be substantiated by Holbrook et al.[9] In this first stage described by the latter investigators, the zygotes, which first differentiate into kinetes, 10 to 14 μm in length and 4 to 6 μm in width, are already present. These kinetes once again infect the gut cells and form large spherical bodies of 12 to 14 μm diameter, which then develop into second-generation kinetes. These kinetes have generally an 8 to 12 μm length by 2 to 4 μm width and are somewhat smaller than those of the first generation. The kinetes penetrate the gut wall and attach to the cells of various organs, especially the ovaries of the tick. In these cells a multiple division occurs again with the formation of a third generation of kinetes, which at this stage penetrate

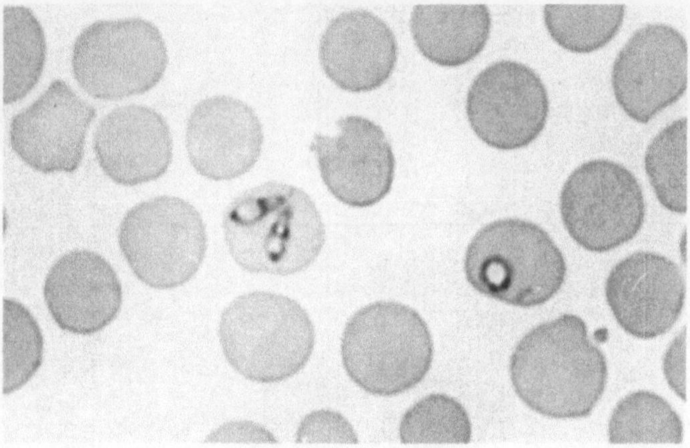

FIGURE 1. *Babesia caballi* (Maroc strain); trophozoite (right) and two merozoites (left) in stained blood smear. (Giemsa; magnification × 2000.)

into the eggs of the tick. Afterwards, a further multiple division of the tick egg cells occurs. After the larvae have hatched and are attached to the host, during the suction phase, some of the kinetes penetrate the salivary glands. The remaining kinetes penetrate other cells and infect the salivary glands first in the nymph or adult stage. The last phase of multiplication takes place in the salivary glands in which a large number of small spherical or pear-shaped, 2.5 to 3-μm-sized sporozoites evolve. During the attachment and feeding of the tick, the sporozoites contained in the saliva are inoculated into the equine host. They infect only the erythrocytes of the equine host. This transovarial transmission is a typical characteristic of the other babesias as well as of *B. caballi*. In addition, *B. caballi* can be transmitted transtadially, i.e., they can be transmitted from the larvae to the nymph or from the nymph to the adult tick. Because of the multifarious propagation of *B. caballi,* the tick vectors represent an infectious reservoir and can remain infectious for horses over many generations.

III. MORPHOLOGY AND LIFE CYCLE OF *BABESIA EQUI*

A. In Equidae

In contrast to *B. caballi, B. equi* multiply initially in the lymphocytes and later in the erythrocytes.[7] After inoculation of the sporozoites by the tick vector, the lymphocytes are initially invaded. In these lymphocytes, characteristic schizonts develop in the form of macroschizonts which later differentiate to microschizonts. The schizonts of *B. equi* are morphologically similar to those of the *Theileria* and can be identified in the lymph nodes 12 to 14 days after attachment of the ticks. The microschizonts contain a large number of merozoites which, after destruction of the host cell, are released and penetrate the erythrocytes. This development was also demonstrated in vitro.[7,13]

Babesia equi parasites are generally noticeable in stained blood smears due to their large morphologic variety (Figure 2). In addition to the small (1 to 2 μm), comma-shaped stages, the spheric, and to some extent, amoeboid, 2- to 3-μm-sized stages prevail. A further propagation occurs in the erythrocytes and, in addition to a double segmentation, there is also a division into four new stages. The four merozoites are then arranged into a "Maltese cross" formation which is characteristic of *B. equi* and serves as a differential feature.

B. In the Tick Vector

The developmental cycle of *B. equi* in the tick is only partially understood. In particular,

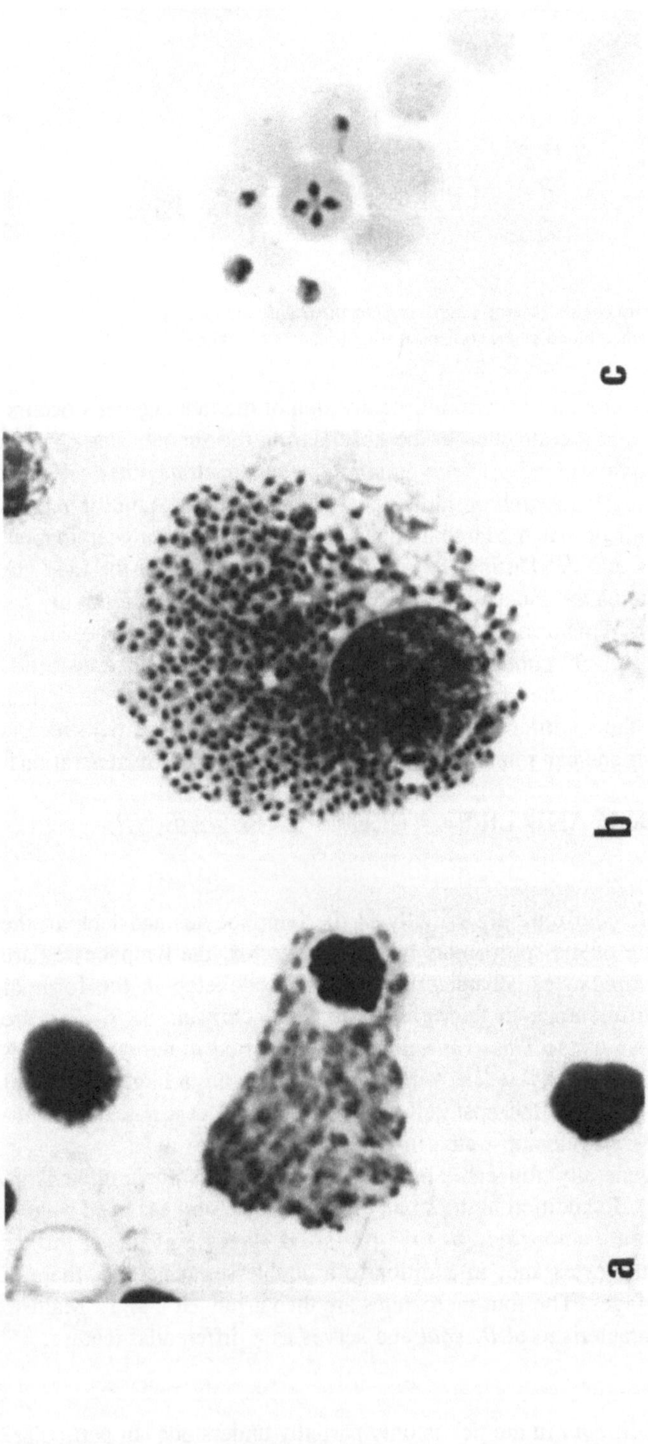

FIGURE 2. *Babesia equi* (Maroc strain); developmental stages in lymphocyte. (a) Macroschizont, 12 days postinfection in lymph node smear. (Magnification × 1600.) (b) Microschizont with numerous merozoites. (Magnification × 1600.) (c) Trophozoites and merozoites in erythrocytes with "Maltese cross" formation. (Magnification × 2000.)

the early developmental stage in the gut of the larvae or nymphs has not yet been precisely examined. Club-shaped kinetes can already be demonstrated at the moment of repletion or 1 day thereafter in the gut of the tick vector *Hyalomma marginatum*. The kinetes have a length of 10 to 12 μm and are approximately 3 to 4 μm in diameter. They appear to multiply through a single multiple division which leads to production of more kinetes. Afterwards, they in turn penetrate the gut wall and attack the cells of the salivary glands.[65] By 8 days after the repletion of the nymphs, they can be detected in the alveolar cells of the juvenile salivary glands.[14] They are rounded off to sporontes in the salivary glands. The sporontes develop into polymorphic bodies with numerous chromatin bodies which are located in a blueish cytoplasm and contain vacuoles. At the same time, the host cell is transformed to a megakaryocytic growth. These megakaryocytic cells displace other cells of the alveola and increase the size of the infected alveolar cells (up to 300 μm), in comparison to normal alveolar cells. During the suction phase, the differentiation to the sporozoites occurs as the next developmental stage. For *B. equi*, only a transstadial transmission is possible, i.e., the infected larva or nymph transmit the organism to the next developmental stage.

IV. GEOGRAPHICAL DISTRIBUTION

Due to the wide distribution of the various tick vectors, equine babesiosis presents a global problem for equestrian sports and trade. *B. caballi* and *B. equi* exist together in most regions. Both species are usually transmitted by the common tick vector; however, *B. equi* infections are more frequently diagnosed than those due to *B. caballi*.

The geographical distribution of these parasites was comprehensively reviewed by Friedhoff.[15] In southern Europe, both *Babesia* species are endemic. Therefore, the entire Balkan Peninsula, as well as Italy, France, and the Iberian Peninsula, are considered endemic regions. In the southern and southeastern parts of the U.S.S.R., both species are prevalent. In the Byelorussian Forest Ridge area, the only vector of *B. caballi* observed is *D. reticulatus*.

In western Europe, the distribution area appears to extend to Belgium,[18] whereas the Netherlands, Ireland, and Great Britain are free of equine babesiosis. North of the Alps, in Switzerland, Austria, and West Germany, there have been no reports of autochthonous cases of babesiosis in spite of the fact that tick vectors *D. reticulatus* and *D. marginatus* are found there. In Czechoslovakia and Poland, data of the distribution pattern are not available.

Throughout Asia, with the exception of the U.S.S.R., babesiosis of the horse is probably an endemic disease. An exception to this is Japan, where to this date no cases of equine babesiosis have been reported.[19] In Turkey and the Near and Middle East, as well as in India where *Hyalomma* and *Rhipicephalus* ticks are widespread, there is a high infection rate among horses and donkeys.[20]

Throughout all of Africa, including Madagascar, equine babesiosis is enzootic.[21] In wild horses and donkeys, high infection rates have been confirmed.[22]

In 1976, equine babesiosis was diagnosed for the first time in imported horses in Australia.[23] This disease did not spread, presumably because of lack of a suitable tick vector.[24]

In South and Central America, equine babesiosis is also quite prevalent. Rudolph et al.[25] reported clinical cases of babesiosis in Chile. They were also able to substantiate, through serological investigations, that 40% of the examined horses had a positive antibody titer.

Both species of *Babesia* were presumably brought into the U.S. via imported horses from Cuba. In 1961, the first clinical cases of babesiosis (*B. caballi*) were diagnosed. Since the tick vector *D. nitens* is widely spread throughout Florida and the southern part of the U.S., *B. caballi* is now endemic in some areas of Florida.[26,27] Further spreading of equine babesiosis was prevented by rigid control measures. However, a complete eradication of the imported epidemic was not achieved. Likewise, in 1965, *B. equi* appeared for the first time in Florida. The vector, which is accountable for this *Babesia* species in the U.S., remains unknown.[28-29]

V. EPIZOOTOLOGY

All horse stock, the donkey, and its crossbreeds, as well as the zebra, are susceptible to *B. caballi* and *B. equi*. Equine babesia, like other babesia, are transmitted under natural conditions only by ticks. Ticks of the species *Dermacentor, Hyalomma,* and *Rhipicephalus* are known as vectors for both species of *Babesia.*[30] In enzootic regions, horses are regularly infested by ticks and are already infected as foals, but are under the protection of maternal antibodies. Through reinfections, they acquire a strong immune status.[21] In enzootic regions, clinical cases of babesiosis are relatively rare.[31] In the case of *B. equi*, the parasitemia persists for many years (probably for life), and such horses constitute a major reservoir. Ticks can also be infected during blood-feeding on horses which are dormantly infected and show only an insignificant parasitemia.[32] Unlike *B. equi*, reservoir is the sole vector of *B. caballi.* Due to the transovarial transmission, the parasites exist for many generations in the tick and can infect horses over a period of years.[33] Hence, for North America and Australia, there is a danger of introducing these parasites through importation of infected horses. However, if also *Hyalomma, Dermacentor,* and *Rhipicephalus* ticks are introduced into these countries, it must be regarded as a much greater danger. In the vast regions of these countries, there are suitable biotopes for these vectors.[15]

Such potential tick vectors as *D. reticulatus* and *D. marginatus* exist in Great Britian and Germany. However, these populations appear to be so scarce that no enzootic infection foci develop.

In rare cases, transmission of the parasites occurs via the use of unsterile syringes and needles.[34]

VI. COURSE OF INFECTION AND CLINICAL MANIFESTATIONS

The infectious stages of the sporozoites are inoculated into the mammalian host 2 to 5 days after the infestation with the infected ticks. The larvae of the *B. caballi* are infectious; mainly the nymphs and adult ticks transmit the infection. In the case of *B. equi,* only nymphs and adult ticks are efficient vectors due to the transstadial mode of transmission.

After infestation with the ticks, the incubation period is 12 to 30 days for *B. caballi*[35,36] and 12 to 15 days for *B. equi.*[7,16] After experimental needle infection, the incubation period is 5 to 15 days for both *Babesia* species,[37] depending on the dose of the inoculum.

The mortality rate depends upon the general immune status of the affected animals and the virulence of the pathogenic organism.[38] In Florida, where the disease has been endemic since 1961, the mortality rate is nearly 10%.[39] The mortality rate in South Africa ranged between 10 and 50%. A high mortality rate occurs during the initial infection of older horses from *Babesia*-free areas introduced into endemic regions.[40]

Sudden death 24 to 48 hr postinfection is relatively rare; usually the horses die 8 to 11 days after the appearance of first clinical signs.[41] Rarely is a parasitemia greater than 10% observed in *B. caballi* infections.[21,37,42] As a rule, there are low levels of parasitemia (1 to 5%) during *B. equi* infections,[41] but there are cases where a high parasitemia of up to 80% has been reported.[42] In splenectomized animals, there are regularly high levels of *B. equi* parasitemias reaching up to 100%, and often these cases are fatal.

Fever coincides with the presence of *B. caballi* merozoites in the blood. In the case of *B. equi* infection, the body temperature increases 1 to 2 days prior to parasitemia. The acute course of the disease lasts for 7 to 12 days.[40] In chronic cases, the clinical signs can persist for an average of 22 days.[41] Initially, in both diseases, the body temperature rises to 39 to 42°C, and in *B. equi* infections, intermittent bouts of fever are frequently observed. However, in the case of *B. caballi* infections, the fever persists for a long time.[29] The pulse rate increases to 80 to 100 beats per minute,[37] and in some cases, there is a noticeable jugular pulsation.[41]

During the acute phase of the disease, there are often petechial hemorrhages of the mucous membranes, which assume a yellowish brown color.[43] During a more chronic progression, especially in the case of *B. equi*, palpebral and extensive subcutaneous edema of the abdomen, breast, genital organs, and extremities can be observed.[40,44] Further, frequent resting in the semiprone position, unsuccessful or weak attempts to stand up, lowered head position, increasing listlessness to complete apathy and anorexia, and extensive weight loss are commonly observed. In the case of *B. caballi* infection, central nervous system involvement has been reported.[40,45]

A wide spectrum of other clinical signs has been described. The incidence, frequency, and expression of some of them are not always characteristic of equine babesiosis. Gastrointestinal disturbances, such as colic and diarrhea, are observed as well as lacrimation rhinorrhoea, bronchitis, and pneumonia (pulmonary edema).[44]

Abortions are common, and fetal erythrocytes are parasitized.[41,46,47] Intrauterine transmission of *B. equi* appears to be relatively common.[48,49] The fetus can be infected by both *Babesia* species during the various stages of pregnancy without the mare showing clinical symptoms.[50]

In addition to the various external manifestations, a pronounced anemia which is not proportional to the parasitemia is observed. Roberts[37] noticed a drop in the hematocrit down to 20% 1 week after infection with *B. caballi*. Rudolph et al.[25] and Mahoney et al.,[24] who obtained similar results, demonstrated a substantial decrease in the hematocrit during equine babesia infections. Horses usually died during experimental infections when the packed cell volume (PCV) dropped below 10%.[65] The phagocytosis of the infected and normal erythrocytes is a possible explanation for this phenomena. Metabolic antigens which are produced by the babesia protozoa adhere to the membranes of normal erythrocytes and are indiscriminately eliminated by the immune defense mechanisms of the host.[51]

Hemoglobinuria, which frequently appears during babesiosis of other animals, is observed only in a few cases during *B. equi* infections of the horse.[29] In case the infected animals survive the acute phase of the disease, the infection will assume a subclinical chronic course. Normocytic and normochromic anemia is seen during the recovery phase, which indicates increased erythropoesis. *Babesia caballi* can still be detected in the blood 12 to 42 months after the initial infection.[9] Infections with *B. equi* persist for the entire life span of the horse, and recrudesence often occurs after many years. Noticeable changes in the leukocyte count can be observed 2 weeks after patency. Leukocytosis with a nuclear shift to the left, monocytosis, and neutrocytophilia indicate extensive activation of the cellular immune system.[25,40,41,52]

VII. PATHOGENESIS AND PATHOLOGY OF EQUINE BABESIOSIS

During *B. caballi* infections, the major pathological effects are targeted to the destruction of the erythrocytes. In addition, there is capillary occlusion leading to dysfunction of the affected organs, especially the lungs, liver, and central nervous system. A kallikrein shock results from the activation of kallikreinogen, which is accompanied by a severe fall in blood pressure, leading to dyspnea and sudden death. These changes are similar to those observed in *B. bovis* infection of cattle.

In contrast, during *B. equi* infection there is a more pronounced anemia resulting from the massive parasitemia and destruction of the erythrocytes (see Chapter 6).

At autopsy of horses which die of acute *B. caballi* infection there is subcutaneous edema in the regions of the sternum, lower abdomen, and extremities, and serous exudates in all body cavities as well as in the pericardium. Pronounced jaundice of the serous membranes is quite characteristic. Hepatosplenomegaly and glomerulonephropathy, as well as pulmonary edema, occur regularly. Petechial hemorrhages are observed on the seromucosal membranes.

Allen et al.[53,54] observed erythrophagocytosis in the lymph nodes. The liver exhibits centrolobular necrosis, enlarged sinusoids with leukocytic infiltration, and activated "Kupffer's cells" containing phagocytized erythrocytes. The authors concluded that anemia resulted from an increased phagocytic activity of the stimulated and proliferating reticuloendothelial cell system.

Similar findings have also been compiled in acute cases of *B. equi* infections. In the foreground stand extensive icteroid discolorations of the tissues, serous fluids in the body cavities, subendocardial hemorrhages, enlarged hemorrhagic lymph nodes as well as hepato, spleno-, and renomegaly. Pulmonary edema is present only in isolated cases.

A fibrin deposition in the blood vessels, a proliferation of the reticuloendothelial cells, and macrophages containing phagocytized erythrocytes were observed. There is extensive impairment of the kidneys with degeneration of the tubular epithelium and deposition of hemoglobin in the renal tubules. In addition to the activated alveolar macrophages of the lungs, thrombi were observed in the larger lymphatic vessels. In the blood vessels of the central nervous system, erythrocytes containing parasites were found. The latter change was not associated with capillary or any other types of thrombi.

VIII. DIAGNOSIS AND SEROLOGY

Clinical diagnosis can be made on the basis of typical clinical signs, evidence of exposure to infected ticks, or history of blood transfusion. Definitive diagnosis depends on the identification of *Babesia* organisms in blood smears stained with Giemsa or acridine orange. Direct parasitologic verification of chronic *B. caballi* infection is almost impossible, but is occasionally successful with *B. equi*. In general, chronic or inapparent infections can only be confirmed after transfusion of approximately 500 mℓ of blood into a susceptible animal.[42] The differential diagnosis of babesiosis must exclude trypanosomiasis, equine infectious anemia, leptospirosis, and poisoning.

Serodiagnosis utilizing the complement fixation test (CFT) has been approved and appraised.[42] This is the official test in the U.S., Canada, Austria, and Japan for screening animals imported from endemic regions. This test is also highly rated by other *Babesia*-free countries. Other serologic methods, such as the indirect fluorescent antibody (IFA) test, the agar gel diffusion test, enzyme immunoassay, etc., are thoroughly discussed by Weiland and Reiter in Chapter 9.

IX. IMMUNITY

In the course of *B. equi* or *B. caballi* infection, Equidae develop an immunity that protects them against reinfections. After recovering from the acute infection, horses still remain carriers of the parasite for a longer period of time. These animals develop an "infection-immunity"-type of immune status. In 1979, Hourrighan and Knowles[35] claimed to have observed sterile immunity after elimination of the protozoan organisms, but these findings have not been independently confirmed.

Immunity to *B. equi* is thought to involve at least partially cell-mediated immune mechanisms. Zweygarth et al.[55] demonstrated that peripheral blood lymphocytes (PBL) from immune animals inhibited the in vitro growth of lymphoblasts containing schizonts. Banerjee et al.[56] were able to demonstrate delayed cutaneous hypersensitivity in donkeys infected by *B. equi* as well as inhibition of leukocytic migration. The authors state that there is a correlation between cell-mediated immunity and protective immunity. During the development of *B. equi,* schizont stages similar to those of *Theileria* are present in the vertebrate host. Carrying the analogy further, it is conceivable that immunity is only achieved after *B.*

equi schizonts invade immunocompetent cells. In *Theileria parva* infection, such an immune phenomenon is thought to be required before infected cattle can acquire immunity.[57]

There is no cross immunity between *B. equi* and *B. caballi*.[58] In addition to the cellular immune defense mechanisms, there is a humoral response which is activated by *B. caballi* infection (see Chapter 7).

In enzootic areas, constant superinfections induce stable immunity. The foals of inapparently infected mares are already infected during the first postnatal year. Under protection of maternal antibodies, they usually go through an inapparent infection. Hence, clinical diseases are rather seldom in enzootic regions (enzootic stability). A breakdown of the immune defense and a new flare-up of the infection only occur under extremely stressful situations.[29]

X. CHEMOTHERAPY AND CONTROL

In combating babesiosis, one must differentiate between two primary goals of treatment: the therapy for the acutely diseased horse and the elimination of inapparent infections.[15]

In the treatment of acute babesial infections, there are some babesicidic agents available. Basically, *B. caballi* can be handled better with chemotherapeutic drugs than *B. equi*.

The following babesicidic drugs can be used in the treatment of equine babesiosis.

Diminazen (Berenil)* — In most cases the application of 4 to 5 mg/kg of body weight administered daily leads to clinical recovery. During *B. equi* infection, a repetition of the treatment with the same dosage may be required. However, complete elimination of the parasite is not possible.[60]

Imidocarb (Imazil)** — Two doses of 2 to 3 mg/kg of body weight of Imazil every 24 hr leads to clinical recovery from *B. equi* infections and a complete elimination of *B. caballi*. This dose, however, can cause toxic side effects. The application of higher dosages with 4 × 4 mg/kg of body weight every 72 hr causes such drastic side effects that the animals may die of drug toxicosis unless atropine sulfate is administered.[61,62]

Amicarbalid (Diaprom)*** — A single injection of 9 to 10 mg/kg of body weight is sufficient for clinical recovery from *B. caballi* and *B. equi* infections. Higher dosages cause pronounced side effects without eliminating *B. equi*.

Quironium (Acaprin)† — The recommended therapeutic dosage, which is 1.2 mℓ of a 5% solution per 100 kilograms of body weight should be administered in two injections, 6 hr apart. These injections are to be administered only subcutaneously. Due to the relatively frequent side effects on horses, quironium sulfate should only be administered in exceptional cases. After the appearance of side effects (restlessness, muscle tremor, salivation, defecation, and dyspnea), which occur approximately 10 to 15 min after injection, atropine sulfate immediately is indicated as an antidote.

A "sterilsatio magna" of *B. caballi* can be achieved by the application of high but still subtoxic doses of a babesicide. Through therapeutic measures, only clinical recovery of *B. equi* is achieved, but complete elimination of the parasites is seldom or impossible even after repeated administration of high dosages. The treated horses develop a preimmunity. They remain carriers presumably for life and present a permanent source of infection for the tick vectors (see Chapter 14).

Utilization of chemotherapy in the battle against equine babesiosis is limited and indeed very difficult. In the U.S., there have been attempts to wipe out equine babesiosis in the following manner: in addition to parasitologic and serologic surveillance of improved Equi-

* Hoechst AG, Frankfurt, West Germany.

** Wellcome Foundation, London, U.K.

***May and Baker, Dagenham, U.K.

† Bayer AG, Leverkusen, West Germany.

dae, those horses in endemic areas (e.g., Florida) are likewise examined. If found positive, they are quarantined and treated with chemotherapeutics. The therapy procedure begins with administration of imidocarb against *B. caballi* and at the same time with acaricides against the tick vectors. In spite of this, the eradication of equine babesiosis in Florida has to this day not been accomplished.

XI. CONCLUSION

Most of the horse and donkey population of the world (approximately 120 million) is exposed to equine babesiosis and lives in endemic areas.[63] Only about 10% of the entire world horse population inhabits regions which are free of babesiosis. There is perpetual danger that babesiosis might be introduced into previously noninfested areas and spread in such regions by intensive horse trading, and especially by equestrian sports. Valuable race and tournament horses of high quality breeds from industrial countries are especially endangered. The transportation of horses by airplane, in which enormous distances can be bridged in relatively short periods of time, conceals a great danger that the tick vector might be unknowingly introduced. Climatic prerequisites and suitable biotopes for the ticks exist in previous babesia-free regions. Therefore, countries like the U.S., Australia, and others have taken precautionary measures to prevent the introduction of babesiosis. Serological monitoring for the detection of parasite carriers, acaricide treatment, and quarantine are necessary protective measures taken by the import countries.

Our knowledge concerning equine babesiosis is still quite incomplete and requires further intensive research. A drawback to research is the fact that both pathogenic organisms are parasitic only in Equidae and cannot be transmitted to laboratory rodents.[64] Such experimental investigations on horses require a great deal of effort and are quite expensive. Suitable chemotherapeutics are at the moment not available to combat equine babesiosis. The development of an effective and compatible chemotherapeutic drug which eliminates *B. equi* infections would certainly be a great advancement.[15] A vaccination program for tournament horses with either attenuated merozoites or inactivated antigens would be desirable. However, vaccinated animals would ultimately become seropositive and, as such, could not be imported into countries with strict serological surveillance. To what extent an attenuated vaccine from the schizonts, similar to that of *T. annulata*, can protect against *B. equi* infections, warrants vigorous research.

Furthermore, an elucidation of some of the stages in the developmental cycle of *B. caballi* and *B. equi* in the tick vectors is required. On the basis of the developmental cycle of the babesia in the equine host and in the tick vector, the taxonomic position of *B. equi* should be reexamined and revised in light of the striking similarity between this organism and *Theileria* species.[7]

REFERENCES

1. **Nuttall, G. H. F. and Strickland, C.,** Die Parasiten der Pferdepiroplasmose des biliary fever, *Zentralbl. Bakteriol. Mikrobiol. Hyg. B*, 56, 524, 1910.
2. **Laveran, A.,** Contribution a l'etude de Piroplasma equi, *C. R. Soc. Biol.*, 12, 385, 1901.
3. **Nuttall, G. H. F. and Strickland, C.,** On the occurrence of two species of parasites in equine piroplasmosis or biliary fever, *Parasitology*, 12, 65, 1912.
4. **Peice, M. A.,** *Nuttallia franca* 1909 (Babesiidae) preoccupied by *Nuttallia dall*, 1989 (Psammobiidae): a re-appraisal of the taxonomic position of the avian piroplasms, *Int. J. Parasitol.*, 5, 285, 1975.
5. **Levine, N. D.,** A newly revised classification of the Protozoa, *J. Protozool.*, 27, 37, 1980.
6. **Levine, N. D.,** *Veterinary Protozoology*, Iowa State University Press, Ames, 1985.

7. **Schein, E., Rehbein, G., Voigt, W. P., and Zweygarth, E.,** *Babesia equi* (Laveran 1901). I. Development in horses and in lymphocyte culture, *Tropenmed. Parasitol.*, 32, 223, 1981.
8. **Simpson, C. F., Kirkham, W. W., and Kling, J. M.,** Comparative morphologic features of *Babesia caballi* and *Babesia equi, Am. J. Vet. Res.*, 28, 1693, 1967.
9. **Holbrook, A. A., Anthony, D. W., and Johnson, A. J.,** Observations on the development of *Babesia caballi* in the tropical horse tick *Dermacentor nitens* (Neumann), *J. Protozool.*, 15, 391, 1968.
10. **Koch, R.,** Beitrage zur Entwicklungsgeschichte der Piroplasmen, *Z. Hyg. Infektionskr.*, 54, 1, 1906.
11. **Mehlhorn, H., Schein, E., and Voigt, W. P.,** Light and electron microscopic study on developmental stages of *Babesia canis* within the gut of the tick *Dermacentor reticulatus, J. Parasitol.*, 66(2), 220, 1980.
12. **Friedhoff, K. T. and Buscher, G.,** Rediscovery of Koch's Strahlenkorper of *Babesia bigemina, Z. Parasitenkd.*, 50, 345, 1976.
13. **Rehbein, G., Zweygarth, E., Voigt, W. P., and Schein, E.,** Establishment of *Babesia equi*-infected lymphoblastoid cell lines, *Z. Parasitenkd.*, 67, 125, 1982.
14. **Moltmann, U. G., Mehlhorn, H., Schein, E., Voigt, W. P., and Friedhoff, K. T.,** Ultrastructural study on the development of *Babesia equi* (Coccidia: Piroplasmia) in the salivary glands of its vector ticks, *J. Protozool.*, 30, 218, 1983.
15. **Friedhoff, K. T.,** Die Piroplasmen der Equiden — Bedeutung fur den internationalen Pferdeverkehr, *Berl. Muench. Tieraerztl. Wochenschr.*, 95, 368, 1982.
16. **Enigk, K.,** Die Ubertrager der Pferdepiroplasmose, ihre Verbreitung und Biologie, *Arch. Wiss. Prakt. Tierheilkd.*, 78, 209, 1943.
17. **Enigk, K.,** Zur Epidemiologie der Pferdepiroplasmose, *Dtsch. Tropenmed. Z.*, 47, 333, 1943.
18. **Chevrier, L., Soule, C., and Dorchies, P. H.,** Les Piroplasmoses equines in aparentes, *Bull. Acad. Vet. France*, 52, 37, 1979.
19. **Tatezawa, E.,** Piroplasmosis of imported ponies, *Jpn. Med. Vet. J.*, 54, 712, 1979.
20. **Kalra, I.S., Verma, B. B., and Sinba, B. P.,** Clinical observations on equine babesiosis, *Indian Vet. J.*, 53, 967, 1976.
21. **Retief, G. P.,** A comparison of equine piroplasmosis in South Africa and the United States, *J. Am. Vet. Med. Assoc.*, 145, 912, 1964.
22. **Young, A. S. and Purnell, R. E.,** Observations on *Babesia equi* in the salivary glands of *Rhipicephalus evertsi, Bull. Epizoot. Dis. Afr.*, 21, 377, 1973.
23. **Churchill, R. C.,** Babesiosis of a horse in Australia, *Aust. Vet. J.*, 52, 487, 1976.
24. **Mahoney, D. F., Wright, I. G., and Frerichs, W. M.,** The identification of *Babesia equi* in Australia, *Aust. Vet. J.*, 53, 461, 1977.
25. **Rudolph, W., Correa, J., Zurita, L., and Manley, W.,** Equine piroplasmosis: leukocytic response to *Babesia equi* (Larveran, 1901) infection in Chile, *Br. Vet. J.*, 131, 601, 1975.
26. **Roby, T. O. and Anthony, D. W.,** The transmission of equine piroplasmosis by the tropical horse tick, *Dermacentor nitens* (Neumann), *J. Am. Vet. Med. Assoc.*, 142, 768, 1963.
27. **Roby, T. O., Anthony, D. W., Thorton, C. W., and Holbrook, A. A.,** The hereditary transmission of *Babesia caballi* in the tropical horse tick, *Dermacentor nitens* (Neumann), *Am. J. Vet. Res.*, 25, 495, 1964.
28. **Strickland, R. K. and Gerrish, R. R.,** Distribution of the tropical horse tick in the United States, with notes on associated cases of equine piroplasmosis, *J. Am. Vet. Med. Assoc.*, 144, 875, 1964.
29. **Taylor, W. M., Bryant, J. E., Anderson, J. B., and Willers, K. H.,** Equine piroplasmosis in the United States — a review, *J. Am. Vet. Med. Assoc.*, 155, 915, 1969.
30. **Klinckmann, G.,** Sporozoitenstabilate von *Babesia equi* aus *Hyalomma anatolicum anatolicum* und *Rhipicephalus turanicus*, Ph.D. thesis, Tierarztliche Hochschule, Hannover, West Germany, 1981.
31. **Littlejohn, A. and Walker, E. M.,** Some aspects of the epidemiology of equine babesiosis, *J. S. Afr. Vet. Assoc.*, 50, 308, 1979.
32. **De Buhr, H.,** Versuche zur Verbesserung von Babesicidprufungen, Ph.D. thesis, Tierarztliche Hochschule, Hannover, West Germany, 1981.
33. **Abramov, I. V.,** About the length of retention of the agent of piroplasmosis of horses *Piroplasma caballi* in ticks *Hyalomma plumbeum* Panzer, 1796, *Veterinariya (Moscow)*, 32, (3), 42, 1955.
34. **Atkinson, R.,** Importation in shambles, *Equus*, 2, 14, 1979.
35. **Hourrighan, J. L. and Knowles, A. M.,** Equine piroplasmosis, *Am. Assoc. Equine Pract. Newsl.*, 1, 119, 1979.
36. **Holbrook, A. A.,** The biology of *Babesia*: development of the *Babesia* of horses, in *Proc. 2nd Int. Conf. Equine Infectious Diseases*, S. Karger, Basel, 1969, 249.
37. **Roberts, E. D.,** Equine piroplasmosis, *J. Am. Vet. Med. Assoc.*, 141, 1324, 1962.
38. **Maurer, F. D.,** Equine piroplasmosis — another emerging disease, *J. Am. Vet. Med. Assoc.*, 141, 699, 1962.
39. **Ristic, M.,** Equine babesiosis and trypanosomiasis, *1st Int. Symp. Equine Hematology*, 1975, 393.
40. **Sippel, W. L., Cooperrider, D. E., Gainer, J. H., Allen, R. W., Mouw, J. E. B., and Teigland, M. B.,** Equine piroplasmosis in the United States, *J. Am. Vet. Med. Assoc.*, 694, 1962.

41. **Gautam, O. P. and Dwivedi, S. K.**, Equine babesiosis: a severe outbreak in a stud farm at Hissar, *Indian Vet. J.*, 53, 546, 1976.
42. **Holbrook, A. A., Frerichs, W. M., and Allen, P. C.**, Laboratory diagnosis of equine piroplasmosis, *Proc. 3rd Int. Conf. Equine Infectious Diseases*, Vol. 3, S. Karger, Basel, 1972, 467.
43. **Weber, H.**, Untersuchungen uber den Verlauf experimenteller Infektionen mit *Babesia equi* bei Kleinpferden, Ph.D. thesis, Fachbereich Veterinarmedizin, Berlin, West Germany, 1983.
44. **Gerber, H.**, Krankheiten des Atmungsapparates, in *Krankheiten des Pferdes*, Wintzer, H.-J., Ed., Paul Parey, Berlin, 1982, chap. 1.1.
45. **Meynard, J. A. and Boudichaud, J. A.**, Diagnosis of acute equine piroplasmosis, *Proc. 3rd Int. Conf. Equine Infectious Diseases*, S. Karger, Basel, 1972, 462.
46. **Du Plessis, J. L. and Basson, P. A.**, Babesiosis in aborted equine foetuses: a report on two cases in South Africa, *J. S. Afr. Vet. Med. Assoc.*, 37, 267, 1966.
47. **Meynard, J. A.**, Piroplasmose congenitale chez le cheval, *Rec. Med. Vet.*, 77, 340, 1951.
48. **Enigk, K.**, Weitere Untersuchungen zur Ubertragerfrage der Pferdepiroplasmosen, *Arch. Tierheilkd.*, 79, 59, 1944.
49. **Neitz, W. O.**, Classification, transmission, and biology of piroplasms of domestic animals, *Ann. N.Y. Acad. Sci.*, 64 (2), 56, 1956.
50. **Erbsloh, J. K. E.**, Babesiosis in the newborn foal, *J. Reprod. Fertil. Suppl.*, 23, 725, 1976.
51. **Cox, H. W.**, The spleen in babesiosis, malaria and other red cell infections. La rate dans les babesiosis, malaria et autres infections a erythrocytes, in *Progress in Protozoology*, Universite de Clermont, France, 1973, 97.
52. **Pipano, E.**, Piroplasmosis — a review, *Refu. Vet.*, 23, 47, 1966.
53. **Allen, P.C., Frerichs, W. M., and Holbrook, A. A.**, Experimental acute *Babesia caballi* infections, *Exp. Parasitol.*, 37, 373, 1975.
54. **Allen, P. C., Frerichs, W. M., and Holbrook, A. A.**, Experimental acute *Babesia caballi* infections, *Exp. Parasitol.*, 37, 67, 1975.
55. **Zweygarth, E., Ahmed, J. S., Rehbein, G., and Voigt, W. P.**, Cell-mediated immune response to *Babesia equi*-transformed lymphoblastoid cells *in vitro*, *Zentralbl. Bakteriol. Parasitenkd. Infektionskr. Hyg. Abt. 1 Orig. Reihe A*, 254, 281, 1983.
56. **Banerjee, D. P., Singh, B., Gautam, O. P., and Sarup, S.**, Cell-mediated immune response to equine babesiosis, *Trop. Anim. Health Prod.*, 9, 153, 1977.
57. **Brown, C. G. D., Cunningham, M. P., Joyner, L. P., Purnell, R. E., Branagan, D., Corry, G. L., and Bailey, K. P.**, *Theileria parva*: significance of leukocytes for infected cattle, *Exp. Parasitol.*, 45, 55, 1978.
58. **Du Toit, P. J.**, Experimentelle Studien uber die Perdepiroplasmose, *Arch. Schiffs Trop. Hyg.*, 23, 141, 1919.
59. **Kuttler, K. L.**, Chemotherapy of babesiosis: a review, in *Babesiosis*, Ristic, M. and Kreier, J. P., Eds., Academic Press, New York, 1981, 65.
60. **Bauer, F.**, Uber die Behandlung von Babesien-Infektionen mit Berenil, Sonderdruck aus, *Die blauen Hefte fur den Tierarzt*, Behringwerke AG, West Germany, 1967, 33.
61. **Frerichs, W. M., Allen, P. C., and Holbrook, A. A.**, Equine piroplasmosis (Babesia equi): therapeutic trials of imidocarb dihydrochloride in horses and donkeys, *Vet. Rec.*, 93, 73, 1973.
62. **Frerichs, W. M. and Holbrook, A. A.**, Treatment of equine piroplasmosis (*Babesia caballi*) with imidocarb dipropionate, *Vet. Rec.*, 95, 188, 1974.
63. **Jasiorowski, H. A., Ozawa, Y., Assaad, F. A., Bogel, K., and Blajan, L.**, in *Animal Health Yearbook, 1985*, Animal Production and Health Series, Bellver-Gallent, M., Ed., Food and Agriculture Organization, Rome, 1985, 25.
64. **Frerichs, W. M., Johnson, A. J., and Holbrook, A. A.**, Equine piroplasmosis: attempts to infect laboratory animals with *Babesia equi*, *Am. J. Vet. Res.*, 30, 1333, 1969.
65. **Schein, E.**, unpublished results.

Chapter 13

HUMAN BABESIOSIS

George Healy and Miodrag Ristic

TABLE OF CONTENTS

I. HISTORICAL PERSPECTIVE

Although a report of the first human *Babesia* infection was not published until the 1950s,[1] it is apparent that Wilson and Chowning in 1904[2] saw organisms in the blood films from patients with spotted fever. In their camera lucida drawings of blood films, they illustrated what appeared to be *Babesia* organisms in the erythrocytes (Figure 1). Their studies on "*Pyroplasmosis hominis*" later showed that the disease was due to a Rickettsia (*R. rickettsiie*), the now recognized cause of Rocky Mountain spotted fever.

Skrabalo and Deanovic[1] published the first well-illustrated case of a human *B. bovis* infection in a 33-year-old Yugoslavian tailor who had been splenectomized 12 years previously. Cattle on pastures in the vicinity of the patient's home were known to be infected with *B. bovis*, and *Ixodes ricinus*, the tick vector, was also present.

A period of 11 years elapsed before the next human infection was reported. Scholtens et al. in 1968[3] recorded a case of babesiosis in a 46-year-old male resident of San Francisco, Calif. He had been splenectomized 4 years previously, during treatment for hereditary spherocytosis. The *Babesia* species was not determined, and the patient survived the infection. At about the same time, Fitzpatrick et al.[4,5] reported a fatal *Babesia* infection in a 48-year-old male, deep sea fisherman in Northern Ireland. Details of the case were reviewed by Kennedy.[6] The patient had been splenectomized 4 months previously in the course of surgery for gastroenterostomy and vagotomy. The infection was initially diagnosed as malaria due to *Plasmodium falciparum*. However, after careful examination of stained blood films and epidemiologic evidence, which ruled out malaria (no travel to malarious area or blood transfusions), the organism was identified as *B. divergens*. An expedition to the area where the patient was thought to have acquired the infection while camping[7] disclosed the presence of *Ixodes ricinus* ticks, vectors of *B. divergens*.

In 1970, Western et al.[8] and Ristic et al.[9] in the U.S., reported the first human case of babesiosis caused by the rodent parasite *B. microti* (Figure 2B, C, D). The patient, a 59-year-old nonsplenectomized female resident of Nantucket Island, Mass., survived the infection. Identification of the suspected *Babesia* species was confirmed by Van Peenen and Healy[10] by comparing an isolate obtained from blood of the patient inoculated into mice with a known *B. microti* infection maintained in rodents. Blood from the patient was inoculated into CFW mice, Sherman rats, a rhesus monkey, and a beagle puppy, resulting in transient, low-level parasitemias.[11] Blood films from hamsters inoculated with the patient's blood showed patent heavy parasitemia. Since then, the hamster has been the animal of choice for maintenance of isolates of human *B. microti* and for inoculation of human blood suspected of containing the parasite.[12]

II. GEOGRAPHIC DISTRIBUTION AND SPECIES

The majority of the human *Babesia* infections in North America, specifically the U.S., have been due to *B. microti* and will be discussed below. Six infections, however, have been attributed to other *Babesia* species[3,13-15] (Table 1). The *Babesia* organisms in the erythrocytes of the six human North American cases were not identified as to species, although the morphology of the parasites in the blood of the two splenectomized invididuals in California were thought to resemble *B. equi*, found in horses and other equids (Figure 2A). The individual reported by Bredt et al.[15] did keep dogs and horses on his premises, but none of the animals were ill prior to or at the time of the patient's illness. Although organisms were recovered in the blood of hamsters inoculated with the three Mexican cases,[14] the authors did not designate the species.

To date, eight of ten published cases from Europe and Asia have been associated with bovine species of *Babesia*.[1,5,16-24] All but two of the individuals were splenectomized and

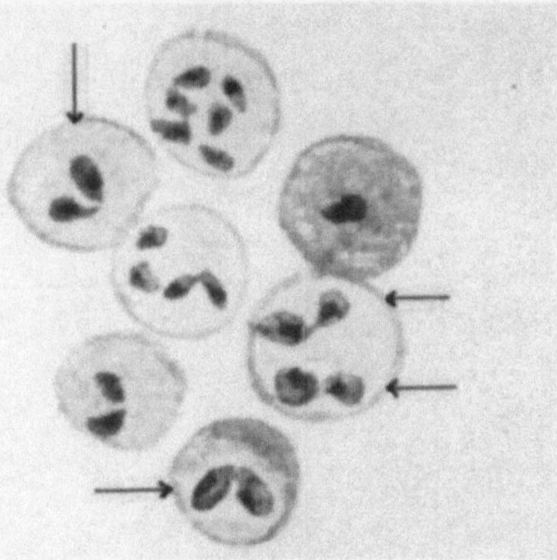

FIGURE 1. Camera lucida drawing of "pyroplasms" (arrows) in blood of spotted fever patient. (From Wilson, L. B. and Chowning, W. M., *J. Infect. Dis.*, 1, 31, 1904.)

seven of ten died, presumably from complications associated with the *Babesia* infection (Table 2). *B. divergens* was the species identified in four cases, *B. bovis* in three, and the species was not identified in three cases. A significant breakthrough occurred with the case from Scotland reported by Entrican et al.[19] Lewis and Williams[25] were able to infect calves with blood from the human and later isolated the strain of parasite in the mongolian gerbil. Further work in 1980 by Lewis et al.[26] characterized the human *B. divergens* isolate in calves and gerbils.

Liddell et al. provided detailed information on the Scottish human isolate in the gerbil model.[27] They were able to propagate the strain through 150 semicontinuous, intravenous passages. The infections were normally fatal, with death occurring as early as 44 hr. By the 150th passage, the average maximum parasitemia was 80%, attesting to the virulence of this isolate.

A detailed analysis of the parasites in the blood films of the first two cases reported in France[16,17] was made by Gorenflot and Piette.[28] In a well-illustrated paper, the authors concluded that the first French case was due to *B. divergens* (Figure 3). They further concluded that the second case was probably an equine species, based on the presence of many paired, pyriform, and Maltese cross forms in the blood films.

The number of human *B. microti* infections in the U.S. is difficult to determine. In recent years, babesiosis has been a reportable disease in Massachusetts and New York. Table 3 lists the number of cases in the two states, giving a total of 143. Several years ago, Dammin et al.[29] in an analysis of the incidence of *B. microti* in three states (Massachusetts, New York, and Rhode Island), reported a total of 99, both clinical and subclinical infections. Ruebush et al., in 1981,[30] in a 1975 epidemiologic survey of persons on Nantucket Island reported that sera of 14 of 394 (3.6%) persons had positive indirect immunofluorescent (IIF) antibody titers[31] (≥64), indicating prior experience with the parasite. These were in addition to the parasitologically confirmed clinical cases in 1969 (one), 1973 (one), 1973 (seven), and 1976 (five). Filstein et al., in 1980,[32] reported results of a serosurvey on specimens collected from residents of Shelter Island, N.Y., another endemic area for *B. microti*. They obtained positive IIF titers (≥64) in 7 of 102 (6.9%) of the Shelter Island sera, while none of 300 control sera obtained from the Greater New York Blood Council were positive. Control sera comprised 150 specimens from residents of New York City and 150 from the New York metropolitan area.

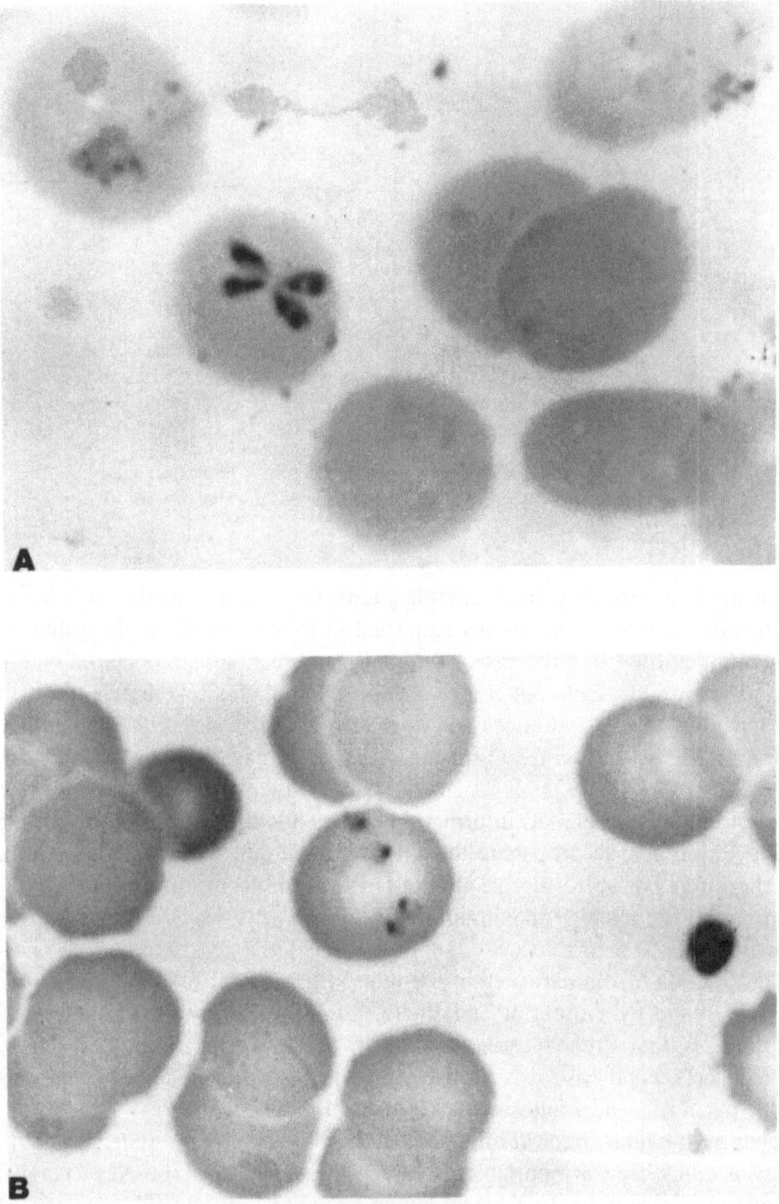

FIGURE 2. *Babesia* parasites in blood films. (A) *B. equi* (?). (Magnification × 950.) (After Bredt, A. B., *JAMA*, 245, 1938, 1981.) (B), (C), (D) *B. microti*. (Magnification × 950.)

Over 22 published reports of human *B. microti* infection have appeared in recent years.[33-54] Most of the cases have been diagnosed in the Massachusetts-New York endemic areas. In addition, Smith and Clevenger[55] published a case report in 1985 concerning a 56-year-old Indiana resident. The woman, however, had recently visited her son on Long Island, N.Y. Demicco and Dolin,[56] in 1985 diagnosed and successfully treated a *B. microti* infection in a 40-year-old male resident of Vermont. The patient, however, had visited Nantucket, Mass. 1 month previous to his hospitalization in Vermont.

A significant finding in the geographic distribution of human *B. microti* infections was

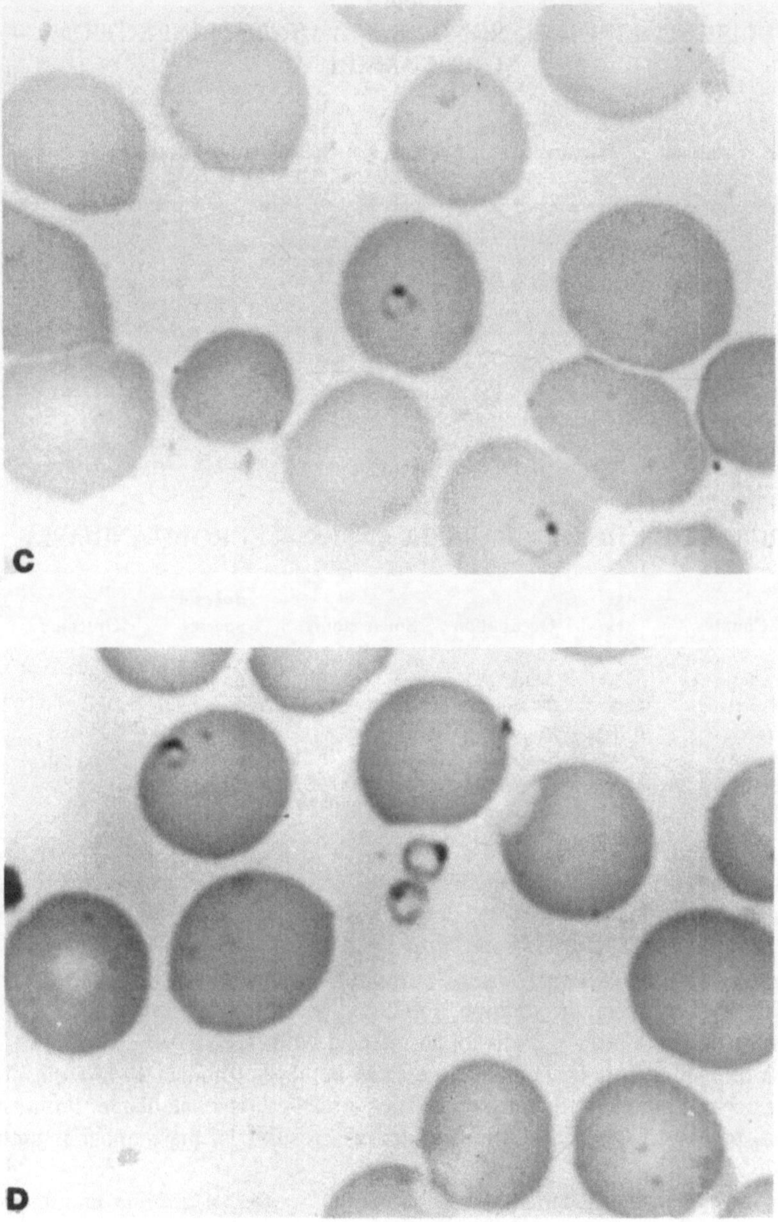

FIGURE 2C and D.

recently made by Steketee et al.[57] They reported an autochthonous case in a splenectomized 54-year-old male resident of Wisconsin. He had no travel history to endemic areas within the previous 30 years and had no history of blood transfusions since the splenectomy carried out 10 years previously for a Billroth II operation for peptic ulcer. The 52-year-old wife of the patient had been ill 2 years previously with myalgia, fever, joint pains, and fatigue. Her illness was subsequently diagnosed as Lyme disease. It is interesting, however, that a serum sample obtained from her at the time of her husband's illness was positive in the IIF test[31] at a titer of 1024. Blood films stained with Giemsa were negative for parasites and hamsters inoculated with her blood did not develop parasitemia. *Ixodes dammini,* the tick vector of

Table 1
PUBLISHED REPORTS, NON-*BABESIA MICROTI* INFECTIONS — NORTH AMERICA

Year	Age/sex	Occupation	Place	Splenectomy	*Babesia* species	Ref.
1968	46/M	Amateur photographer	California	Yes	*B. equi*	3
1976	51/M	Laborer	Georgia	No	?	13
1976	49/M	Construction worker	Mexico	No	?	14
1976	31/M	Animal caretaker	Mexico	No	?	14
1976	29/M	Animal caretaker	Mexico	No	?	14
1981	36/M	Policeman	California	Yes	*B. equi*	15

Table 2
PUBLISHED HUMAN BABESIA CASES — EUROPE AND ASIA

Year	Country	Age/ sex	Occupation	Splenectomy	*Babesia* species	Outcome	Ref.
1957	Yugoslavia	33/M	Tailor	Yes	*B. bovis*	Died	1
1968	N. Ireland	48/M	Fisherman	Yes	*B. divergens*	Died	5
1976	France	61/F	Housewife	Yes	*Babesia* sp.	Survived	16
1976	France	53/M	Farmer	Yes	*B. divergens*	Survived	17
1978	U.S.S.R.	48/F	Farmer	Yes	*B. bovis*	Died	18
1979	Scotland	34/M	Accountant	Yes	*B. divergens*	Died	19,20
1979	France	53/M	Vagabond	Yes	*Babesia* sp.	Died	21
1986	France	70/M	?	Yes	*B. divergens*	Died	22
1986	Spain	55/M	?	No	*B. bovis*	Died	23
1986	Spain	26/M	?	No	*Babesia* sp.	Survived	24

both *B. microti* and *Borrelia burgdorferi,* the spirochetal agent of Lyme disease are known to be present in Wisconsin. In addition, the blood of 1 of 16 *Peromyscus leucopus* mice trapped on the patient's property was found infected with *B. microti.*

From the foregoing, it is obvious that the exact number of human *B. microti* infections, which have occurred in the U.S. since the first case in 1969, is incalculable. From anecdotal information, published reports, and the results of serosurveys, the number is probably in excess of 200.

Previous reviews of the geographic distribution and species of *Babesia* in humans can be found in the papers by Cox,[58] Dammin,[59] Donnelly,[60] Garnham,[61] Hoare,[62] and Ruebush.[63]

III. DIAGNOSIS

A. Direct Detection — Blood Films

Diagnosis of human *Babesia* infections has been generally accomplished by the microscopic examination of Giemsa- or Wright-stained peripheral blood smears. Photomicrographs of human blood films containing the bovine, equine, and unknown species of *Babesia* have been published by Skrabalo and Deanovic,[1] Scholtens et al.,[3] Fitzpatrick et al.,[4] Healy et al.,[13] Bredt et al.,[15] Gorenflot et al.,[17,28] and Calvo de Mora et al.[23] Photomicrographs of human *B. microti* organisms in blood films have been illustrated by Anderson et al.,[64] Healy,[65] Parry et al.,[46] Smith and Clevenger,[55] Sun et al.,[50] and Western et al.[8] Color

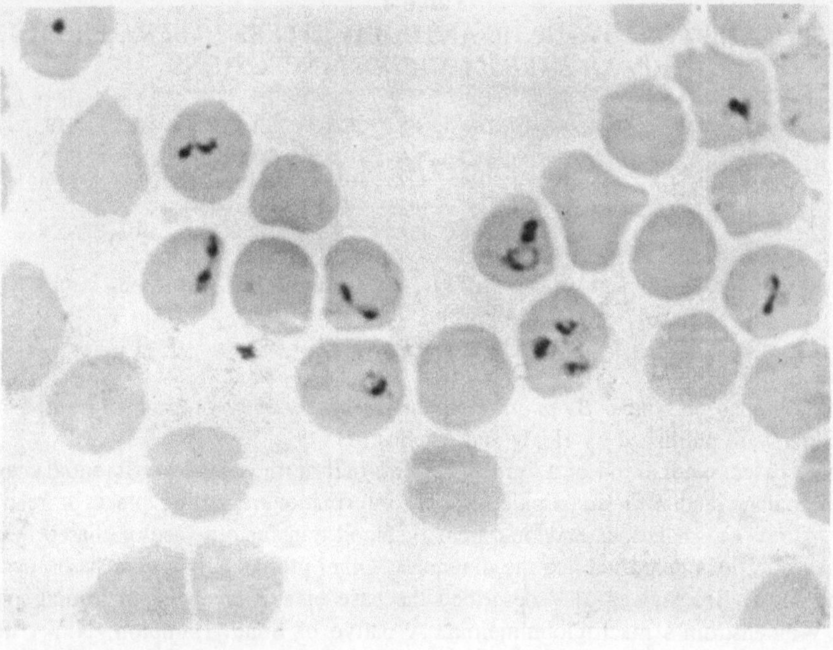

FIGURE 3. *Babesia divergens*. (Magnification × 950.) (From Gorenflot, A. and Piette, M., *Med. Malad. Infect.*, 11, 334, 1981. With permission.)

Table 3
HUMAN *BABESIA MICROTI* CASES — STATES REQUIRING REPORTING

Calendar year	Massachusetts[a]	New York[b]
1969—1976	12	↓
1977	0	
1978	1	
1979	1	
1980	5	21 Total
1981	4	2
1982	5	10
1983	4	13
1984	1	14
1985	5	18
1986	9	18
Total	47	96

[a] Cases confirmed clinically or serologically — courtesy Dr. Lynn Mofenson.

[b] Cases confirmed by parasitologic diagnosis of blood films, or inoculation of hamsters; all serologically reactive — courtesy Dr. Jorge Benach.

Table 4
ACUTE AND FOLLOW-UP IIF ANTIBODY TITERS IN SERA FROM 12 HUMAN *BABESIA MICROTI* INFECTIONS

Week	NG	LW	JD	MD	DF	FE	GR	WB	NJ	MT	HB	PA
0—2	256	1024	256	1024	1024	4096	1024	1024	4096	4096	1024	1024
3—4	1024		1024			1024	256		4096			1024
5—8		64	1024		256	1024	256	1024		1024	1024	
Month												
5—7		64		64	64	64	64		64	64	1024	
Year												
1	64		64	256	16	64	64	64				256

photomicrographs of typical *B. microti* organisms in a variety of Giemsa-stained human blood films were published by Healy and Ruebush.[66]

In some cases, recognition of a human *Babesia* parasite involves a differential diagnosis between malaria and *Babesia*. Epidemiologic information regarding place of residence, history of travel to malarious endemic areas or blood transfusions, use of shared syringes, or exposure to ticks help establish the diagnosis. Other problems in diagnosis do occur. In a recent paper, Bruckner et al.[34] described the case of a 76-year-old male diagnosed as having Waldenstrom's macroglobulinemia. A native of South Hampton, N.Y., he was hospitalized while in Los Angeles, Calif., with persistent fever and sweating. Peripheral blood films examined by an autoanalyzer failed to indicate any abnormal erythrocytes. With symptoms persisting, a detailed review of the patient's history was carried out. Questioning elicited the information that the patient had a spontaneous rupture of the spleen 2 years previously, as well as that he owned marshlands on Long Island, N.Y., where he had been bitten by ticks on numerous occasions. Manual microscopic examinations of a blood film disclosed *B. microti* organisms at a level of 1 parasitized erythrocyte per 15 to 20 oil objective fields. Reexamination of the films, tested by the autoanalyzer 11 days previously, disclosed by manual examination a similar parasitemia. The patient was treated with quinine and clindamycin and made an uneventful recovery. It was believed that the diagnosis of Waldenstrom's macroglobulinemia was incorrect. The inability of an autoanalyzer to discriminate between *B. microti* infected cells and normal erythrocytes points to the need for human microscopic examination of blood films from individuals whose symptoms or history might suggest presence of a blood-borne parasite such as *Babesia*, malaria, or even trypanosomiasis.

B. Indirect Detection — Serodiagnosis

The use of serologic tests for the diagnosis of human babesiosis has been pursued primarily in cases of *B. microti* infections. Following the reported cases of human *B. microti* infections on Nantucket Island,[8,64] Chisholm et al.[31] developed an IIF test for use in diagnosis, persistence of antibody in sera from confirmed cases,[67] and for epidemiologic studies.[30] Table 4 indicates IIF titers obtained by testing sera from 12 cases of clinical babesiosis. Most of the acute sera had antibody titers $\geq$1024, which dropped to 64, some as early as 7 months and most by the end of 1 year. Chisholm et al.[31] concluded that the IIF test was sensitive, specific, and reproducible. Occasional positive reactions ($>$16) were noted with sera from slide-proven malaria cases. Sera from a control group (New York City residents) and persons exposed to ticks in Montana and diagnosed as suspected or confirmed Colorado tick fever were not reactive in the IIF test.

The authors extended their initial evaluation of the antigenic specificity of the test.[68] They found that human antisera to *B. microti* antigen showed cross reactions with antigens of other *Babesia* species, but at lower dilutions and during the acute phase of the disease. Cross reactions with antibodies from cases of malaria, other parasitic diseases, and Colorado tick fever were uncommon. Acute and convalescent sera from four patients with Rocky Mountain spotted fever showed a rise in titer to *B. microti* antigen. These latter findings recall the illustrations of Wilson and Chowning[2] (Figure 1). However, human *Babesia* sp. infections have yet to be diagnosed in Oklahoma, Tennessee, Maryland, North Carolina, or South Carolina, the origin of the 1005 spotted fever sera tested. The authors also reported 6 of 185 sera, from North Carolina Cherokee Indian children, positive at IIF titers of $\geq$256, a level considered indicative of the recent *B. microti* infection.

IV. EPIDEMIOLOGY

A. Tick Transmission

The acquisition of a *Babesia* infection by animals (and humans) has been known since the 1893 pioneer studies by Smith and Kilbourne[69] on Texas cattle fever caused by *B. bigemina* and transmitted by *Boophilus annulatus*. In the human *Babesia* infections of European bovine and/or equine origin, *Ixodes ricinus* has been recognized as an important vector, and in some cases, infected cattle and ticks were epidemiologically related to the acquisition of human infections.[1,7] In most of the European cases, however, although the *Babesia* species was identified, there was little information about the epidemiology of the infection, or attempts to collect ticks, except for the *B. divergens* infection in Northern Ireland.[7]

The epidemiology of human *B. microti* infections in North America has been the subject of a decade of ongoing investigations. Spielman and co-workers, recently summarized results of these investigations in an excellent review[70] on the ecology of *Ixodes dammini*-borne human babesiosis and Lyme disease. Following the cluster of human *B. microti* infections on Nantucket Island in the early 1970s, Healy et al.[71] found natural infections in 5 field mice (*Microtus pennsylvanicus*) and 31 of 39 white-footed or deer mice (*Peromyscus leucopus*).

Additional field studies by Spielman et al.,[72] collecting blood specimens from Nantucket and several other locations in New England, resulted in positive findings in deer mice, field mice, rats (*Rattus norvegicus*), cotton-tailed rabbits (*Sylvilagus floridanus*), and chipmunks (*Tamias striatus*). The white footed mouse was most heavily infected on Nantucket, with infection rate ranging between 42 to 78%, depending on the season of the year, as determined by hamster inoculation of mouse blood.[73] Spielman et al.[74] identified the tick vector and determined that it was a new species, *Ixodes dammini*.

Piesman and Spielman[75] presented field and laboratory evidence for the nymphal stage of the tick as being important in transmission to humans, although Piesman et al.[76] recently did show the *B. microti* sporozoites in salivary glands of adult female ticks were infectious to hamsters. Piesman et al.[77] also recently found *I. dammini* nymphs simultaneously infected with *B. microti* and *Borellia burgdorferi*, the spirochetal agent of human Lyme disease. Both organisms have caused concurrent infections in severe human cases in recent years.[42,78]

Deer are inextricably a part of the epidemiology of human *B. microti* infections.[79] The vector ticks, as larvae, feed on rodents such as white footed mice; nymphs and adults feed on deer, which are not infected with the parasite, but provide a blood meal. In the midsummer and fall, nymphal *I. dammini* will attach to humans who venture into areas inhabited by deer. The increase in human infections on Nantucket and other islands in Massachusetts, the Long Island, N.Y., complex, and even Wisconsin[57] has been linked to the reintroduction or increased population of deer in those sites. Hoogstraal,[80] in a review of the changing patterns of tick-borne diseases in modern society, called attention to the babesiosis problem in the U.S. by stating, "the simple act of introducing deer and the deer tick into a favorable rodent-inhabited biotope, used by numerous visitors for summer holidays, appears to have been responsible for another emerging disease during the past decade."

B. Transfusion Babesiosis

Although human babesiosis, especially the bovine and equine species, have been characterized by symptomatic and sometimes fatal infections, there is evidence of the occurrence of mild, self-limited, symptomatic or even asymptomatic infections. Filstein et al.,[32] Ruebush et al.,[30,81] and Osorno et al.[14] reported etiologic and/or serologic evidence of present or recent asymptomatic babesial parasitemia. It is therefore likely that such individuals post a risk if their blood is used as a donor source in transfusions. This is part of the general subject of transfusion-associated parasitic infections recently reviewed by Bruce-Chwatt.[82]

Four cases of transfusion-acquired *B. microti* infections have been reported in recent years. Jacoby et al.[40] reported disease in a 70-year-old splenectomized male who received platelets from a 17-donor source. Serum from one of the donors, who was asymptomatic, had an IIF titer of 256. Organisms were not found in blood smears, nor did hamsters inoculated with the donor's blood become parasitemic.

Grabowski et al.[38] reported the case of a transfusion-dependent, thalassemia major, 23-year-old splenectomized woman. Over a period of 6 years, she had received only frozen-thawed erythrocytes (3 U/month). The serum from one of six donors of the previous 3 months had an IIF titer of 256. No parasites were seen on blood films.

Wittner et al.[52] reported babesiosis in a newborn, premature female. She received 11 blood transfusions; blood films and sera were obtained from ten donors. Sera from a 35-year-old resident of Long Island had an IIF titer of 64 and subsequent blood film examination and hamster inoculation showed *B. microti* organisms.

Marcus et al.[41] described a fatal case in a 79-year-old woman who had received two units of blood during surgery for sigmoidectomy and drainage of an abdominal abscess. Blood smears and serum obtained from one of the donors, a 34-year-old male, were tested; blood films were negative, but hamsters inoculated with the blood developed heavy parasitemia. The donor's serum was positive at an IIF titer of $\geq$4096. He had been asymptomatic from the time of his presumed acquisition of the parasite, while camping on Cape Cod 7 days before donating blood. He was still asymptomatic, 4 to 7 weeks after the presumed time of his tick exposure when the hamster inoculations were done.

The longevity of *B. microti* infections in asymptomatic individuals has not been prospectively studied. Evidence from transfusion cases, however, does indicate that parasitemia may persist for several weeks. Ruebush et al.[83] followed the course of *B. microti* infections in seven rhesus monkeys infected by inoculation of infected hamster blood. In five monkeys, patent low-level parasitemia was evident for a 90-day observation period. In two other monkeys observed over a 600 day period, one animal was positive by hamster subinoculation for 135 days, while blood films were negative after 72 days. The second rhesus was positive for *B. microti* by hamster subinoculation for 559 days after infection, while blood films were negative after 159 days.

V. CLINICAL ASPECTS

Human babesiosis, like other infectious diseases, runs the gamut from the asymptomatic to the severe, even fatal, presentation. Asymptomatic infections have been detected in serosurveys and investigations of donors in transfusion-associated infections. The ability to detect low-level parasitemia, not demonstrable by thin or even thick blood film examination, by inoculation of patient blood into candidate gerbils for the bovine species[27] and hamster for the rodent forms[12] has amply demonstrated the existence of asymptomatic infections.

Symptoms associated with the majority of cases caused by the bovine/equine *Babesia* spp. have been the most severe. The association of splenectomy with the *Babesia* infection has magnified the symptoms which include fever, chills, hemolytic anemia, sweating, myalgia, arthralgia, nausea, vomiting, and prostration. In earlier cases described, the hemolytic

Table 5
REPORTED *BABESIA MICROTI* INFECTIONS — U.S. 1983—1986

Patient #	Age/sex	Location infection acquired	Reason for splenectomy	Interval of splenectomy to infection	Outcome	Ref.
1	54/M	Fire Island, N.Y.	0	—	Recovered	33
2	74/F	Shelter Island, N.Y.	0	—	Recovered	36
3	55/M	Shelter Island, N.Y.	0	—	Recovered	36
4	66/M	Shelter Island, N.Y.	0	—	Recovered	36
5	47/M	Shelter Island, N.Y.	Idiopathic cytothrombo-penic purpura	?	Recovered	36
6	78/M	Shelter Island, N.Y.	0	—	Recovered	36
7	66/F	Shelter Island, N.Y.	0	—	Recovered	36
8	76/M	Long Island, N.Y.	0	—	Recovered	34
9	79/F	Pittsfield, Mass.	0	—	Died[a]	37
10	54/F	Shelter Island, N.Y.	0	—	Recovered	39
11	66/M	Nantucket, Mass.	0	—	Died[b]	42
12	14/M	Barnstable, Mass.	Hereditary spherocytosis	10 years	Recovered	43
13	30/F	Montauk, N.Y.	0	—	Recovered	47
14	25/F	Montauk, N.Y.	Trauma	14 years	Recovered	48
15	47/M	Long Island, N.Y.	Idiopathic cytothrombo-penic purpura	6 years	Recovered	50
16	Newborn/M	New York, N.Y.	0	—	Recovered[a]	52
17	69/M	Montauk, N.Y.	0	—	Recovered	54
18	56/F	Long Island, N.Y.	0	—	Recovered	55
19	40/M	Nantucket, Mass.	0	—	Recovered	56
20	54/M	Wisconsin	Billroth II operation peptic ulcer disease	10 years	Recovered	57
21	36/M	Shelter Island, N.Y.	Car accident	4 years	Recovered[b]	78

[a] Transfusion acquired.
[b] Concurrent *Borrelia burgdoferi* (Lyme Disease) infection.

anemia presented much the same as "blackwater fever", a hallmark of severe *Plasmodium falciparum* malaria. In point of fact, the early bovine-source human babesiosis cases were diagnosed as malaria.[1,5]

The classic symptoms associated with human *B. microti* infections have been presented by Western et al.,[8] Ruebush et al.,[84] Benach and Habicht,[85] Gombert et al.,[36] and several other authors citing individual case reports. Patients have presented with intermittent fever, chills, drenching sweats, arthralgia, myalgia, malaise, anorexia, and weight loss. The spectrum of symptoms and severity has, for the most part, been associated with older age groups, the level of parasitemia, and the absence of a spleen. Rosner et al.,[86] in 1984, reviewed 22 published cases of babesiosis in splenectomized humans.

Table 5 lists the reported cases of human *B. microti* infections in the U.S. published since 1983. Several features of the infections are unique since they offer new information on the clinical importance of *B. microti* infections. Only two of the patients in Table 5 died, and neither was without a spleen. Patient 9 was a 79-year-old female who expired from acute respiratory distress syndrome as a complication of her *Babesia* infection; patient 11 was a 66-year-old male who succumbed to a fatal pancarditis while hospitalized with classic symptoms and an etiologic diagnosis of *B. microti* infection. The availability of serologic tests for Lyme disease stimulated a test for antibody in antemortem serum, and the results were positive. The autopsy, reported by Marcus et al.,[42] disclosed spirochetes of *Borrelia burgdorferi* in the patient's myocardium.

The finding by Piesman et al.[77] that nymphal *I. dammini* ticks may be concurrently infected with *B. microti* and *B. burgdorferi* adds another dimension to the possible severity of *B.*

microti infections coupled with tick-acquired Lyme disease. Grunwaldt et al.[78] called attention to the evidence of the simultaneous occurrence of babesiosis and Lyme disease in a 36-year-old splenectomized male from Shelter Island, N.Y. (patient 21, Table 5), who recovered from the infection(s). More recently, Benach et al.,[87] in 1985, showed serologic evidence of simultaneous occurrence of antibody to both diseases. They found that sera of 54% of patients with babesiosis (*B. microti*) tested, had IgG and IgM antibodies to *B. burgdorferi* antigens; 66% of sera from selected patients with Lyme disease from areas where babesiosis is thought not to occur, also had IgG and IgM antibodies to *B. microti* antigens. Cross reactivity does not seem to be a problem, since laboratory animals infected with *B. microti* do not have antibody to Lyme disease, and animals immunized with Lyme disease spirochetes do not develop antibodies to *B. microti*.

Although all but two of the patients in Table 5 recovered, even those who were splenectomized (#5, #12, #14, #15, #20, and #21), certain recent observations on the associated clinical findings in *B. microti* bear mention. Auerbach et al.[33] described a case of severe hemophagocytosis in a 54-year-old male with *B. microti*. The authors cautioned that babesiosis should be considered in the differential diagnosis of hemophagocytosis. The patient recovered and the condition disappeared after therapy with clindamycin and quinine. Mathewson et al.[43] described a self-limited *B. microti* infection in a 14-year-old splenectomized boy (#12, Table 5) whose association with Nantucket Island involved a 4-hr visit for a tennis match, 6 weeks prior to onset of symptoms! The presence of immune complexes in babesiosis has been recorded. Ortiz et al.,[45] in 1982, reported retinal nerve fiber layer infarcts in a 34-year-old splenectomized woman with babesiosis, whose condition improved following therapy for the *B. microti* infection. Sun et al.[50] reported *B. microti* infection in a 47-year-old splenectomized male (case #15, Table 5) whose parasitemia approached 85% of erythrocytes. Two exchange transfusions of 7 and 6 units helped decrease the parasitemia to <5%, along with chloroquine, quinine, and pentamidine therapy. Low levels of infected red cells were demonstrable by hamster inoculation for 7 months following initial hospitalization. Wolf et al.[53] reported finding in 3 of 30 human *B. microti* infections an autoimmune response to red blood cells. The patients had positive direct antiglobulin tests with noncomplement-binding IgG antibodies coating their red blood cells. Eluates prepared from the patients' red blood cells reacted with cells taken from healthy individuals. The authors concluded that the autoimmune response was thus aimed at integral structures of the red blood cell membrane, possibly associated with Rh protein, and did not appear to be directed against unique antigens of parasite origin. Other autoimmune phenomena in babesiosis have been described by Cahill et al.[35] and Grunwaldt.[39] Woessner et al.,[24] in a brief communication, noted a severe *Babesia* sp. infection in a 26-year-old splenectomized male with an imbalance of T cell subset population.

Raucher et al.[47] reported a *B. microti* infection in a nonsplenectomized pregnant female in the 19th week of gestation. Clinically, the patient presented with spiking fevers to 40.5°C and pancytopenia. A 4.3% parasitemia was recorded. She was transfused with 3 units of packed red blood cells, and no other therapeutic modalities were administered. She made an uneventful recovery and delivered a 3100 g infant girl. No etiologic or serologic evidence of infection with the child was noted, and she remained well during a 7-month follow-up, without anemia or illness.

Benach et al.[87] examined the immunoresponsiveness in acute and follow-up *B. microti* patients' sera. In general, they found that responses to nonspecific mitogens were suppressed during acute babesiosis. Levels of serum IgG and IgM and C1Q binding were significantly increased, whereas C3 and C4 levels were suppressed in acute-phase sera.

The host parasite relationship of human-derived *Babesia* sp. in animal models has been noted above.[11,12,25,83] Additional studies were published by Garnham and Bray,[89] Lykins et al.,[91] Benach et al.,[90] and Moore and Kuntz.[92] The latter authors noted different susceptibility of primates to infection with *B. microti*. Two African baboons, two cynomologus macaques, and two squirrel monkeys (one of each pair splenectomized) survived intraperitoneal inoculation of hamster blood containing the Gray strain of *B. microti*. All animals recovered from the infection, and parasites were no longer demonstrated in Giemsa-stained blood films 20 days after inoculation. In contrast, all of six spleen-intact Capuchin monkeys died by the 5th day postinoculation. Relevant aspects of human and animal babesiosis have been thoroughly reviewed by Ristic and Healy.[93]

VI. THERAPY

The treatment of human *Babesia* infections has involved the use of antiparasitic agents and supportive therapy. Experience has shown that most animal babesiacidal agents are effective only at highly toxic doses. Antimalarial compounds on the other hand have, until recently, been effective in relieving symptoms, but have shown poor antibabesial activity. Recent case reports for the treatment of *B. microti* infections have indicated the effectiveness of quinine and clindamycin.[33,43,50,52,56] It is of historic interest that Wilson and Chowning,[2] in their 1904 paper on "pyroplasmosis hominis", reported success in treating several of their patients with quinine.

Only three of the ten European human cases of infection with bovine/equine *Babesia* spp. survived (Table 2). It is therefore difficult to assess the effectiveness of the treatment in the European cases. The first surviving French case[16] received ampicillin, gentamycin, chloroquine, quinine, and blood transfusion. The other French patient[17] survived following chloroquine, peritoneal and hemodialysis, and blood transfusion. The surviving patient in Spain[24] was asymptomatic. The species of *Babesia* was not identified, and the patient was treated with unspecified drugs.

In the North American cases of non-*B. microti* infections (Table 1), the latest reported case[15] received chloroquine, quinine sulfate, and pyramethamine. Because of hemolytic anemia and severe thrombocytopenia, he also required erythrocyte and platelet transfusions. The first non-*B. microti* infected patient[3] was treated with chloroquine and made an uneventful recovery.

The therapy of human *B. microti* infections has included a variety of antiparasitic agents. While most *B. microti* infected persons have survived, the more severe clinical presentation in older persons and the deleterious effect of splenectomy have necessitated various supportive measures such as platelet and blood transfusions and red cell exchange in cases of high parasitemia and severe hemolytic anemia. Chloroquine had been given to numerous patients with resolution of symptoms, but persistence of parasitemia in many instances.[44] Francioli et al.[94] reported success in treating three patients with pentamidine and discussed the current status of chemotherapy. Francioli reported that the three nonsplenectomized patients improved clinically after 5 days of therapy, but parasites persisted for at least 5 weeks, as demonstrated by hamster inoculation of the patients' blood. Pain at the injection site was a major side effect, and one of the patients required two blood transfusions because of a low hemoglobin level (8.6 g/dℓ).

As first reported by Wittner et al.[52] the combination of quinine sulfate and clindamycin seems a useful antibabesial therapy regimen, at least in cases of *B. microti*, and is now recommended.[95] Smith et al.,[96] however, recently reported the failure of quinine-clindamycin therapy in a case of *B. microti* infection in a 72-year-old male in New Hampshire. The infection was thought to have been transfusion acquired. The patient died of acute myocardial infarction, complicated by cardiogenic shock.

REFERENCES

1. **Skrabalo, Z. and Deanovic, Z.**, Piroplasmosis in man. Report on a case, *Doc. Med. Geogr. Trop.*, 9, 11, 1957.
2. **Wilson, L. B. and Chowning, W. M.**, Studies in pyroplasmosis hominis ("spotted fever" or "tick fever") of the Rocky Mountains, *J. Infect. Dis.*, 1, 31, 1904.
3. **Scholtens, R. G., Braff, E. H., Healy, G. R., and Gleason, N. N.**, A case of babesiosis in man in the United States, *Am. J. Trop. Hyg.*, 17, 810, 1968.
4. **Fitzpatrick, J. E. P., Kennedy, C. C., McGeown, M. G., Oreopoulos, D. G., Robertson, J. H., and Soyannwo, M. A.**, Further details of a third recorded case of redwater (babesiosis) in man, *Br. Med. J.*, 4, 770, 1969.
5. **Fitzpatrick, J. E. P., Kennedy, C. C., McGeown, M. G., Oreopoulos, D. G., Robertson, J. H., and Soyannwo, M. A.**, Human case of piroplasmosis (Babesiosis), *Nature (London)*, 217, 861, 1968.
6. **Kennedy, C. C.**, Human babesiosis: a summary of a case in Ireland, *Trans. R. Soc. Trop. Med. Hyg.*, 74, 156, 1980.
7. **Garnham, P. C. C., Donnelly, J., Hoogstraal, H., Kennedy, C. C., and Watson, G. A.**, Human babesiosis in Ireland: further observations and the medical significance of this infection, *Br. Med. J.*, 4, 768, 1969.
8. **Western, K. A., Benson, G. D., Gleason, N. N., Healy, G. R., and Schultz, M. G.**, Babesiosis in a Massachusetts resident, *N. Engl. J. Med.*, 283, 854, 1970.
9. **Ristic, M., Conroy, J. D., Siwe, S., Healy, G. R., Smith, A. R., and Huxsoll, D. L.**, *Babesia* species isolated from a woman with clinical babesiosis, *Am. J. Trop. Med. Hyg.*, 20, 14, 1971.
10. **Van Peenen, P. F. D. and Healy, G. R.**, Infection of *Microtus ochrogaster* with piroplasms isolated from man, *J. Parasitol.*, 56, 1029, 1970.
11. **Gleason, N. N., Healy, G. R., Western, K. A., Benson, G. D., and Schultz, M.**, The "Gray" strain of *Babesia microti* from a human case established in laboratory animals, *J. Parasitol.*, 56, 1256, 1970.
12. **Brandt, F., Healy, G. R., and Welch, M.**, Human Babesiosis: the isolation of *Babesia microti* in golden hamsters, *J. Parasitol.*, 63, 934, 1977.
13. **Healy, G. R., Walzer, P. D., and Sulzer, A. J.**, A case of asymptomatic babesiosis in Georgia, *Am. J. Trop. Med. Hyg.*, 25, 376, 1979.
14. **Osorno, B. M., Vega, C., Ristic, M., Robles, C., and Ibarra, S.**, Isolation of *Babesia* spp. from asymptomatic human beings, *Vet. Parasitol.*, 2, 111, 1977.
15. **Bredt, A. B., Weinstein, W. M., and Cohen, S.**, Treatment of babesiosis in asplenic patients, *JAMA*, 245, 1938, 1981.
16. **Bazin, C., Lamy, C., Piette, M., Gorenflot, A., Duhamel, C., and Vala, A.**, Un nouveau cas de babesiose humaine, *Nouv. Presse Med.*, 5, 799, 1976.
17. **Gorenflot, A., Piette, M., and Marchand, A.**, Babesioses animales et santé humaine. Premier cas de babesiose humaine observe en France, *Rec. Med. Vet.*, 152, 289, 1976.
18. **Rabinovich, S. A., Voronina, Z. L., Stepanova, N. I., Marvashvili, G. M., Bakradze, T., Odisarija, M. S., and Gvasalija, N. I.**, First detection of human babesiosis in USSR and a brief analysis of the cases recorded in the literature, *Med. Parazitol. Parazit. Bolezni.*, 47, 97, 1978.
19. **Entrican, J. H., Williams, H., Cook, J. A., Lancaster, W. M., Clark, J. C., Joyner, L. P., and Lewis, D.**, Babesiosis in man: a case from Scotland, *Br. Med. J.*, p. 474, 1979.
20. **Williams, H.**, Human babesiosis, *Trans. R. Soc. Trop. Med. Hyg.*, 74, 157, 1980.
21. **Bayle, J., Dumon, H., Verdot, J. J., and Muratore, R.**, Piroplasmose humaine. Une observation, *Nouv. Presse Med.*, 8, 44, 1979.
22. **Varache, C., Delille, J., Gorenflot, A., Marmonier, A., and Haas, C.**, Un nouveau cas de babésiose humaine au Mans, *Med. Malad. Infect.*, 2, 93, 1986.
23. **Calvo de Mora, A., Garcia Castellano, J. M., Herrera, C., and Jimenez-Alonso, J.**, Human babesiosis: report of a case with fatal outcome, *Med. Clin. (Barcelona)*, 85, 515, 1985.
24. **Woessner, S., Drobnic, L., Lafuente, R., Verdaguer, A., Florensa, L., and Sans-Sabrafen, J.**, Asymptomatic babesiosis in a Polish immigrant with an imbalance in T subpopulations, *Med. Clin. (Barcelona)*, 82, 284, 1984.
25. **Lewis, D. and Williams, H.**, Infection of the Mongolian gerbil with the cattle piroplasm *Babesia divergens*, *Nature (London)*, 278, 179, 1979.
26. **Lewis, D., Purnell, R. E., Shaw, S. R., and Bevington, J. P.**, The isolation and characterization of human and bovine strains of *Babesia divergens* from Drumnadrochit Scotland, *Parasitology*, 81, 145, 1980.
27. **Liddell, K. G., Lucas, S. B., and Williams, H.**, *Babesia divergens* infections in the Mongolian gerbil: characteristics of a human strain, *Parasitology*, 82, 205, 1980.
28. **Gorenflot, A. and Piette, M.**, Aspects microscopiques observés au cours des deux permiers cas français de Babésiose humaine, *Med. Malad. Infect.*, 11, 334, 1981.

29. **Dammin, G. J., Spielman, A., Benach, J. L., and Piesman, J.,** The rising incidence of clinical *Babesia microti* infection, *Hum. Pathol.*, 12, 398, 1981.
30. **Reubush, T. K., Juranek, D. D., Spielman, A., Piesman, J., and Healy, G. R.,** Epidemiology of human babesiosis on Nantucket Island, *Am. J. Trop. Med. Hyg.*, 30, 937, 1981.
31. **Chisholm, E. S., Ruebush, T. K., Sulzer, A. J., and Healy, G. R.,** *Babesia microti* infection in man: evaluation of an indirect immunofluorescent antibody test, *Am. J. Trop. Med. Hyg.*, 27, 14, 1978.
32. **Filstein, M. R., Benach, J. L., White, D. J., Brody, B. A., Goldman, W. D., Bakal, C. W., and Schwartz, R. S.,** Serosurvey for human babesiosis in New York, *J. Infect. Dis.*, 141, 518, 1980.
33. **Auerbach, M., Haubenstock, A., and Soloman, G.,** Systemic babesiosis. Another cause of the hemophagocytic syndrome, *Am. J. Med.*, 80, 301, 1986.
34. **Bruckner, D. A., Garcia, L. S., Shimizu, R. Y., Goldstein, E. J., Murray, P. M., and Lazar, G. S.,** Babesiosis. Problems in diagnosis using autoanalyzers, *Am. J. Clin. Pathol.*, 83, 520, 1985.
35. **Cahil, K. M., Benach, J. L., Reich, L. M., Bilmes, E., Zins, J. H., Siegel, F. P., and Hochweiss, S.,** Red cell exchange: treatment of babesiosis in a splenectomized patient, *Transfusion*, 21, 193, 1981.
36. **Gombert, M. E., Goldstein, E. J., Benach, J. L., Tenenbaum, M. J., Grunwaldt, E., Kaplan, M. H., and Eveland, L. K.,** Human babesiosis: clinical and therapeutic considerations, *JAMA*, 248, 3005, 1982.
37. **Gordon, S., Cordon, R. A., Mazdzer, E. J., Valigorsky, J. M., Blagg, N. A., and Barnes, S. J.,** Adult respiratory distress syndrome in babesiosis, *Chest*, 86, 633, 1984.
38. **Grabowski, E. F., Giardina, P. J., Goldberg, D., Masur, H., Read, S. E., Hirsh, R. L., and Benach, J. L.,** Babesiosis transmitted by a transfusion of frozen-thawed blood, *Ann. Intern. Med.*, 96, 466, 1982.
39. **Grunwaldt, E.,** Babesiosis on Shelter Island, *N.Y. State J. Med.*, 77, 1320, 1977.
40. **Jacoby, G. A., Hunt, J. V., Kosinski, K. S., Demirjian, Z. N., Huggins, C., Etkind, P., Marcus, L. C., and Spielman, A.,** Treatment of transfusion-transmitted babesiosis by exchange transfusion, *N. Engl. J. Med.*, 303, 1098, 1980.
41. **Marcus, L. C., Valigorsky, J. M., Fanning, W. L., Joseph, T., and Glick, B.,** A case report of transfusion induced babesiosis, *JAMA*, 248, 465, 1982.
42. **Marcus, L. C., Steere, A. C., Duray, P. H., Anderson, A. E., and Mahoney, E. B.,** Fatal pancarditis in a patient with coexistent Lyme disease and babesiosis. Demonstration of spirochetes in the myocardium, *Ann. Intern. Med.*, 103, 374, 1985.
43. **Mathewson, H. O., Anderson, A. E., and Hazard, G. W.,** Self-limited babesiosis in a splenectomized child, *Pediatr. Infect. Dis.*, 3, 148, 1984.
44. **Miller, L. H., Neva, F. A., and Gill, F.,** Failure of chloroquine in human babesiosis (*Babesia microti*). Case report and chemotherapeutic trials in hamsters, *Ann. Intern. Med.*, 88, 200, 1978.
45. **Ortiz, J. M. and Eagle, R. C.,** Ocular findings in human babesiosis (Nantucket fever), *Am. J. Ophthalmol.*, 93, 307, 1982.
46. **Parry, M. F., Fox, M., Burka, S. A., and Richar, W. J.,** *Babesia microti* infection in man, *JAMA*, 238, 1282, 1977.
47. **Raucher, H. S., Jaffin, H., and Glass, J. L.,** Babesiosis in pregnancy, *Obstet. Gynecol.*, 63, 7S, 1984.
48. **Rowin, K. S., Tanowitz, H. B., Rubenstein, A., Kunkel, M., and Wittner, M.,** Babesiosis in asplenic hosts, *Trans. R. Soc. Trop. Med. Hyg.*, 78, 442, 1984.
49. **Scharfman, W. B. and Taft, E. G.,** Nantucket fever. An additional case of babesiosis, *JAMA*, 328, 1281, 1977.
50. **Sun, T., Tenenbaum, M. J., Greenspan, J., Teichberg, S., Wang, R. T., Degnan, T., and Kaplan, M. H.,** Morphologic and clinical observations in human infection with *Babesia microti*, *J. Infect. Dis.*, 148, 239, 1983.
51. **Teutsch, S. M., Etkind, P., Burwell, E. L., Sato, K., Dana, M. M., Fleishman, P. R., and Juranek, D. D.,** Babesiosis in post-splenectomy hosts, *Am. J. Trop. Med. Hyg.*, 29, 738, 1980.
52. **Wittner, M., Rowin, K. S., Tanowitz, H. B., Hobbs, J. F., Saltzman, S., Weng, B., Hirsch, R., Chisholm, E., and Healy, G. R.,** Successful chemotherapy of transfusion babesiosis, *Ann. Intern. Med.*, 96, 601, 1982.
53. **Wolf, C. F., Resnick, G., Marsh, W. L., Benach, J., and Habicht, G.,** Autoimmunity to red blood cells in babesiosis, *Transfusion*, 22, 538, 1982.
54. **Zaino, E. C. and Amelkin, S.,** Babesiosis: a potentially grave disease for the immunosuppressed, *N.Y. State J. Med.*, 81, 384, 1981.
55. **Smith, D. R. and Clevenger, R. R.,** *Babesia microti* in an Indiana woman, *Indiana Med.*, 78, 212, 1985.
56. **Demicco, D. D. and Dolin, R.,** Personal communication, 1984.
57. **Steketee, R. W., Eckman, M. R., Burgess, E. C., Kuritsky, J. N., Dickerson, J., Schell, W. L., Godsey, M. S., and Davis, J. P.,** Babesiosis in Wisconsin. A new focus of disease transmission, *JAMA*, 253, 2675, 1985.
58. **Cox, F. E.,** Human babesiosis, *Nature (London)*, 285, 535, 1980.

59. **Dammin, G. J.,** Babesiosis, *Semin. Infect. Dis.*, 7, 169, 1978.
60. **Donnelly, J.,** Human babesiosis, *Trans. R. Soc. Trop. Med. Hyg.*, 74, 158, 1980.
61. **Garnham, P. C. C.,** Human babesiosis: European aspects, *Trans. R. Soc. Trop. Med. Hyg.*, 74, 153, 1980.
62. **Hoare, C. A.,** Comparative aspects of human babesiosis, *Trans. R. Soc. Med. Hyg.*, 74, 153, 1980.
63. **Ruebush, T. K.,** Human babesiosis in North America, *Trans. R. Soc. Trop. Med. Hyg.*, 74, 149, 1980.
64. **Anderson, A. E., Cassaday, P. B., and Healy, G. R.,** Babesiosis in man. Sixth documented case, *Am. J. Clin. Pathol.*, 62, 612, 1974.
65. **Healy, G. R.,** *Babesia* infections in man, *Hosp. Pract.*, p. 107, 1979.
66. **Healy, G. R. and Ruebush, T. K.,** Morphology of *Babesia microti* in human blood smears, *Am. J. Clin. Pathol.*, 73, 107, 1980.
67. **Ruebush, T. K., Chisholm, E. S., Sulzer, A., and Healy, G. R.,** Development and persistence of antibody in persons infected with *Babesia microti*, *Am. J. Trop. Med. Hyg.*, 30, 291, 1981.
68. **Chisholm, E. S., Sulzer, A. J., and Ruebush, T. K.,** Indirect immunofluorescence test for human *Babesia microti* infection-antigenic specificity, *Am. J. Trop. Med. Hyg.*, 35, 921, 1986.
69. **Smith, T. and Kilbourne, F. L.,** Investigations into the nature, causation and prevention of Texas or southern cattle fever, *U.S. Dep. Agric. Bur. Anim. Ind. Bull.*, 1, 1, 1893.
70. **Spielman, A., Wilson, M. L., Levine, J. F., and Piesman, J.,** Ecology of *Ixodes dammini*-borne human babesiosis and Lyme disease, *Annu. Rev. Entomol.*, 30, 439, 1985.
71. **Healy, G. R., Spielman, A., and Gleason, N.,** Human Babesiosis: reservoir of infection on Nantucket Island, *Science*, 192, 479, 1976.
72. **Spielman, A., Etkind, P., Piesman, J., Ruebush, T. K., Juranek, D. D., and Jacobs, M. S.,** Reservoir hosts of human babesiosis on Nantucket Island, *Am. J. Trop. Med. Hyg.*, 30, 560, 1981.
73. **Etkind, P., Piesman, J., Ruebush, T. K., Spielman, A., and Juranek, D. D.,** Methods for detecting *Babesia microti* in wild rodents, *J. Parasitol.*, 66, 107, 1980.
74. **Spielman, A., Clifford, C. M., Piesman, J., and Corwin, M. D.,** Human babesiosis on Nantucket Island, U.S.A. Description of the vector, *Ixodes* (Ixodes) *dammini*, n. sp. (Acarina: Ixodidae), *J. Med. Entomol.*, 15, 218, 1979.
75. **Piesman, J. and Spielman, A.,** Human babesiosis on Nantucket Island: prevalence of *Babesia microti* in ticks, *Am. J. Trop. Med. Hyg.*, 29, 742, 1980.
76. **Piesman, J., Karakashian, S. J., Lewengrub, S., Rudinska, M. A., and Spielman, A.,** Development of *Babesia microti* sporozoites in adult *Ixodes dammini*, *Int. J. Parasitol.*, 16, 381, 1986.
77. **Piesman, J., Mather, T. N., Telford, S. R., and Spielman, A.,** Concurrent *Borrelia burgdorferi* and *Babesia microti* infection in nymphal *Ixodes dammini*, *J. Clin. Microbiol.*, 24, 446, 1986.
78. **Grunwaldt, E., Barbour, A. G., and Benach, J. L.,** Simultaneous occurrence of babesiosis and Lyme disease, *N. Engl. J. Med.*, 308, 1155, 1983.
79. **Piesman, J., Spielman, A., Etkind, P., Ruebush, T. K., and Juranek, D.,** Role of deer in the epizootiology of *Babesia microti* in Massachusetts, USA, *J. Med. Entomol.*, 15, 537, 1979.
80. **Hoogstraal, H.,** Changing patterns of tickborne diseases in modern society, *Annu. Rev. Entomol.*, 26, 75, 1981.
81. **Ruebush, T. K., Juranek, D. D., Chisholm, E. S., Snow, P. C., Healy, G. R., and Sulzer, A. J.,** Human babesiosis on Nantucket Island. Evidence for self-limited and subclinical infections, *N. Engl. J. Med.*, 297, 825, 1977.
82. **Bruce-Chwatt, L. J.,** Transfusion associated parasitic infections, *Proc. Clin. Biol. Res.*, 182, 101, 1985.
83. **Ruebush, T. K., Collins, W. E., Healy, G. R., and Warren, M.,** Experimental *Babesia microti* infections in non-splenectomized *Macca mulatta*, *J. Parasitol.*, 65, 144, 1979.
84. **Ruebush, T. K., Cassaday, P. B., Marsh, H. J., Lisker, S. A., Vorhees, D. B., Mahoney, E. B., and Healy, G. R.,** Human babesiosis on Nantucket Island. Clinical features, *Ann. Intern. Med.*, 86, 6, 1977.
85. **Benach, J. L. and Habicht, G. S.,** Clinical characteristics of human babesiosis, *J. Infect. Dis.*, 144, 481, 1981.
86. **Rosner, F., Zarrabi, M. H., Benach, J. L., and Habicht, G. S.,** Babesiosis in splenectomized adults. Review of 22 reported cases, *Am. J. Med.*, 76, 696, 1984.
87. **Benach, J. L., Coleman, J. L., Habicht, G. S., MacDonald, A., Grunwaldt, E., and Giron, J. A.,** Serological evidence for simultaneous occurrence of Lyme disease and babesiosis, *J. Infect. Dis.*, 152, 473, 1985.
88. **Benach, J. L., Habicht, G. S., and Hamburger, M. I.,** Immunoresponsiveness in acute babesiosis in humans, *J. Infect. Dis.*, 146, 369, 1982.
89. **Garnham, P. C. C. and Bray, R. S.,** The susceptibility of the higher primates to piroplasms, *J. Protozool.*, 6, 352, 1959.

90. **Benach, J. L., White, D. J., and McGovern, J. P.**, Babesiosis in Long Island, host-parasite relationships of rodent and human derived *Babesia microti* isolates in hamsters, *Am. J. Trop. Med. Hyg.*, 27, 1073, 1978.
91. **Lykins, J. D., Ristic, M., and Weisiger, R. M.**, *Babesia microti:* pathogenesis of parasite of human origin in the hamster, *Exp. Parasitol.*, 37, 388, 1975.
92. **Moore, J. A. and Kuntz, R. E.**, *Babesia microti* infections in nonhuman primates, *J. Parasitol.*, 67, 454, 1981.
93. **Ristic, M. and Healy, G. R.**, Babesiosos, in *Handbook Series in Zoonoses*, Vol. 1, Steele, J., Ed., CRC Press, Boca Raton, Fla., 1982, 151.
94. **Francioli, P. B., Keithly, J. S., Jones, T. C., Brandstetter, R. D., and Wolf, F. P.**, Response of babesiosis to pentamidine therapy, *Ann. Intern. Med.*, 94, 326, 1981.
95. Medical Letter on Drugs and Therapeutics—Drugs for Parasitic Infections, Vol. 28, January 31, 1986.
96. **Smith, R. P., Evans, A. T., Popovsky, M., Mills, L., and Spielman, A.**, Transfusion-acquired babesiosis and failure of antibiotic treatment, *JAMA*, 256, 2726, 1986.

Chapter 14

CHEMOTHERAPY OF BABESIOSIS*

Kenneth L. Kuttler

TABLE OF CONTENTS

* Mention of a trademark or proprietary product does not constitute a guarantee or warranty of the product by the U.S. Department of Agriculture and does not imply its approval to the exclusion of other products that also may be suitable.

I. INTRODUCTION

Babesiosis is an infectious, noncontagious disease of vertebrates, often characterized by fever, anemia, and hemoglobinuria, caused by any one of numerous tick-transmitted blood protozoa belonging to the genus *Babesia.*[1,2] These babesias tend to be host-specific, but exceptions do occur as noted in Chapter 1. Different babesias have unique morphology, virulence, pathogenesis, and arthropod vectors, as well as differences in response to treatment. There appears to be a continuing increase in *Babesia* distribution. At one time, these hemoparasites were thought to be largely restricted to the tropics and subtropics, but northern Europe, North America, and Asia have been in recent years the source of many reports describing babesiosis in horses, cattle, dogs, sheep, wildlife, and even man.[2] There appear to be numerous tick species adapted to northern latitudes which have proven to be efficient *Babesia* vectors.

The ever broadening scope of babesiosis in man and animals has focused attention on the development of appropriate treatment to moderate clinical signs in acutely infected individuals, and to eliminate or prevent infection.

Economically, babesiosis is probably of greatest importance in cattle. In the U.S., treatment of cattle for babesiosis has assumed a secondary role because of a successful *Boophilus* tick eradication program which was followed by the disappearance of cattle babesiosis in the U.S. In most other areas of the world, however, ticks have not been eradicated, and the disease persists and is controlled by a variety of measures including chemotherapy and chemoprophylaxis.

In recent years, a number of human cases of babesiosis have been reported, primarily involving *Babesia microti* (a rodent *Babesia*) in the U.S. and *B. divergens* (a cattle *Babesia*) in northern Europe, transmitted by *Ixodes dammini* and *I. ricinus,* respectively.[3]

Babesiosis is emerging as an increasingly important disease in dogs both in Europe and the U.S. where it has been introduced from endemic zones in the tropics and subtropics.[4-6] Babesiosis in sheep and swine has been recognized in recent years as the cause of losses in these animals.[2,7] Babesiosis in cats has been described in South Africa; it is the only *Babesia* that responds favorably to antimalarial therapy.[8] Equine babesiosis is widespread and occasionally produces clinical disease associated with death. A major concern with this infection is the restriction it imposes on the movement of horses and the requirement that carrier horses be cleared of infection by chemotherapy before being imported into areas where the disease does not occur.

A variety of chemical compounds have been successfully used in the treatment of babesiosis. A list of some of the more commonly used compounds is presented in Table 1. In addition to the use of the appropriate babesiacidal compound, supportive treatment should not be overlooked. Generally, if specific and effective treatment is given early in the course of infection, before the onset of severe anemia or nervous system disorders, recovery is the rule. If delayed, however, supportive treatment, including blood transfusions, fluids, and vitamins, contributes to success in treatment. The exertion and excitement associated with the restraint of some animals being treated may actually contribute to death. Judgment as to when and how extensively an animal is to be handled and treated must take into account the temperament of the animal, the location, and the stage of infection. It is possible that in some instances no treatment is indicated, even through infection is advanced.[9]

Treatment of acute babesiosis is usually concerned with moderating clinical signs, characterized by fever and the anemia associated with parasitemia. Some of the babesiacidal compounds are quite effective, and one injection will result in a rapid drop and possible elimination of the parasitemia. When and if the parasite is eliminated, a reservoir of infection is also eliminated. While this is often a desirable goal, in some instances it may be contraindicated if the animal is to be kept in an endemic zone where reexposure will occur.

Table 1
COMPOUNDS FREQUENTLY USED TO TREAT BABESIOSIS IN DOMESTIC ANIMALS

Compound or compound group	Proprietary name	Chemical description[a]
Acridine derivative		
Acriflavine hydrochloride (euflavine, trypaflavine)	Gonacrine[b]	A mixture of 3,6-diamino-10 methylacridinium chloride and 3,6-diaminoacridine
Azo-naphthalene dyes		
Trypan blue	Congo blue Niagara blue	3,3′-{3,3′-Dimethyl{1,1′-biphenyl}-4,4′-diyl bis(azo)} bis{5-amino-4 hydroxy-2,7-naphthalenedisulfonic acid} tetrasodium salt
Diamidine derivatives		
Aromatic: Diminazene diaceturate	Berenil[c] Ganaseg[d]	4,4′-Diamidinodiazoaminobenzene diaceturate
Pentamidine isethionate	Lomidine[b]	4,4′-Diamidinodiphenoxypentane di(beta-hydroxyethane sulfonate)
Phenamidine isethionate	Lomadine[b]	4,4′-Diamidinodiphenyl ether di(beta-hydroxyethane sulfonate)
Carbanilide		
Amicarbalide	Diampron[b]	3,3′-Diamidinocarbanilide
Imidocarb dipropionate	Imizol[e]	3,3′-Bis(2-imidazolin-2yl) carbanilide dipropionate
Phenanthridine compounds		
Isometamidium	Samorin[b]	8-{(*m*-Amidinophenylazo)amino}-3-amino-5 ethyl-6-phenylphenanthridinium chloride
Quinoline derivatives		
Quinuronium sulfate	Acaprin[f] Akiron Pirevan Piroplasmin	6,6′-Ureylenebis(1-methylquinolinium)bis(methosulfate)
	Babesan[g]	N,N′-di-6-quinolinylurea
8-Aminoquinolines		
Primaquine phosphate	Primaquine[g]	N[4]-(6-methoxy-8-quinolinyl)-1,4-pentanediamine
Tetracyclines		
Oxytetracycline	Terramycin LA/200[h]	4-(Dimethylamino)-1,4,4a,5,5a,6,11,12a-octahydro-3,5,6,10,12,12a-hexahydroxy-6 methyl-1,11-dioxo-2 naphthacenecarboxamide
Chlortetracycline	Aueromycin[i]	7-Chloro-4 dimethylamino-1,4,4a,5,5a,6,11,12a-octahydro-3,6,10,12,12a-pentahydroxy-6 methyl-1,11-dioxo-2-naphthacenebarboxamide

[a] Obtained from the *Merck Index*, 9th ed., Merck & Co., Rahway, N.J., 1976.
[b] May & Baker Ltd., Dangenham, England.
[c] Farbwerke-Hoechst AG, Frankfurt, West Germany.
[d] Squibb Mathieson, E. R., Squibb & Sons de Mexico, Mexico City, Mexico.
[e] Burroughs Wellcome & Co. Ltd., London, England.
[f] Ludabel Farbenfabriken, Bayer, Leverkusen, West Germany.
[g] Imperial Chemical Industries Ltd., Macclesfield, Cheshire, England.
[h] Pfizer, Inc., New York, N.Y.
[i] American Cyanamide, Princeton, New Jersey.

Elimination of *Babesia* by treatment eventually renders the animal susceptible to reinfection.[10] If treatment eliminates infection early in the course of the disease, the level of sterile immunity is less than if treatment follows a high parasitemia. Reexposure, when sterile immunity is strong, may not result in clinical disease, and a premunizing infection is acquired. However, if reinfection is delayed and sterile immunity has waned, then infection may be associated with clinical signs and even death.

The establishment of drug efficacy in the treatment of babesiosis (see Table 2) is complicated by the large number of babesias and vertebrate hosts, with differing degrees of drug susceptibility and tolerance. *Babesia rodhaini* infection in mice has been used successfully to screen potential babesiacidal compounds.[11,12] Caution is warranted because parallel action is not consistent and results may be misleading. An example is that diminazene, while highly effective against most bovine babesias, is not equally as effective against *B. rodhaini.* The susceptibility of the Mongolian gerbil to *B. divergens* from cattle provides an additional laboratory model for babesiacidal drug screening.[13] Irvin and Young have shown that the in vitro uptake of tritiated purines, particularly hypoxanthine, by *Babesia* could be inhibited by babesiacidal drugs in proportion to their efficacy in vivo.[14,15] This screening procedure was applicable to *B. microti* and *B. rodhaini* of mice and *B. divergens* and *B. major* of cattle. An in vitro chemotherapeutic screen could probably be developed using culture techniques recently reported for the growth of *B. bovis* and *B. divergens* in bovine erythrocytes.[16-18] The growth of *B. equi* lymphocytic schizonts in tissue culture has been adapted to evaluate various chemical compounds.[19]

The mode of action for most compounds is not entirely clear, but the acridine and diamidine derivatives are thought to combine with nucleic acids.[20,21] There is an intercalation of the acriflavine molecule between the coils of the DNA molecule, with the amine groups of acriflavine forming bonds with adjacent thymine-containing base pairs, causing partial uncoiling and denaturation of the DNA double helix. Studies of the actions of diminazene on *B. herpailuri* show a dilation of membrane-bound organelles and the perinuclear space, dissolution of the cytoplasm, and destruction of the nucleus.[22]

Babesia equi in erythrocytes previously exposed to imidocarb were swollen and vacuolated, except for an area just beneath the plasma membrane, where the compressed nucleus was marginated, elongated, and clumpy.[23] Electron microscopy revealed four stages of degeneration of *B. equi* in erythrocytes of ponies treated with imidocarb. The first stage showed dilation of the nuclear cisterna and clumping of chromatin. The second stage was characterized by dissolution of the nuclear envelope and passage of nuclear matrix into cytoplasm. The third stage was associated with vacuolation of the cytoplasm and compression of the chromatin to the periphery of the cell. Degeneration continued until the organism consisted of a vacuole, and chromatin was condensed into an irregular mass of clumped material.[23] Similar ultrastructural changes were observed with *B. herpailuri* in the erythrocytes of cats after treatment with imidocarb, except that nuclear damage was not observed.[24]

Drug-resistant babesias can be developed experimentally and probably occur.[25,26] This problem has not yet evolved as a major constraint in chemotherapy, but techniques of using some compounds as prophylactic agents with sublethal, low blood levels of active drug in constant contact with the babesias could contribute to the evolution of drug-resistant organisms.[26] A recent report has described a field isolate of *B. bovis* in Israel that is two or three times more resistant to amicarbilide than other similar isolates.[27]

The small babesias are generally more resistant to treatment. *B. bigemina, B. caballi,* and *B. canis* usually respond more readily to treatment than their small counterparts, *B. bovis, B. equi,* or *B. gibsoni.*

Table 2
SUBJECTIVE ESTIMATE OF RELATIVE EFFICACY OF SOME OF THE MORE COMMONLY USED BABESIACIDAL COMPOUNDS

	Diminazene	Imidocarb	Amicarbalide	Phenamidine	Quinuronium	Trypan blue	Pentamidine	Oxytetracycline	Primaquin
Cattle									
Babesia bigemina	++++	++++	++++	++	+++	++	++	+	
B. bovis	+++	+++	++	++	++	—		+	
B. divergens	++	+++	++		+	—		+	
Horses									
B. caballi	+++	++++	+++	+++	++	++			
B. equi	++	+++	+	++	—	—			
Dogs									
B. canis	+++	+++	++	+++	++	+	++		
B. gibsoni	++			++	—	—	++		
Cats									
B. herpailuri	+++	+++							
B. felis		—		—	—	—			+++
Sheep									
B. motasi	+++		+++		+	+			
B. ovis	+	++			±	—			
Swine									
B. trautmanni				++	++	++			
B. perroncitoi	+++				++	—			

Note: + to ++++ = Relative effectiveness; — = ineffective.

II. BOVINE BABESIOSIS

Trypan blue was probably the first specific drug used to successfully treat bovine babesiosis. Theiler reported cures of *B. bigemina* with this compound by the intravenous (i.v.) injection of 2 to 3 mg/kg;[28] however, it was later discovered that it was not effective against *B. bovis.*[29] Trypan blue has the disadvantages of producing discolorations of the flesh and body secretions (including milk) of the animal, and causing severe tissue sloughing if injected extravascularly. For this reason, and because more effective drugs have been developed, it is now rarely used to treat bovine babesiosis.

A. Quinoline Derivatives

For many years, the quinoline derivatives were the drugs of choice in the treatment of bovine babesiosis.[30,31] These drugs have a low therapeutic index and may produce transient signs of discomfort such as excessive salivation, panting, frequent urination, and general uneasiness, which are associated with its anticholinesterase activity. Such reactions usually disappear in about 1 hr and may be alleviated or reduced by the administration of atropine or epinephrine.[32,33] A fatal, toxic reaction to quinuronium was observed following three daily injections of 0.5 mg/kg intramuscularly (i.m.).[32] Postmortem examination revealed a volvulus at the jejunal-ilial junction. It was postulated that quinuronium was responsible for intestinal hyperperistalsis which resulted in the volvulus.

Quinuronium is still in use, however, and one or two treatments given 24 hr apart at the rate of 1 mg/kg subcutaneously (s.c.) is usually sufficient in treating either *B. bigemina* or *B. bovis.*[34] At 1 mg/kg, quinuronium was therapeutically effective in treating splenectomized calves showing clinical signs of *B. divergens* infection, but was not useful prophylactically[35] when given 1 day before exposure. A *B. divergens* recrudescence may occur 2 to 3 weeks after quinuronium therapy. A relatively insoluble salt, quinuronium 5,5′-methylene-bis-salicylate, given as a 1-g implant, has been shown to prevent *B. bovis*-induced babesiosis for 21 days.[36] Similar insoluble salts injected s.c. at the rate of 25 to 50 mg/kg are reported to induce resistance to *B. divergens* challenge for up to 3 to 4 months.[37]

B. Acridine Derivatives

Acriflavin and related compounds are reported to be effective against both *B. bigemina* and *B. bovis* when given i.v. as a 5% solution using 4.4 mℓ/100 kg body weight.[9,38] Even though they are effective, the acridine derivatives appear to have given way to the diamidine derivatives which have become widely used in the last 15 to 20 years.

C. Diamidine Derivatives

These compounds have proven effective and safe in the treatment of bovine babesiosis. There are numerous such diamidines, but probably the most commonly used compounds are diminazene, imidocarb, amicarbalide, phenamidine, and pentamidine.[9,25,34] Amicarbalide administered i.m. has been successfully used at the rate of 5 to 10 mg/kg against *B. divergens, B. bovis,* and *B. bigemina.* Phenamidine isethionate, at 8 to 13.5 mg/kg s.c., has been reported effective against *B. bigemina.*[9,34] Diminazene is usually given i.m. at the rate of 3 to 5 mg/kg, but titrations show it to be effective against *B. bigemina* at much lower levels.[9,34] As little as 0.25 mg/kg was effective in moderating the course of *B. bigemina* infections in splenectomized calves, but a level of 0.5 mg/kg is probably the lowest safe dosage that should be used in clinical cases.[9]

Pentamidine, at the level of 0.5 to 2.0 mg/kg, administered s.c. to calves in the acute phase of *B. bigemina* infection, usually led to clinical recovery. As much as 5 mg/kg administered during patency did not destroy the carrier state in intact calves. For this reason, pentamidine has proven useful in chemoimmunization because it effectively moderates clinical signs of infection while allowing the carrier state to persist.[39]

The addition of 0.25 mg/kg of betamethasone to diminazene (3.5 mg/kg) and to quinuronium (0.9 mg/kg) in the treatment of cattle in the acute phase of a *B. bigemina* infection failed to produce any enhancing effects.[40] Diminazene has been successfully used with vaccine programs in South Africa and South America.[10,41] Diminazene is administered 7 or 10 days after vaccination with an attenuated *B. bovis* or with a combination of *B. bigemina* and *B. bovis*. The drug moderates the effects of the primary infection, but allows a state of premunity to develop, hence a high level of protection. When *B. bigemina* is given concurrently with the *B. bovis* vaccine, revaccination with *B. bigemina* 2 to 3 weeks later is advisable to establish solid immunity. Even though diminazene has not been shown to produce extended prophylaxis, workers in South Africa have shown it to influence the onset of *B. bovis* infections for up to 2 weeks and up to 4 weeks with *B. bigemina*.[10] If, therefore, vaccination with a live agent is to follow diminazene treatment, a delay beyond the period of maximum drug action is recommended.

Babesia ovata, an immunologically distinct species, has recently been described as being pathogenic to cattle in Japan.[42] This *Babesia* morphologically resembles *B. major* and *B. bigemina* and, like these two babesias, is sensitive to treatment with diminazene and amicarbalide.

Imidocarb, a diamidine of the carbanilide series, is one of the more recent babesiacidal compounds found effective in the treatment and prevention of bovine babesiosis.[9,43,44] Early formulations utilized the dihydrochloride salt, but the dipropionate salt is now more commonly used. This preparation appears to be less irritating to the tissues, although both are soluble and highly active therapeutically. Imidocarb is safe and effective when used at the recommended levels of 1 to 3 mg/kg s.c. or i.m. for infections due to *B. bigemina, B. bovis,* and *B. divergens*.[9,25,44,45] Imidocarb is not recommended for i.v. use.[9]

Imidocarb used at 2.4 mg/kg was effective against *B. divergens* infections in splenectomized calves. When given 1 or 2 weeks before infection, it had the effect of reducing the severity while not preventing infection; treatment on the day of infection inhibited parasitemia, and no immunity occurred. *Babesia* infection was eliminated when imidocarb was given 7, 14, and 21 days after exposure, and sterile immunity was observed at a level corresponding to the length and severity of initial infection.[9] A second related study showed that 3.6 mg/kg imidocarb given 14 days before *B. divergens* exposure of splenectomized calves allowed the development of immunity while reducing the severity of infection.[46] The administration of 6.0 mg/kg given 14 days before exposure prevented the infection and no immunity was seen. Under field conditions in France, imidocarb administered at the rate of 2 mg/kg was found effective in preventing clinical babesiosis in cattle for 6 to 8 weeks.[47] It also controlled an outbreak of babesiosis due to *B. divergens* in Ireland when used at 1 to 2 mg/kg on 682 animals.[48]

Imidocarb has been used in connection with some vaccine programs. The simultaneous administration of 0.15 mg/kg imidocarb and live attenuated *B. bovis* vaccine has given satisfactory results. When the vaccine was given 7 days prior to imidocarb treatment, a dose of between 0.15 and 0.6 mg/kg imidocarb was required for effective control.[45] A combined *B. bovis* and *B. bigemina* vaccine given 21 and again 61 days after imidocarb at 3 mg/kg allowed the development of adequate premunity to both parasites.

Treatment of intact calves with 5 mg imidocarb per kilogram 14 days before and 14 days after exposure to *B. bovis*-infected *Boophilus microplus* larvae not only prevented *Babesia bovis* infection, but rendered the next generation of larvae incapable of transmitting this infection.[49] These results were not, however, duplicated using either *Boophilus decoloratus* or *B. microplus* ticks infected with *Babesia bigemina*. Splenectomized calves treated with imidocarb at 5 mg/kg 2 weeks before tick infestation, and with 3 mg/kg on the day of tick infestation, prevented patent infections in the calves, but did not eliminate *B. bigemina* from *Boophilus decoloratus* ticks.[50] *Boophilus microplus* tick infection rates were suppressed by imidocarb dipropionate, quinuonium, and amicarbalide, but these compounds failed to eliminate tick forms of *Babesia bigemina* or to prevent subsequent transovarial transmission.[51] A possible explanation of these apparently different findings is that *B. bovis* is not thought to persist in infective forms beyond the larval stage of its *Boophilus* vector, implying that engorging nymphs or adult females must acquire the infection in each generation from an infected bovine to complete their life cycle and, in turn, infect the larvae of the next generation.[52,53] Imidocarb will effectively eliminate a *Babesia* parasitemia in the vertebrate host for an extended time, thus interrupting the *B. bovis* cycle in ticks, whereas it is unlikely that similar treatment would have any significant effect on *B. bigemina* in the tick vector, where it resides during all tick stages, or on its ability to be transmitted.

Imidocarb has been utilized to prevent acute babesiosis in cattle being shipped from *Babesia*-free areas to endemic areas of the tropics. In one such trial, 44 young Charolais cattle were moved from Texas to Haiti.[54] They were treated s.c. with 2.8 mg/kg body weight before being exposed to infected *Boophilus* ticks. After 130 days in a tick environment, over 70% of the cattle showed serologic evidence of babesiosis, but there were no deaths due to acute babesiosis. It is probable that exposure of the cattle to *Boophilus* ticks resulted in inapparent *Babesia* infection, followed by premunition, without acute infection or death losses.

A recent unique innovation for the administration of imidocarb has been carried out in which the drug was encapsulated in bovine carrier erythrocytes with no ill effects on the carrier cells.[55] The drug-loaded carrier cells had an in vivo half-life of 28 to 32 days, and a dose equivalent to 0.1 mg/kg body weight imidocarb was thought to have provided a level of protection against *Babesia* infections for at least 35 days.

Imidocarb is a cholinesterase inhibitor, and under some conditions may be toxic, notwithstanding a high therapeutic index for *Babesia*. Toxic reactions vary, largely due to the wide range of individual animal tolerance to the drug. Imidocarb treatment on 469 adult cows at the level of 5 mg/kg i.m., given two times at a 2-week interval, was accompanied by a 5% mortality, some, but not all, of which were shown to be drug related.[56] A study was made in calves in which five animals in each of three groups were inoculated with 5, 10, and 20 mg/kg imidocarb two times at a 14-day interval.[57] All five calves which received the 20 mg dosage died. Four died before the second injection on days 7, 10, 11, and 11, respectively; the last calf died 4 days after the second injection. Mortality did not occur in the animals that received the 5- and 10-mg levels. All calves showed excessive lacrimation, serous nasal discharge, and increased frequency of defecation for 30 to 40 min after injection. Persistent reactions, including diarrhea, dyspnea, and depression, occurred in one of five calves receiving the 5-mg level and in two of five calves receiving the 10-mg level. Dose-dependent injection site reactions were characterized by areas of myositis, focal areas of necrosis, edema, and granulation tissue.[57] Fatal toxicosis was associated with renal hyperemia and enlargement, pulmonary congestion, edema, hydrothorax, and hydroperitoneum. Renal lesions were mainly seen as severe cellular necrosis in the proximal convoluted tubules.

Imidocarb is very slowly metabolized and eliminated.[58,59] The persisting tissue residues restrict its use in food animals in some countries. Tissue assays for drug residues have been used in sheep and cattle to give some indication of drug retention. The injection of 4.5 mg/kg i.m. in sheep resulted in peak plasma levels of 7.9 μg/mℓ within 4 hr, followed by rapid decline within the next 2 hr to 4.6 μg/mℓ, and then a slow decline for several weeks.[58,59] Average tissue levels of imidocarb 28 days after the inoculation of 1 to 2 mg/kg in three cattle were 2.03 μg/g in liver, 2.13 μg/g in kidney, 0.13 μg/g in muscle, and 0.02 μg/g in fat.[60]

D. Tetracyclines

With the introduction of the long-acting, 200 mg/mℓ formulation of oxytetracycline (LA/200), renewed interest in the use of this antibiotic for the treatment of babesiosis has occurred. The relative lack of toxicity, absence of side effects, and general safety of the tetracyclines would strongly recommend their use if efficacy was comparable to other classes of compounds, As early as 1953, chlortetracycline was reported effective in retarding the growth of *B. equi* in splenectomized donkeys.[61] The author has observed[106] that splenectomized calves fed 2 to 11 mg/kg chlortetracycline daily were refractory to *B. bigemina* exposure.

Oxtetracycline LA/200 had a detectable prophylactic effect when given four times (20 mg/kg i.m.) at weekly intervals to *Babesia*-susceptible cattle introduced to an area known to be heavily contaminated with *B. bigemina*-infected *Boophilus* ticks.[62] Infection apparently occurred in the absence of clinical manifestations, and the cattle subsequently became immune.

A similar study with *B. bovis* has suggested that treatment with LA/200 14 to 28 days before exposure is related to a modified clinical response.[63] The mode of action is not understood since blood drug levels after 4 weeks would likely be insignificant. Treatment of cattle with 20 mg/kg LA/200 7 to 15 days after vaccination with an attenuated *B. bovis* strain produced a detectable moderating influence on the subsequent *B. bovis* reaction without preventing a replicating infection and the establishment of immunity.[64]

The injection of LA/200 at 20 mg/kg every 4 days prevented a *B. divergens* parasitemia in a group of cattle exposed to field infection for a 6-week period, whereas all of 21 controls developed patent parasitemias.[65] Using a controlled *B. divergens* exposure, cattle were inoculated with 5, 10, and 20 mg/kg LA/200 every 4 days for 3 weeks after exposure. *B. divergens* infection was prevented at the 20-mg/kg level, and infections were significantly moderated at the 5- and 10-mg/kg level.[65]

These reports do not indicate therapeutic efficacy at levels comparable to other babesiacidal compounds, but clearly suggest an activity that might have a useful function in moderating cattle babesiosis.

In practice, treatment of bovine babesiosis is frequently limited to the use of specific babesiacidal drugs, but when the disease is advanced, or in severe cases, additional supportive treatment is indicated. Blood transfusions at the rate of 1 ℓ of blood for each 45 kg of body weight up to a maximum of 8 ℓ is recommended.[66] Good success has been achieved with this technique when accompanied by specific babesiacidal drug therapy.

III. EQUINE BABESIOSIS

DeKock published a report in which he unsuccessfully attempted to treat nutalliosis (*B. equi*) in horses with a large number of chemical compounds including inorganic salts, organic arsenicals, various dyes, hyalin, formalin, quinine, hydrobromide, camphor, turpentine, etc.[67] Included was trypan blue which proved effective against *B. caballi*, but not *B. equi*.

Acute equine babesiosis caused by *B. caballi* responds to many of the same drugs that are effective in cattle; however, *B. equi* is more difficult to treat. The most commonly used drugs in treating equine babesiosis are amicarbalide (8 mg/kg), imidocarb (2 to 5 mg/kg), and diminazene (5 to 10 mg/kg).[9] One injection of 2 mg/kg imidocarb is usually sufficient to bring about a clinical remission of infection due to *B. caballi*. Quinuronium sulfate was described as being ineffective in treating *B. equi* infections in splenectomized donkeys.[68] Diminazene was successfully used to treat acute *B. equi* infections in donkeys at the rate of 12 mg/kg.[69] The addition of a tetracycline at 10 mg/kg to the diminazene treatment failed to produce any potentiating effects.[68] These workers also found that imidocarb administered at the rate of 2 mg/kg was ineffective in treating *B. equi* infection in donkeys, but was effective when 5 mg/kg was given two times at a 48-hr interval. This latter dosage appeared effective in eliminating *B. equi* infection in these splenectomized donkeys, based on subinoculation trials.[69]

Elimination of the carrier infection has assumed importance, particularly in animals being shipped from endemic zones to areas where infection is limited or does not exist. The following treatments appear effective in eliminating *B. cabelli*:[70,71]

1. Diminazene, 5 mg/kg, two times, 24-hr interval
2. Phenamidine isethionate, 8.8 mg/kg, two times, 24-hr interval
3. Amicarbalide, 8.8 mg/kg, two times, 24-hr interval
4. Imidocarb, 1 to 2 mg/kg, two times, 24-hr interval

Treatment of *B. equi* infection with amicarbalide, at the rate of 11 mg/kg, four times at 24- and 48-hr intervals, was unsuccessful; and only four of eight horses became free of infection when they received seven daily injections of amicarbalide at 22 mg/kg which approaches toxic levels.[9] Imidocarb, given 4 mg/kg four times at 72-hr intervals, was shown to remove *B. equi* infections in 13 of 14 horses.[72] It is recommended that this treatment (4 mg/kg) be divided into two equal doses given 30 min apart to minimize the undesirable acute side effects associated with imidocarb treatment.

The apparent success of these procedures must be considered along with other reports in which similar and even more drastic treatments with imidocarb and other babesiacidal compounds failed to clear *B. equi*.[73] The author's personal experience[106] has been that treatment of nine *B. equi*-infected horses with imidocarb at the level of 4 mg/kg, four times at 72-hour intervals, was uniformly unsuccessful as determined by subinoculation trials approximately 90 days after treatment. These trials were conducted on horses infected with *B. equi* of European origin. The successful treatment reported by Frerichs et al.[72] was on horses infected with *B. equi* of Western Hemisphere origin. The possibility of strain differences in susceptibility to therapy should be considered.

Recent in vitro studies comparing drug activity on *B. equi* schizonts in lymphocytic cell cultures showed that the antitheileria compounds, halofuginone and Wellcome® 993 C, suppressed thymidine incorporation by more than 80%, whereas diminazene showed no noticeable effect.[19]

In horses, imidocarb at therapeutic levels is reasonably safe, but in larger amounts it can be toxic and responsible for serious adverse side effects. Eight *B. equi*-infected donkeys died following treatment with 4 mg/kg given four times at 72-hr intervals.[72] The LD_{50} of imidocarb in horses was determined to be 16 mg/kg given two times at a 24-hr interval.[74] Although horses treated with 2 mg/kg showed no adverse reactions, all horses receiving 4 mg/kg or more manifested signs compatible with excessive stimulation of the cranial and sacral portions of the parasympathetic division of the autonomic nervous system, characterized by sweating, extreme restlessness, and abdominal pain. There were increased levels of blood urea nitrogen, serum aspartate, aminotransferase, serum sorbitol dehydrogenase, serum creatine phosphokinase, and increasing severity of renal, hepatic, and pulmonary lesions.[74] Mortalities which occurred within 6 days following the first injection were attributed to acute cortical tubular renal necrosis and acute periportal hepatic necrosis.

Chlortetracycline has been reported to be effective against *B. equi* if given in repeated i.v. doses early in the course of infection.[61] Six daily injections at three-dose levels, 0.5 mg/kg, 2.5 mg/kg, and 2.6 mg/kg, were described as effective in retarding the multiplication of *B. equi*.

IV. CANINE BABESIOSIS

Treatment of canine babesiosis has included a wide variety of ineffectual drugs such as ammonia, extract of belladonna, quinine, benzoate of soda, carbolic acid, and calomel.[75] Currently, treatment for canine babesiosis includes specific babesiacidal agents coupled with supportive treatment, including blood transfusions, fluid therapy, alkalinizing agents, liver protectants, diuretics, restricted exercise, and improved nutrition.[76] Trypan blue was successfully used in 1909 for the treatment of *B. canis*.[77] This treatment consisted of 1 mℓ/kg of a 1% solution (10 mg/kg) given s.c. Quinuronium is more active than trypan blue, but has a low therapeutic index and must be used with some caution. The recommended s.c. dose of 0.25 mg/kg of quinuronium is effective, but does not eliminate infection.[75] At 1 mg/kg, toxic signs may be observed. While trypan blue and quinuronium are effective against *B. canis*, they are not recommended for *B. gibsoni*.[78]

One of the first diamidines to be recommended for the treatment of canine babesiosis was phenamidine at a s.c. dose of 10 mg/kg. Although effective against both *B. canis* and *B. gibsoni*, and less toxic than quinuronium, it did not eliminate infection and relapses occurred.[79] The dose should be repeated after an interval of 24 hr in *B. gibsoni* infection or after 6 to 7 days in *B. canis* infection.[80,81] In addition to phenamidine, Klinefelter found diminazene at 7 mg/kg to be effective in treating *B. canis*.[82] Although effective against both *B. canis* and *B. gibsoni*, diminazene, when used at 7.5 to 10.0 mg/kg as a single dose, may produce some abnormal CNS involvement such as nystagmus, ataxia, extensor rigidity, opisthotonus, coma, and even death.[75,83,84] This can be avoided by using 3.0 mg/kg diminazene two times on consecutive days. As little as 2.5 to 3.5 mg/kg diminazene aceturate has been successful in treating *B. canis*, but 5 to 7 mg/kg was required for *B. gibsoni*.[5,75,84] Diminazene is not recommended for use at levels over 7 mg/kg because of possible CNS changes. *B. gibsoni* was successfully treated with diminazene, totaling 11 mg/kg, given in two doses 5 days apart, but chloroquine phosphate, primaquine phosphate, sodium arsenamide, and chlorguanide hydrochloride were not effective.[85] Pentamidine at the level of 16.5 mg/kg given twice on consecutive days was reported clinically effective against *B. gibsoni* and *B. canis*, but relapses commonly occurred with *B. gibsoni*.[83] Pentamidine was responsible for pain at the i.m. injection site, hypotension, tachycardia, nausea, and vomiting. The antitrypanosomal agent, isometamidium, has recently been used successfully in the treatment of *B. canis* at 2 mg/kg i.m. and compares favorably with amicarbalide and diminazene.[86]

Imidocarb at 5 mg/kg has been found to be very effective against *B. canis*.[87] Following treatment, there was a rapid reduction in parasitemia and a corresponding increase in PCV. Fewer than 4% of the animals so treated relapsed with *Babesia* over a 3-month period, notwithstanding the probability of reinfection. In Nigeria the treatment of 260 dogs diagnosed as having babesiosis (*B. canis*) with 5 mg/kg imidocarb has resulted in a 95.8% recovery rate.[88] Another study reported a favorable response in *B. canis*-infected dogs following imidocarb treatment with 3 mg/kg.[89] When the drug was given at the time of *B. canis* exposure, carrier premunition occurred without evidence of clinical infection.[89] Field experience suggests chemoprophylaxis for at least 4 weeks in 75% of dogs treated with 4.8 to 6.4 mg/kg imidocarb.[90] Another study suggests that the prophylactic properties of imidocarb were limited to moderating the severity of infection.[91]

The use of a single therapeutic regime for babesiosis in dogs is not recommended, but should be influenced by the severity of infection and the age and condition of the dog. Intensive supportive treatment is often required to counteract shock and disseminated intravascular coagulation, a common sequel in canine babesiosis. This phenomenon has also been seen in malaria. Either diminazene or imidocarb are usually the babesiacidal drugs of choice. Trypan blue may be used in patients that are categorized as severely infected, because it does not stimulate the parasympathetic nervous system and thus aggravate the medical shock that is frequently present in those critically ill patients, nor does it possess the potentially toxic effects on the CNS sometimes seen with the diamidines. In the severe category, i.v. fluid therapy is recommended immediately, and the babesiacidal drug trypan blue infused i.v. at the rate of 10 mg/kg ina 1% solution, followed by blood transfusion, antibiotics, and vitamin complexes. Diminazene, at the rate of 3.5 mg/kg in a 7% solution, given s.c., plus antiboitics and vitamins, is frequently adequate in mild cases. In those cases that responded, improvement was noted within a few hours after the initial treatment.[75]

V. FELINE BABESIOSIS

A large number of *Babesia* species occurs in different domestic and feral cats. *Babesia felis* (a small *Babesia*) appears to be of importance among domestic cats.[92] A recent article described a large *Babesia* in domestic cats in South Africa, which was tentatively identified as *B. herpailuri.*[93] *B. felis* is unique in its response to chemotherapy in that the usual babesiacidal drugs fail to produce a remission of infection.[8] Primaquine phosphate, an antimalarial compound, administered *per os* or as an i.m. injection, was found to be effective and was the drug of choice in the treatment of *B. felis*. The dosage used was 0.5 mg/kg body weight of primaquine base. Repeated treatment was well tolerated, but single doses in excess of 1 mg/kg are known to cause mortality in cats. Chloroquine, diminazene, phenamidine, quinuronium, euflavin, trypan blue, imidocarb, and oxytetracycline were all ineffective against *B. felis.*[8] *B. herpailuri,* on the other hand, responds to either imidocarb or diminazene, at 5 mg/kg.[22,24,93]

VI. HUMAN BABESIOSIS

Babesiosis in man has reached the point when it should no longer be considered a benign medical rarity.[94] For a time, the accepted treatment of human babesiosis consisted for the most part of antimalarial drugs and supportive treatment.[95,96] Quinine sulfate and chloroquine are generally ineffective in reducing *Babesia* parasitemias. In the U.S., the principal cause of babesiosis in man is *B. microti;*[3] in Europe, *B. divergens* is more commonly involved.[3,96] In one instance, a 65-year-old man infected with *B. microti* failed to respond to supportive and antimalarial treatment, but was successfully treated with diminazene, and the *Babesia* parasitemia disappeared. Some time after his recovery from babesiosis, the patient developed acute idiopathic polyneuritis (Landry-Guillian-Barre syndrome). A splenectomized man infected with *B. divergens* died after treatment with chloroquine and pyrimethamine, plus supportive treatment.[96] In two cases of *B. microti* infection, treatment with pentamidine was successful in eliminating parasitemia.[97] Low-level parasitemias, detected only by hamster inoculations, persisted following pentamidine therapy.[97] There was considerable discomfort at the site of the pentamidine injection.

Clindamycin has recently been used successfully for the treatment of *B. microti* infections in hamsters.[98] It has also been used successfully in a near-fatal case of babesiosis in man, which occurred following blood transfusion from a *Babesia* carrier.[99] Because of enterocolitis which may accompany treatment with clindamycin alone, it has been necessary to use it in combination with vancomycin. The addition of oral quinine to the clindamycin/vancomycin therapy enhances the efficacy of this treatment.

Many cases of human babesiosis can be effectively handled with symptomatic and supportive care, but these treatment regimes generally fail to produce results comparable to those seen in animals when specific babesiacidal compounds are used.[100]

VII. OVINE AND PORCINE BABESIOSIS

Sheep babesias (*B. motasi* and *B. ovis*) and swine babesias (*B. trautmanni* and *B. perroncitoi*) are thought to be susceptible to treatment using the compounds previously mentioned.[9] *Babesia ovis* was shown to be relatively resistant to a single injection of quinuronium and imidocarb, but treatment with 2 mg/kg imidocarb s.c., three times at 24-hr intervals, was successful.[101,102] The large *B. motasi* has been reported in northern Europe and northern Wales.[103,104] Babesiosis occurring as a result of this infection can be successfully treated with amicarbalide at the rate of 5 mg/kg and also diminazene at 5.5 mg/kg.[104]

An outbreak of porcine babesiosis in Senegal caused by *B. perroncitoi* resulted in 318 deaths.[7] Over 4000 swine were treated i.m. with diminazene at 5 mg/kg body weight.[7] This was followed by decreased death rate and clinical cures, although parasitemia was still detectable in some animals.

VIII. CONCLUSIONS

Except for *B. felis* and a European isolate of *B. equi,* most babesias are susceptible to one or more of the recognized babesiacidal drugs. Instances of drug resistance have been reported, but this has not to date evolved as a major problem. Greater emphasis is needed in areas of chemoprophylaxis and the study of the so-called ''sterile immunity'' which may occur and persist for differing periods following radical cures.

Treatment before or after premunizing vaccines is common and provides an element of safety and avoids the occasional severe response. Chemoprophylaxis alone might be feasible when used to prevent fatal tick-acquired infections in susceptible cattle introduced into endemic zones where exposure occurs on a continuing basis. Theoretically, a point will be reached when drug levels are sufficiently low enough to allow infection to occur, but still high enough to prevent a fulminating infection, clinical signs, and death. Such animals may then develop carrier or premune status without undergoing severe or fatal reactions.

The availability of highly specific and effective chemotherapeutic agents has contributed to the practice of excessive reliance on these drugs in the treatment of infection to the extent that little attention has been paid to the need for supportive symptomatic treatment. With the increasing number of babesiosis cases occurring in man and companion animals, greater emphasis is being given to supportive and symptomatic treatment. The extrapolation of these techniques to large, food-producing animals is probably indicated in selected cases.

Additional pharmacokinetic studies are indicated on the more commonly used drugs to determine safe and effective doses and establish the effect of treatment intervals in cases where multiple doses are indicated. Such studies might also facilitate clearance of babesiacidal compounds for use in the U.S. if the need should ever occur. It is fortunate that babesiosis in food-producing animals does not occur in the U.S. because if an outbreak occurred, there is not at present a single, approved babesiacidal compound available that could be legally used to prevent death losses.

Recent studies involving *B. equi* have revealed a number of unique characteristics which have placed some doubt as to the validity of its classification as a *Babesia*.[105] The presence of lymphocytic schizonts, and the absence of transovarial tick transmission, are characteristics which closely resemble those of *Theileria*.[73] The marked difference of *B. equi* in response to treatment, plus the apparent unique response to antitheilerial compounds, further contribute to the concept that this *Babesia* is related to the theilerial organisms.[19] The failure of *B. felis* to respond to traditional babesiacidal compounds suggests the need to further study the taxonomic classification of this hemoparasite. A careful evaluation of the large and small babesias relative to their response to therapy is needed and may well contribute to our future understanding of these organisms.

REFERENCES

1. **Smith, T. and Kilbourne, F. L.,** Investigations into the nature, causation and prevention of Texas or Southern cattle fever, *U.S. Dep. Agric. Bur. Anim. Ind. Bull.*, 1, 1, 1893.
2. **Purnell, R. E.,** Babesiosis in various hosts, in *Babesiosis*, Ristic, M. and Kreier, J. P., Eds., Academic Press, New York, 1981, 25.
3. **Brocklesby, D. W.,** Human babesiosis, *J. S. Afr. Vet. Assoc.*, 50, 302, 1979.
4. **Anderson, J. F. and Magnerelli, L. A.,** Canine *Babesia*, new to North America, *Science*, 204, 1431, 1979.
5. **Breitschwerdt, E. B., Malone, J. B., MacWilliams, P., Levy, M. G., Qualls, W., and Prudich, M. J.,** Babesiosis in the greyhound, *J. Am. Vet. Med. Assoc.*, 182, 978, 1983.
6. **Uilenberg, G., Top, P. D. J., Arends, P. J., Kool, P. J., Van Dirk, J. E., Van Schieveen, P. B. E., and Zwart, D.,** Indigenous canine babesiosis in the Netherlands, *Tijdschr. Diergeneeskd.*, 110, 93, 1985.
7. **Vercruysse, J. and Parent, R.,** Report of an account of porcine babesiosis due to *Babesia perroncitoi* in Senegal, *Afr. Med.*, 20, 435, 1981.
8. **Potgieter, F. T.,** Chemotherapy of *Babesia felis* infection: efficacy of certain drugs, *J. S. Afr. Vet. Assoc.*, 52, 289, 1981.
9. **Kuttler, K. L.,** Chemotherapy of Babesioisis: a review, in *Babesiosis*, Ristic, M. and Kreier, J. P., Eds., Academic Press, New York, 1981, 65.
10. **DeVos, A. J.,** Epidemiology and control of bovine babesiosis in South Africa, *J. S. Afr. Vet. Assoc.*, 50, 357, 1979.
11. **Beveridge, E.,** *Babesia rodhaini*. A useful organism for the testing of drugs designed for the treatment of piroplasmosis, *Ann. Trop. Med. Parasitol.*, 47, 134, 1953.
12. **Lucas, J. M. S.,** The chemotherapy of experimental babesiosis in mice and splenectomized calves, *Res. Vet. Sci.*, 1, 218, 1960.
13. **Lewis, D. and Williams, H.,** Infection of the Mongolian gerbil with the cattle piroplasm, *Babesia divergens*, Nature (London), 278, 170, 1979.
14. **Irvin, A. D. and Young, E. R.,** Possible *in vitro* test for screening drugs for activity against *Babesia* and other blood protozoa, *Nature (London)*, 169, 407, 1977.
15. **Irvin, A. D. and Young, E. R.,** Further studies on the uptake of tritiated nucleic acid precursors by *Babesia* spp. of cattle and mice, *Int. J. Parasitol.*, 9, 109, 1979.
16. **Erp, E. E., Gravely, S. M., Smith, R. D., Ristic, M., Osorno, B. M., and Carson, C. A.,** Growth of *Babesia bovis* in bovine erythrocyte cultures, *Am. J. Trop. Med. Hyg.*, 27, 1061, 1978.
17. **Levy, M. G. and Ristic, M.,** Continuous cultivation in a microaerophilous stationary phase culture, *Science*, 207, 1218, 1980.
18. **Varynen, R. and Thomi, J.,** Continuous *in vitro* cultivation of *Babesia divergens*, *Acta Vet. Scand.*, 23, 471, 1982.
19. **Zweygarth, E., Ahmed, J. S., and Rehbein, G.,** The effect of halofuginone, Wellcome 993 C, oxytetracycline, and diminazene diaceturate on *Babesia equi* infected lymphoblastoid cell cultures, *J. Parasitol.*, 79, 542, 1984.
20. **Newton, B. A.,** Interaction of berenil with deoxyribonucleic acid and some characteristics of the berenil-deoxyribonucleic acid complex, *Biochem. J.*, 105, 50, 1967.
21. **Neville, D. M.,** The interaction of acridine dyes with DNA, An X-ray diffraction and optical investigation, *J. Molec. Biol.*, 17, 57, 1966.
22. **Hebel, R. and Denning, H. K.,** Licht und Eletronenmikroskopische Untersuchungen an *Babesia herpailuri* nach Behandlund mit 4,4′ Diamidinodiazoamino-benzol (Berenil), *Z. Parasitenkd.*, 33, 1, 1969.
23. **Simpson, C. F. and Neal, F. C.,** Ultrastructure of *Babesia equi* in ponies treated with imidocarb, *Am. J. Vet. Res.*, 41, 267, 1980.
24. **Gobel, E. and Denning, H. K.,** Die Ultrastruktur der Trophozoiten und Merozoiten von *Babesia herpailuri* in Erythrozyten der Hauskatze nach Behandlung mit Imidocarb (3,3′-bis-2-imidazolin-2-yl) — carbanilid, *Z. Parasitenkd.*, 49, 97, 1976.
25. **Joyner, L. P.,** The chemotherapy of protozoal infections of veterinary importance, *J. Protozool.*, 28, 17, 1981.
26. **Dalgliesh, R. J. and Stewart, N. P.,** Tolerance to imidocarb induced experimentally in tick-transmitted *Babesia argentina*, *Aust. Vet. J.*, 53, 176, 1977.
27. **Yeruham, I., Pipano, E., and Davidson, M.,** A field strain of *Babesia bovis* apparently resistant to amicarbalide isethionate, *Trop. Anim. Health Prod.*, 17, 29, 1985.
28. **Theiler, A.,** The treatment of redwater in cattle with trypan blue, *Vet. J.*, 68, 64, 1912.
29. **Rees, C. W.,** Characteristics of the piroplasms *Babesia argentina* and *Babesia bigemina* in the United States, *J. Agric. Res.*, 48, (5), 427, 1934.
30. **Legg, J.,** Treatment of piroplasmosis (*P. bigeminum*) with akiron, *Aust. Vet. J.*, 12, 227, 1936.

31. **Legg, J.,** The treatment of babesiellosis (*B. argentina*) with scaprin, *Aust. Vet. J.*, 15, 121, 1939.
32. **Vanzini, V. R., Gomez, M. R., Hadani, A., Bermudez, A. C., Luciani, C. A., Guglielmone, A. A., and Angold, A. J.,** A case of intestinal torsion in a cow, apparently caused by administration of acaprin as a babesiacide, *Rev. Med. Vet. Argent.*, 6, 503, 1981.
33. **Eyre, P.,** Some pharmacodynamic effects of the babesicidal agents quinuronium and amicarbalide, *J. Pharm. Pharmacol.*, 19, 509, 1967.
34. **Barnett, S. F.,** The chemotherapy of *Babesia bigemina* infection in cattle, *Res. Vet. Sci.*, 6, 397, 1965.
35. **Purnell, R. E., Lewis, D., and Young, E. R.,** Quinuronium sulphate for the treatment of *Babesia divergens* infections of splenectomized calves, *Vet. Rec.*, 108, 538, 1981.
36. **Newton, L. G. and O'Sullivan, P. J.,** Chemoprophylaxis in *Babesia argentina* infection in cattle, *Aust. Vet. J.*, 45, 404, 1969.
37. **Ryley, J. F.,** A chemoprophylactic approach to babesiosis, *Res. Vet. Sci.*, 5, 411, 1964.
38. **Riek, R. F.,** Babesiosis, in *Infectious Blood Disease of Man and Animals*, Weinman, D. and Ristic, M., Eds., Academic Press, New York, 1968, 219.
39. **Pipano, E., Jeruhan, I., and Frank, M.,** Pentamidine in chemoimmunisation of cattle against *Babesia bigemina* infections, *Trop. Anim. Health Prod.*, 11, 13, 1979.
40. **Lohr, F. K., Otieno, P. S., and Gacanga, W.,** Therapy trial using betamethasone in addition to specific treatment with a diamidine and quinuronium derivate in cattle experimentally infected with *Babesia bigemina*, *Bull. Anim. Health. Prod.*, 24, 199, 1976.
41. **Thompson, K. C., Todorovic, R. A., Mateus, C., and Adams, L. G.,** Methods to improve the health of cattle in the tropics: immunisation and chemoprophylaxis against haemoparasitic infections, *Trop. Anim. Health Prod.*, 10, 75, 1978.
42. **Minami, T. and Ishihara, T.,** *Babesia ovata* sp.n. isolated from cattle in Japan, *Natl. Inst. Anim. Health Q.*, 20, 101, 1980.
43. **Todorovic, R. A., Vizcaino, O. G., Gonzalez, E. F., and Adams, L.G.,** Chemoprophylaxis (Imidocarb) against *Babesia bigemina* and *Babesia argentina* infections, *Am. J. Vet. Res.*, 34, 1153, 1973.
44. **Callow, L. L. and McGregor, W.,** The effect of imidocarb against *Babesia argentina* and *Babesia bigemina* infections of cattle, *Aust. Vet. J.*, 46, 195, 1970.
45. **Taylor, R. J. and McHardy, N.,** Preliminary observations on the combined use of imidocarb and *Babesia* blood vaccine in cattle, *J. S. Afr. Vet. Assoc.*, 50, 326, 1979.
46. **Purnell, R. E., Lewis, D., and Young, E. R.,** Investigations on the prophylactic effect of treatment with imidocarb diproprionate on *Babesia divergens* infections in splenectomized calves, *Br. Vet. J.*, 136, 452, 1980.
47. **Euzeby, J., Rancien, P., and Simon, P. H.,** Outbreak of *Babesia* infection in the Charolles region of France. III. Chemoprophylaxis by imidocarb, *Bull. Soc. Sci. Vet. Med. Comp. Lyon*, 76 (6), 423, 1974.
48. **Haigh, A. J. B. and Hagan, D. H.,** Evaluation of imidocarb dihydrochloride against redwater disease in cattle in Eire, *Vet. Rec.*, 94, 56, 1974.
49. **Kuttler, K. L., Graham, O. H., and Trevino, J. L.,** The effect of imidocarb treatment on *Babesia* in the bovine and the tick (*Boophilus microplus*), *Res. Vet. Sci.*, 18, 198, 1975.
50. **Gray, J. S. and Potgieter, F. T.,** The retention of *Babesia bigemina* infection by *Boophilus decoloratus* exposed to imidocarb dipropionate during engorgement, *Onderstepoort J. Vet. Res.*, 48, 225, 1981.
51. **DeVos, A. J., Stewart, N. P., and Dalgliesh, R. J.,** The effect of chemotherapy on *Babesia bigemina* in the tick vector *Boophilus microplus*, *Int. J. Parasitol.*, 14, 249, 1984.
52. **Mahoney, D. F. and Mirre, G. B.,** A note on the transmission of *Babesia bovis* (syn. *B. argentina*) by the one host tick, *Boophilus microplus*, *Res. Vet. Sci.*, 26, 253, 1979.
53. **Potgieter, F. T.,** The life cycle of *Babesia bovis* and *Babesia bigemina* in Ticks and in Cattle in South Africa, Ph.D. thesis, Rand Afrikaans University, Johannesburg, South Africa, 1977.
54. **Day, W. C. and Kuttler, K. L.,** Animal health considerations involved in the movement of U.S. cattle to Haiti, *Southwest. Vet.*, 28, 229, 1975.
55. **DeLoach, J. R., Wagner, G. G., and Craig, T. M.,** Imidocarb dipropionate encapsulation and binding to reseal carrier bovine erythrocytes for potential babesiosis chemotherapy, *J. Appl. Biochem.*, 3, 254, 1981.
56. **Kuttler, K. L.,** Use of imidocarb to control anaplasmosis, *Southwest. Vet.*, 28, 47, 1975.
57. **Adams, L. G., Corrier, D. E., and Williams, J. D.,** A study of the toxicity of imidocarb dipropionate in cattle, *Res. Vet. Sci.*, 28, 172, 1980.
58. **Aliu, Y. O.,** Absorption, Distribution, and Excretion of Imidocarb Dipropionate (3,3′-bis-(2-imidazolin-2-yl) Carbanilide) in Sheep, Ph.D. thesis, Texas A & M University, College Station, 1974.
59. **Aliu, Y. O., Davis, R. H., Camp, B. J., and Kuttler, K. L.,** Absorption, distribution and excretion of imidocarb dipropionate in sheep, *Am. J. Vet. Res.*, 38, 2001, 1977.
60. **Wood, J. C.,** The activity of imidocarb against *Babesia* infections of cattle, *Ir. Vet. J.*, 25, 254, 1971.
61. **Jansen, B. C.,** The parasiticidal effect of Aureomycin (Lederle) on *Babesia equi* (La Veran 1899) in splenectomized donkeys, *Onderstepoort J. Vet. Res.*, 26, 175, 1953.

62. **Purnell, R. E. and Schroeder, J.**, Herd prophylaxis of tick-borne diseases in susceptible cattle in South Africa using terramycin/LA, *Proc. 13th World Congr. on Diseases of Cattle,* Durban, Hoescht, Johannesburg, South Africa, 1984. 1.
63. **Pipano, E., Krigel, Y., and Markovics, A.**, Oxytetracycline-induced resistance to *Babesia bovis* infection in splenectomized calves, *Trop. Anim. Health Prod.*, 17, 153, 1985.
64. **Pipano, E., Krigel, Y., Markovics, A., Rubinstein, E., and Frank, M.**, Mitigation of the response of Freisian calves to live *Babesia bovis* vaccine by treatment with long acting oxytetracycline, *Vet. Rec.*, 117, 413, 1985.
65. **Taylor, S. M., Elliott, C. T., and Kenny, J.**, Inhibition of *Babesia divergens* in cattle by oxytetracycline, *Vet. Rec.*, 118, 98, 1986.
66. **O'Neill, A. R.**, Blood transfusions in cattle with particular reference to "Redwater" (Babesiosis), *Ir. Vet. J.*, 33, 1, 1979.
67. **DeKock, G. D. W.**, Drug treatment in *nuttalliosis,* Union of S.A., Dept. of Agric. 7th and 8th Reports, *Dir. Vet. Res.*, April, 639, 1918.
68. **Malhotra, D. V., Gautam, O. P., and Banerjee, D. P.**, A note on chemotherapeutic trials against *Babesia equi* infection in donkeys, *Indian J. Anim. Sci.*, 49, 75, 1979.
69. **Singh, B., Banerjee, D. P., and Gautam, O. P.**, Comparative efficacy of diminazine diaceturate and imidocarb dipropionate against *B. equi* infections in donkeys, *Vet. Parasitol.*, 7, 173, 1980.
70. **Frerichs, W. M. and Holbrook, A. A.**, Treatment of equine piroplasmosis (*B. caballi*) with imidocarb dipropionate, *Vet. Rec.*, 95, 188, 1974.
71. **Kirkham, W. W.**, The treatment of equine babesiosis, *J. Am. Vet. Med. Assoc.*, 155, 457, 1969.
72. **Frerichs, W. M., Allen, P. C., and Holbrook, A. A.**, Equine piroplasmosis (*Babesia equi*): therapeutic trials of imidocarb dihydrochloride in horses and donkeys, *Vet. Rec.*, 93, 73, 1973.
73. **Friedhoff, K. T.**, The piroplasms of equidae-significance for international commerce, *Berl. Muench. Tieraerztl. Wochenschr.*, 95, 368, 1982.
74. **Adams, L. G.**, Clinicopathological aspects of imidocarb dipropionate toxicity in horses, *Res. Vet. Sci.*, 31, 54, 1981.
75. **Moore, D. J.**, Therapeutic implications of *Babesia canis* infection in dogs, *J. S. Afr. Vet. Assoc.*, 50, 346, 1979.
76. **Button, C.**, Metabolic and electrolyte disturbances in acute canine babesiosis, *J. Am. Vet. Med. Assoc.*, 175, 475, 1979.
77. **Nutall, G. H. F. and Hadwen, S.**, The successful drug treatment of canine piroplasmosis, together with observation upon the effect of drugs on *Piroplasma* canis, *Parasitology,* 2, 156, 1909.
78. **Groves, M. G. and Dennis, G. L.**, *Babesia gibsoni:* field and laboratory studies of canine infections, *Exp. Parasitol.*, 31, 153, 1972.
79. **Groves, M. G. and Vanniasingham, J. A.**, Treatment of *Babesia gibsoni* infections with phenamidine isethionate, *Vet. Rec.*, 86, 8, 1970.
80. **Ruff, M. D., Fowler, J. L., Fernau, R. C., and Matsuda, K.**, Action of certain antiprotozoal compounds against *Babesia gibsoni* in dogs, *Am. J. Vet. Res.*, 34, 641, 1973.
81. **Joyner, L. P., Davies, S. F. M., and Kendall, S. B.**, Chemotherapy of Babesiosis, in *Experimental Chemotherapy,* Vol. 1, Schnitzer, R. J. and Hawking, F., Eds., Academic Press, New York, 1963, 603.
82. **Klinefelter, M. R.**, Cause, diagnosis, and treatment of canine piroplasmosis, *Vet. Med. Small Anim. Clin.*, 77, 1505, 1982.
83. **Farwell, G. E., LeGrand, E. K., and Cobb, C. C.**, Clinical observations on *Babesia gibsoni* and *Babesia canis* infections in dogs, *J. Am. Vet. Med. Assoc.*, 180, 507, 1982.
84. **Dennig, Von, H. K., Centurier, C., Gobel, E., and Weiland, G.**, Ein Beitrag zur Babesiose des Hundes und ihrer Bedeutung in der Bundesrepublic Deutschland und Berlin-West, *Berl. Muench. Tieraerztl. Wochenschr.*, 93, 373, 1980.
85. **Fowler, J. L., Ruff, M. D., Fernau, R. C., and Furusho, Y.**, *Babesia gibsoni,* chemotherapy in dogs, *Am. J. Vet. Res.*, 33, 1109, 1972.
86. **Stewart, C. G.**, A comparison of the efficacy of isometamidium, amicarbalide, and diminazene against *Babesia canis* in dogs and the effect on subsequent immunity, *J. S. Afr. Vet. Assoc.*, 54, 47, 1983.
87. **Ogunkoya, A. B., Adeyanju, J. B., and Aliu, Y. O.**, Experiences with the use of imizol in treating canine blood parasites in Nigeria, *J. Small Anim. Pract.*, 22, 775, 1981.
88. **Adeyaniju, B. J. and Aliu, Y. O.**, Chemotherapy of canine ehrlichiosis and babesiosis with imidocarb dipropionate, *J. Am. Anim. Hosp. Assoc.*, 18, 827, 1982.
89. **Euzeby, J., Moreau, Y., Chauve, C., Gevrey, J., and Gauthey, M.**, Experimentation des proprietes antipiroplasmiques de l'Imidocarb sur *Babesia canis,* agent de la piroplasmose canine en Europe, *Bull. Acad. Vet. France,* 53, 475, 1980.
90. **Coll, J. L., Descamps, H., Fayette, J. P., Feraud, J. P., and Villemin, P.**, Essai de chimo-prevention de la Babesiose canine avec l'imidocarbe, *Rec. Med. Vet.*, 158, 791, 1982.

91. **Uilenberg, G., Verdiesen, P. A., and Zwart, D.,** Imidocarb: a chemoprophylactic experiment with *Babesia canis, Tijdschr. Diergeneeskd.*, 106, 118, 1981.
92. **Futter, G. J. and Belonje, P. C.,** Studies on feline babesiosis. I. Historical review, *J. S. Afr. Vet. Assoc.*, 50, 105, 1980.
93. **Stewart, C. G., Hackett, K. J. W., and Collett, M. G.,** An unidentified *Babesia* of the domestic cat *(Felis domesticus), J. S. Afr. Vet. Assoc.*, 51, 219, 1980.
94. **Zaino, E. C. and Amelkin, S.,** Babesiosis: a potentially grave disease for the immunosuppressed, *N.Y. State J. Med.*, 81, 384, 1981.
95. **Ruebush, T. K., II, Rubin, R. H., Wolpow, E. R., Cassaday, P. B., and Schultz, M. G.,** Neurologic complications following the treatment of human *Babesia microti* infection with diminazene aceturate, *Am. J. Trop. Med. Hyg.*, 28, 184, 1979.
96. **Entrican, J. H., Williams, H., Cook, I. A., Lancaster, W. M., Clark, J. C., Joyner, L. P., and Lewis, D.,** Babesiosis in man: report of a case from Scotland with observations on the infecting strain, *J. Infect.*, 1, 227, 1979.
97. **Francioli, P. B., Keithly, J. S., Jones, T. C., Brandstetter, R. D., and Wolf, D. J.,** Response of babesiosis to pentamidine therapy, *Ann. Intern. Med.*, 94, 426, 1981.
98. **Rowin, K. S., Tanowitz, H. B., and Wittner, M.,** Therapy of experimental babesiosis, *Ann. Intern. Med.*, 97, 556, 1982.
99. **Wittner, M., Rowin, K. S., Tanowitz, H. B., Hobbs, J. F., Saltzman, S., Wenz, B., Hirsch, R., Chisholm, E., and Healy, G. R.,** Successful chemotherapy of transfusion babesiosis, *Ann. Intern. Med.*, 96, 601, 1982.
100. **Gombert, M. E., Goldstein, E. J. C., Benach, J. L., Tenebaum, M. J., Grunwaldt, E., Kaplan, M. H., and Eveland, L. K.,** Human babesiosis, clinical and therapeutic considerations, *J. Am. Vet. Med. Assoc.*, 248, 3005, 1982.
101. **Michael, S. A. and Refaii, A. H.,** The effect of imidocarb dipropionate on *Babesia ovis* infection in sheep, *Trop. Anim. Health Prod.*, 14, 1, 1982.
102. **Hashemi-Fesharki, R.,** Studies on imidocarb dihydrochloride in experimental *Babesia ovis* infection in splenectomized lambs, *Br. Vet. J.*, 133, 609, 1977.
103. **Lewis, D. and Herbert, I.,** A large *Babesia* of sheep from north Wales, *Vet. Rec.*, 107, 352, 1980.
104. **Uilenberg, G., Rombach, M. C., Perie, N. M., and Zwart, D.,** Blood parasites of sheep in the Netherlands. II. *Babesia motasi (Sporozoa babesiidae), Vet. Q.*, 2, 3, 1980.
105. **Schein, E., Rehbein, G., Voight, W. P., and Zweygarth, E.,** *Babesia equi* (Laveran 1901). I. Development in horses and in lymphocyte culture, *Tropenmed. Parasitol.*, 32, 223, 1981.
106. **Kuttler, K. L.,** Unpublished data.

INDEX

A

B

C

D

Q

R

S

T

U

V

W

Z